Don MacLean

Biological habitat reconstruction

Biological habitat reconstruction

Edited by

G. P. BUCKLEY

Belhaven Press
A division of Pinter Publishers
London and New York

First published in Great Britain in 1989 and reprinted in 1992 by
Belhaven Press (a division of Pinter Publishers),
25 Floral Street, London WC2E 9DS

British Library Cataloguing in Publication Data
A CIP catalogue record for this book is available from the British Library

ISBN 1 85293 058 6

Library of Congress Cataloging in Publication Data
Biological habitat reconstruction.
Includes bibliographies and index.
1. Restoration ecology. I. Buckley, G.P.
QH541.15.R45B56 1989 333.7′153 89-7035
ISBN 1-85293-058-6

Typeset by Book Ens, Saffron Walden, Essex
Printed and bound in Great Britain
by SRP Ltd, Exeter

Contents

List of contributors

M.A. Anderson, Forestry Commission Research Station, Wrecclesham, Farnham, Surrey, GU10 4LH

Penny Anderson, 52 Lower Lane, Chinley, Stockport, Cheshire, SK12 6BD

J.C. Baines, PO Box 32, Stourbridge, West Midlands, DY9 0NQ

C.J. Bickmore, Travers Morgan Consulting Group, 136 Long Acre, London, WC2E 9AE

N.D. Boatman, The Cereals and Gamebirds Research Project, The Game Conservancy, Fordingbridge, Hampshire, SP6 1DF

A.D. Bradshaw, Department of Environmental and Evolutionary Biology, University of Liverpool, Liverpool, L69 3BX

R.J. Brown, Johnsons Seeds, London Road, Boston, Lincolnshire, PE21 8AD

G.P. Buckley, Wye College, University of London, Wye, Ashford, Kent, TN25 5AH

R.G.H. Bunce, Institute of Terrestrial Ecology, Merlewood Station, Grange-over-Sands, Cumbria, LA11 6JU

C.P. Burnham, Wye College, University of London, Wye, Ashford, Kent, TN25 5AH

R.P. Coppeard, Essex County Council, New Street, Chelmsford, Essex

S.E. Cowgill, Biology Department, Building 44, The University, Highfield, Southampton, Hampshire, S09 5NH

Ruth Cox, Institute of Terrestrial Ecology, Monks Wood Experimental Station, Abbots Ripton, Huntingdon, Cambridgeshire, PE17 2LS

B.N.K. Davis, Institute of Terrestrial Ecology, Monks Wood Experimental Station, Abbots Ripton, Huntingdon, Cambridgeshire, PE17 2LS

C.G. Down, Department of Mineral Resources Engineering, Imperial College, London, SW7 2BP

J.W. Dover, The Cereals and Gamebirds Research Project, The Game Conservancy, Fordingbridge, Hampshire, SP6 1DF

A. Frost, Institute of Terrestrial Ecology, Monks Wood Experimental Station, Abbots Ripton, Huntingdon, Cambridgeshire, PE17 2LS

K.S. Funnell, formerly Principal Landscape Architect with Kent County Council, Springfield, Maidstone, Kent, ME14 2LT

M.W. Gough, Institute of Terrestrial Ecology, Monks Wood Experimental Station, Abbots Ripton, Huntingdon, Cambridgeshire, PE17 2LS

B.H. Green, Wye College, University of London, Wye, Ashford, Kent, TN25 5AH

D.R. Helliwell, Yokecliffe House, West End, Wirksworth, Derbyshire, DE4 4EG

D.A. Hill, British Trust for Ornithology, Beech Grove, Station Road, Tring, Hertfordshire HP23 5NR

J.G. Hodgson, Unit of Comparative Plant Ecology, (NERC), Department of Plant Sciences, The University, Sheffield, S10 2TN

N.R. Jenkins, Institute of Terrestrial Ecology, Merlewood Station, Grange-over-Sands, Cumbria, LA11 6JU

W.R. Jordan III, The University of Wisconsin-Madison Arboretum, 1207 Seminole Highway, Madison, WI 53711, USA

G. Kaule, Institute für Landschaftsplanung, Universität Stuttgart, 7000 Stuttgart 1, Keplerstrasse 11, West Germany

D.G. Knight, Wye College, University of London, Wye, Ashford, Kent, TN25 5AH

S. Krebs, Institute für Landschaftsplanung, Universität Stuttgart, 7000 Stuttgart 1, Keplerstrasse 11, West Germany

P.J. Larard, Travers Morgan Consulting Group, 136 Long Acre, London, WC2E 9AE

I.C. Ludolf, 22 Queens Avenue, Andover, Hampshire, SP10 3HZ

R.H. Marrs, Institute of Terrestrial Ecology, Monks Wood Experimental Station, Abbots Ripton, Huntingdon, Cambridgeshire, PE17 2LS

A.J. Morton, Department of Pure and Applied Biology, Imperial College, Silwood Park, Ascot, Berkshire, SL5 7T4

C. Newbold, Nature Conservancy Council, Northminster House, Peterborough, PE1 1UA

Steven Packard, The Nature Conservancy of Illinois, 79 W. Monroe Street, Chicago, Il 6063, USA

D.G. Park, Steetley Quarry Products Ltd, Thrislington Works, West Cornforth, Ferryhill, Co. Durham

P.A. Robertson, The Pheasants and Woodlands Project, Game Conservancy, Fordingbridge, Hampshire, SP6 1EF

Jane Smart, Conservation Director, London Wildlife Trust, 80 York Way, London, N1 9AG

M.B. Thorns, Biology Department, Building 44, The University, Highfield, Southampton, Hampshire, SO9 5NH

J.M. Way, Ministry of Agriculture, Fisheries and Food, Nobel House, 17 Smith Square, London, SW1P 3HX

T.C.E. Wells, Institute of Terrestrial Ecology, Monks Wood Experimental Station, Abbots Ripton, Huntingdon, Cambridgeshire, PE17 2LS

P.J. Wilson, The Cereals and Gamebirds Research Project, The Game Conservancy, Fordingbridge, Hampshire, SP6 1EF

M.I.A. Woodburn, The Pheasants and Woodlands Project, Game Conservancy, Fordingbridge, Hampshire SP6 1EF

SECTION 1
Introduction

Interest in 'natural' vegetation for landscaping and aesthetic reasons is not a recent development. Landscape architects have been advocating naturalistic landscapes since the 18th century, while ecologists and naturalists have always favoured the most natural and diverse of habitats with their studies. But the essentially horticultural idea of reconstructing complete biological habitats or communities, warts and all, is very much a contemporary phenomenon.

Concern for the environment is one plausible explanation for the popular support now given to habitat 'creation' exercises. These days we are constantly reminded of the part that housing, industry and modern agriculture have played in dramatically reducing and fragmenting the natural and semi-natural environment. It is not surprising, therefore, that this awareness should have fostered the belief that habitat loss can to some extent be compensated for, and the environment be improved, by re-creating 'new' semi-natural areas which resemble the old.

The roots of habitat 'creation' practice lie precisely in those areas where greatest habitat loss has already occured – the built environment. The approach in the Netherlands, a country with strong horticultural traditions, exemplifies the retreat from formal landscaping solutions around new housing areas in favour of ecological or nature-parks, semi-natural vegetation, relaxed mowing regimes and multi-species plantings of trees and shrubs. In the inner city and on old mineral extraction sites, the emphasis too has shifted from more traditional, comprehensive restoration solutions towards minimal reclamation practices, with 'assisted' natural colonisation, and the use of pioneer species, diverse tall herb communities and colourful flowers.

There are many reasons for habitat 'creation' exercises. The aim has generally been to *establish semi-natural vegetation communities which in some way resemble the semi-natural original*, although not necessarily to re-create their full diversity. Nature conservationists, planners, landscape architects, landscape managers and educationalists all have widely differing views on how far this process should be taken. Such mixed objectives include:

- creating visually attractive vegetation
- providing educational and possibly scientific interest
- safeguarding rare species or scarce ecological communities, and
- constructing low maintenance landscapes.

In each case the starting point is the *habitat stereotype* – a search image of the type of community to be reproduced. The closer that this corresponds to the original, the harder and more complicated it will be to achieve. In this age of technology, the idea of assembling an exact vegetation or habitat replica has instant appeal, and presents an almost irresistible ecological challenge. However there are serious dangers, both technical and ethical, in pressing habitat reconstruction to this logical extreme. Abuses can quickly arise if the practioners are landscapers or industrial developers with something to prove. For example it is all too easy for the intrinsically more interesting, rare and attractive species of semi-natural communities to themselves become the real targets of habitat reconstruction, and not the community as a whole.

The semantics of habitat reconstruction are confused and incautious. In practice, it is most unusual for literal attempts to be made at whole habitat reconstruction: most restoration work is done at the level of the single plant community, and often this fails completely to reproduce either the site conditions or the integration of communities found in nature. Secondly, the term 're-creation' implies that we know exactly (when we do not) how the model habitat is constituted, or how it functions ecologically. 'Creation' is even more ambiguous if we consider that we can no more create habitats than we can their component plants and animals. Compromise terms such as habitat 'reconstruction' or 'restoration' are less assertive, and are better at conveying the suggestion that the reassembly of an original habitat model is at best an imperfect and partially unobtainable objective.

The aim of this book is to discuss the assumptions and ecological principles behind habitat reconstruction, and to offer practical experiences of such techniques. These can be summarised as (a) habitat *creation*, in this context used to indicate the *a priori* construction of interesting and attractive ecological communities; (b) habitat *transplantation*, where an original habitat is moved from a donor to a receptor site; and (c) habitat *enhancement* or *diversification*, in which attempts are made to maximise the ecological potential of existing but degraded or impoverished habitats.

Traditionally, the opportunities for habitat reconstruction have been perceived as the repair of land damaged by mineral extraction, or as landscaping following urban renewal. Nature conservation, too, has enhanced habitats, for example in the re-introduction of rare or locally extinct species to reserves. But there is much scope beyond these applications, especially with new possibilities for land use reappraisal being offered by changing countryside policies. The implications of large areas of agricultural land being withdrawn to reduce farming surpluses, together with less intensive farming and forestry practices, have yet to be fully considered. Similarly, in the city and on the rural urban fringe there are equally exciting possibilities for innovative landscaping.

Unfortunately habitat reconstruction is a poorly researched subject which relies more on practical demonstration than science. To be successful, it depends critically upon an understanding of the ecological principles governing the habitat to be reproduced, the characteristics of the construction site, the individual species to be introduced, and their mutual interaction. An attempt has been made here to address these basic points, while drawing attention to the range of habitat manipulation projects ongoing in Britain and elsewhere.

SECTION 2

Philosophies of habitat reconstruction

Philosophies vary on the extent to which habitat reconstruction should anticipate, copy or influence nature. At one extreme is the horticulturalist's wish to create colourful, interesting and attractive habitats for people in the places where they live, while at the other the nature conservationist is committed to protecting good quality semi-natural habitats from all unnatural human influence. But as Baines points out, there is a place for both these positions and even for some reconciliation and compromise. In the urban environment, the cultivation of colourful meadows, informal woodland and ponds in place of traditional parkland has a distinctly educational and propaganda role. In these surroundings the horticultural ethic prevails: there is no pressing need to attempt the construction of a habitat facsimile as long as the result is attractive and stage-managed for effect. Moreover such habitats undeniably reinforce public appreciation and awareness of real countryside.

In contrast, the conservation ethic aims to maintain and defend the small areas of high quality habitat scattered amongst our predominantly cultivated and urbanised landscapes. No genetic contamination or artificial alteration of these areas can be tolerated. Although conservationists have occasionally promoted or re-introduced rare species, attempts to gild the lily are generally deplored. The most that can be allowed is to encourage the land adjacent to these high quality habitats to diversify through natural colonisation and appropriate management. This is the habitat 'duplication' solution offered by Newbold. He suggests a compromise for habitats which are heavily degraded and well beyond the pale, such as intensively farmed or urban areas remote from semi-natural areas. Here it should be possible to re-create more diverse habitats, but only as a supporting network to the genuine article, and provided that certain ground rules are observed. In perhaps the most frequently referenced article in this book, Wells *et al.* (1981) define several criteria for the inclusion of species in seed mixtures for species-rich grassland. These include the avoidance of rare species in favour of those which are uninvasive and widely distributed, so as to prevent possible contamination of genetic resources elsewhere. This is somewhat at odds with the horticultural view – for example Baines's advocation of mixing exotics with native species in the town 'for rapid, recognisable beauty' and with the policies of some amenity seeds suppliers.

A third, more dispassionate view is that of the scientist – Jordan's 'restoration ecologist', who finds the dialogue between habitat reconstruction and ecology a useful way of testing his ecological theories. As ecology is still a relatively young science, owing much to the observation of field practices, habitat reconstruction is a useful indicator for more systematic, experimental work. However, even this is controversial. Although the heuristic value of restoration to science is immense, scientists themselves do not always welcome this reversal of the accepted method of academic enquiry, as Jordan shows. Many ecologists are also conservationists, and take care not to become too closely associated with habitat manipulation.

Reference

Wells, T.C.E., Frost, A., Bell, S.A., 1981, *Creating attractive grasslands using native plant species*. Nature Conservancy Council, Shrewsbury.

1

Choices in habitat re-creation

J. C. Baines

Introduction

In the past forty years or so, the 'green and pleasant' British countryside has to a large extent been sterilised. The intensive use of large machines, the careless application of agrochemicals and the spreading influence of urban growth have all taken their toll. The countryside is still green: thanks to zero-grazing and coniferisation it is greener than ever before. The trouble is, the life has gone out of our landscapes. The habitat for wildlife has been degraded or destroyed, and as a result, most of our native wild species are in decline, and many are under serious threat.

The destruction has taken two forms. Some habitat has disappeared completely. Ponds and marshes have been filled in, streams and ditches have been culverted and whole woods have been grubbed out to make way for wheatfields, warehouses and bypasses. Many more habitats have been degraded but left standing. In the forty years since the war, for instance, almost all of our species-rich hay meadows have been destroyed. Most of them are still grassy fields, but the annual cycle of a late–summer harvest, autumn grazing of the aftermath and a spring flush of flowers and insects has been replaced with an intensive annual cycle of three or four cuts of fertilised grass monoculture. In the case of surviving woodland, much of it has been de-structured – losing shrubs, regenerating saplings, scrambling climbers and carpets of wild flowers to an introduced cash-crop of conifers. The scale of habitat loss has been spectacular (see Newbold in this volume).

There is now, at the eleventh hour, a growing realisation that loss of habitat means loss of heritage – and that we must urgently do something to halt the decline. While nothing can bring back the ecological complexity of a primary oak woodland, with its 8,000 years of accumulated history and its immensely complex species diversity, habitat re-creation can provide a valuable complement to the uncompromising preservation of our few surviving semi-natural areas. It is very important to recognise that there are two contrasting roles for habitat re-creation. There is the need to re-create the most convincing replicas of ancient and complex habitats possible, but there is also a very important role for simpler habitats which have less scientific credibility but more immediate popular appeal. At present these two objectives are sadly confused.

'Political' habitat creation

Nine out of ten people in Britain live in towns. Our towns are, relatively speaking, less polluted than the countryside, and there is far more green land available which is free from productivity pressures. After two generations of alienation from the natural world, the prime objective in creating new habitats in towns should be to stimulate a public passion for wildlife through increased familiarity with immediately appealing species. We spend over £1,000 million each year mowing the permanent pasture of our municipal open spaces. A simple change of management, adopting a hay meadow regime, and introducing immediately attractive wild flowers such as meadow buttercup, (*Ranunculus acris*), Ox-eye daisy (*Leucanthemum vulgare*) and field scabious (*Knautia arvensis*) is much more appropriate than the long struggle to produce the local diversity of a chalk downland. Even this simplified form of meadow will accommodate far more meadow butterflies, grasshoppers, voles and kestrels than the close-shaved alternative. A close-mown margin and properly planned grass paths will satisfy all but the most tidy-minded of critics; and if people can be involved in hay-making parties and the harvest is used productively, perhaps as fodder for the pets' corner, a whole new constituency of meadowphiles will emerge – eager to support the campaign for conservation in the countryside. This kind of 'political' meadow can be achieved using existing parks' machinery, can accommodate the restrictions of the local authority bonus schedules and should strengthen the case for more, rather than less skill amongst municipal managers.

A similar, simplistic level of habitat development is equally applicable to new woodland. At present there are many public open spaces containing groups of mature trees – some of them woodland relics which pre-date the present park. Normally these trees stand in a desert of shaded and close-mown grass. The leaves are swept away and burned each autumn, and no dead branches are allowed to remain on the trees. A withdrawal of the mowing machines, spreading of extra autumn leaves, swept from playing fields, paths and roads, and perhaps the installation of nest-boxes, would stimulate more bird-song almost instantly. Once trespass has been discouraged, through the change from close-mown grass, it may be possible to introduce a new shrub layer. Of course there should be a predominance of native species – hazel (*Corylus avellana*), privet (*Ligustrum vulgaris*), bramble (*Rubus fruticosus* agg.) and dog rose (*Rosa canina*) for example, but public interest should also be courted by the inclusion of some non-aggressive aliens – deciduous azaleas for their spring perfume for example, or *Amelanchier canadensis* for its glorious spring blossom and colourful autumn foliage.

The parks' nursery is probably already growing polyanthus for the town hall window-boxes. These same horticultural skills can be used to propagate primroses (*Primula vulgaris*), pink campion (*Silene dioica*), foxglove (*Digitalis purpurea*) and lesser celandine (*Ranunculus ficaria*) for the woodland ground flora. Again, for rapid, recognisable beauty, boost the native flora with daffodil bulbs (*Narcissus* spp.) and snowdrops (*Galanthus nivalis*), leopardsbane (*Doronicum pardalianches*) and Lenten rose (*Helleborus orientalis*).

The prime function of these political habitats is one of education and propaganda. It is important to stage-manage the visitors' experience of the wildlife. A well-stocked feeding station, preferably within view of a tearoom window or at least a park bench, will bring the wildlife to the people. Seeing is believing. The best time to sell the difficult idea of the need for dead trees in woodland is when the 'public' are enjoying the sight of a feeding woodpecker, and in stage-managed woodland, the bird-table and bags of peanuts are as important as the trees themselves.

Political habitats do present their own peculiar management problems. It is relatively easy, for example, to increase the habitat diversity of a park lake. The concrete edges and steep shelving sides are hostile to wildlife, and by creating shallows, introducing safe, secure nesting islands, and encouraging a fringe of emergent vegetation, the quantity and variety of wildlife can be rapidly enriched. However, the popularity of that resulting wildlife itself creates conflicts. People will visit to feed the ducks. Very quickly the artificial feeding will attract a surplus of wildfowl, which will then turn their attention to more natural food and graze the emergent water-edge habitat out of existence. Restricted feeding, and frequent culling of aggressive wildfowl species in favour of diversity are likely to prove very unpopular with the public – a real challenge to the educational role of the political habitat.

'Ecological' habitat creation

Political habitats can be re-created anywhere – the more public the location, the better: birch forests on ripped-up tarmac playgrounds, poppy fields on redundant allotments, dragonfly ponds in school courtyards; it is the easy access for people that matters. 'Ecological' habitats are different. Here every effort must be made to re-create the most comprehensive replica of the genuine article, and attention to location is of paramount importance. Access to specialist skills, and long-term security of tenure are also important. For the foreseeable future, this kind of habitat re-creation should be associated as intimately as possible with the surviving rich habitats that constitute our nature reserves and sites of special scientific interest (see Newbold in this volume). This policy offers at least four advantages. Firstly, by re-creating a new meadow in the next field to our existing species-rich one, for instance, it will be possible to use nothing but the local population of plants and animals. Flower and grass seeds can be harvested from the established site for introduction to the new one, without fear of adulterating the indigenous gene pool. Secondly, site conditions are more likely to be suitable for the task in hand, with the appropriate soil type and local climate. Thirdly, there will generally be an established pool of knowledge about the site, and very often skilled people available who are already involved with the established site and able to expand their influence to the sensitive development of the new one. Finally, while it is relatively simple to establish new colonies of the larger plants, it is far less simple to achieve successful recolonisation by the smaller species. If the new site is adjacent to an established complex habitat, then there is a greater likelihood of casual colonisation by the more mobile species, and such supplementary techniques as the transfer of leaf litter or the transplanting of turf communities can be undertaken very easily.

Discussion

The development of passable copies of interesting habitats can be achieved very rapidly. A sheet of poppies (*Papaver* spp.) and corn marigolds (*Chrysanthemum segetum*) can be in bloom within two months from a standing start. Dr Miriam Rothschild has established almost one hundred species of wild flowers in a new meadow in less than ten years replacing a former rye grass ley. The new woodlands of Warrington New Town, despite their lack of maturity, are nevertheless alive with bird-song on land which was a hostile munitions factory less than fifteen years ago. The ecological subleties of our long-established habitats will take much, much longer to achieve, and there is a desperate need to record their progress, and to adjust and adapt in the light of new experience. If the County Naturalists' Trusts, the Royal Society for the Protection of Birds, British Butterfly Conservation Society, The Woodland Trust, Game Conservancy, British Association for Shooting and Conservation, The Nature Conservancy Council, National Parks and relevant private landowners all adopted a policy of employing creative conservation, on land immediately adjacent to established nature reserves, then many species at present in danger would be given a new lease of life. The newly-declared policy of agricultural land release, or set-aside, and the relaxation of intensive farming practices, offer a wonderful opportunity to put back the clock, at least with the less historic habitats, most notably arable communities, wetlands and meadows. If that kind of policy is to be given the political backing it needs, there must be much more substantial public support, and that can best be solicited in towns, where most people live. Political habitats, then, are vital both to the survival of our relic habitats, and to the development of more creditable and scientifically sound 'ecological' habitats too.

2

Semi-natural habitats or habitat re-creation: conflict or partnership?

C. Newbold

Introduction

Extensification or 'set aside' in agriculture offers unparalleled opportunities for nature conservation. The Nature Conservancy Council's priority concern, faced with such potential changes in agricultural policy, is to conserve all remaining semi-natural habitats not designated as sites of Special Scientific Interest (SSSIs) and good wildlife habitat found in the British Isles. Habitat re-creation will also have a role and whilst of great value to wildlife, it must be placed in context. It is no substitute for conserving the semi-natural habitat. Guidance manuals on re-creation have been mis-read and have generated the misconception that the re-creation of habitats is an easy option which can faithfully duplicate the semi-natural. This is not so.

A strategy can be put forward which uses as an example those policy changes affecting agriculture. It does not consider urban developments or forestry policy apart from farm woodlands, although features of the 'set aside' strategy could be relevant to both these sectors of land use. The strategy considers:

- the retention of all existing semi-natural habitat;
- the 'duplication' of communities as a 'passive' form of re-creation, by setting aside land around our SSSIs and National Nature Reserves (NNRs); and
- the re-creation of appropriate habitats related to biogeographical regions.

It is then possible to discuss the value of islands of re-created habitat in a network of conservation field margins intermittently supported on a framework of SSSIs and NNRs.

Destruction of semi-natural and good wildlife habitat

Developments in farming and forestry since 1949 have caused the destruction of large areas of semi-natural habitats in Britain. These include the loss of:

95 per cent of lowland unimproved neutral grassland including herb-rich hay meadows;

80 per cent of lowland sheep walks on chalks and limestone;
40 per cent of lowland acidic heaths;
30–50 per cent of ancient lowland woods;
20 per cent of the hedgerow length;
50 per cent of lowland fens valley and basin mires;
60 per cent of lowland raised mires;
30 per cent of upland unimproved grassland, heaths and blanket bogs.

Not all of this is the effect of agriculture; commercial forestry is another important factor and building development has played some part. But the major cause of loss this century has been intensive agriculture (Anon 1984).

Agricultural policy

The Nature Conservancy Council (NCC) considers that any significant change in agricultural policy which is aimed at reducing the area or intensity of management of crops should take account of the scale of previous semi-natural habitat losses. Protecting the remaining areas of wildlife habitat and re-establishing new habitats should be the key elements of this policy change, which should be consciously designed so as to:

- protect existing areas of semi-natural vegetation from any further loss or deterioration in quality;
- similarly protect and where possible enhance existing good wildlife habitat;
- encourage the re-creation of wildlife habitats in areas where they have disappeared, particularly in areas subject to intensive farming.

Naturalness and semi-naturalness

In Britain there are few truly natural habitats, in the sense of those unmodified by man, and the main reservoir of wildlife lies in the 'semi-natural' areas. This term is applied to plant communities which owe their character to some degree of human intervention, but remain composed of native species and have structural features corresponding to those of natural types. Agriculture, as the predominant land use in Great Britain, exerts a strong influence over the development and survival of such communities. In the lowlands, the main semi-natural areas are now relatively small and fragmented; they include ancient woodland, ancient hedgerow, unimproved grassland, undisturbed fen, unreclaimed saltmarsh and sand dunes. To the west and north of the country, where agricultural use is generally less intensive, considerable expanses of semi-natural vegetation still remain on the moors, mountains and peatlands. Lowland and upland areas are in part protected as NNRs or SSSIs, but the remainder must be safeguarded through general countryside policies.

Good wildlife habitat and artificial habitat

Plant communities which include some non-indigenous species but still retain a large semi-natural component can be termed good wildlife habitat. Examples are recent broadleaved plantations, recent hedgerow, and partly-improved grassland which retains elements of the original plant communities. Artificial habitats are those composed mainly of non-indigenous species – arable fields, grass leys, conifer plantations, etc. All these definitions are to some extent arbitrary classifications. In practice the range from 'natural' to 'artificial' is a continuum; and nature conservation is especially concerned with the upper part of the range.

Protecting semi-natural and good wildlife habitat

Generally speaking, these habitats form the least productive part of any farm, so they are unlikely to be the direct focus of agricultural extensification measures. However, any restructuring of the farm business consequent upon entering an extensification scheme is potentially hazardous, (for example, reducing the cereal area on a mixed farm may encourage the farmer to build up the livestock elements of the business and bring semi-natural grassland into more intensive use). This could be guarded against in an extensification scheme by a simple farm development plan which identifies the areas of semi-natural, and good wildlife habitat on which intensive management techniques should be avoided (see Way, this volume).

An important principle consequent on the introduction of an extensification scheme into Britain should be a re-examination of all other agricultural schemes to ensure that they are not damaging to nature conservation interests. There is no justification for any further destruction of semi-natural and good wildlife habitat for the purposes of agricultural production, and grant-aid for activities which can be damaging to wildlife habitats (land drainage grants being a particular case in point) should be phased out. These should be replaced by new schemes which provide farmers with incentives to retain wildlife habitats and manage them sympathetically. This could involve an extension of some of the mechanisms now available to farmers in Environmentally Sensitive Areas (ESAs).

Re-establishing wildlife habitat

The potential for this in the farmed countryside is very large whether the areas withdrawn are whole farms, single fields or strips alongside field margins, and opportunities should now be taken to restore in some measure what has been lost over the years. But it must be emphasised that creating new habitat is not a substitute for conserving what still remains; the new habitats may never be as rich or ecologically varied as existing semi-natural areas. In addition, it is clearly more cost-effective to retain existing habitats than to create new ones. That said, the new habitats are valuable in their own right, and if created on a more considerable scale than hitherto, could contribute greatly to the diversity and visual amenity of the countryside.

Ecological strategies

The above are the potential policy mechanisms. The ecological strategies in set-aside or extensification schemes are now discussed. First of all, the terminology used by ecologists needs to be clarified.

Habitat creation, re-creation and duplication

Confusion may arise between those areas of land which are by definition created solely for amenity purposes and those which are re-created for ecological purposes. Some ecologists lose sight of the real meaning of the pioneering work by Wells *et al.* (1981) which resulted in the publication *Creating attractive grasslands using native plant species*. Essentially it was aimed at creating amenity grassland. It was not intended to re-create the semi-natural habitat but to provide visually attractive areas in biologically impoverished areas. Such creation techniques should be distinguished from habitat re-creation or duplication which defines the imperfect re-creation on 'degraded' land of communities which were recorded there in recent history. The distinction between the two is made more clear in Table 2.1.

The semi-natural habitat

The conservation of the semi-natural habitat is the NCC's priority concern. It is not only concerned with conserving the semi-natural habit in its NNRs or SSSIs but also, through the potential opportunities afforded by extensification or set-aside policies, in conserving all remaining fragments of semi-natural habitat within a farm or any other land use. The second strategy considers setting aside land around or by such fragments, and around existing NNRs or SSSIs, either as whole fields or strips of land. Through appropriate management this will lead to habitat duplication (see Baines, this volume).

Habitat duplication

Specialised techniques could be used for re-establishing a range of habitats in areas where they have disappeared. In general, in such areas the NCC consider that the duplication of habitats will more accurately re-create habitats. This is our second strategy level. Habitat duplication in one sense is a passive form of habitat re-creation. Seed sources and animals come from nearby or more preferably adjacent semi-natural habitats.

Woodland can be re-created by abandonment of land around an area. Clearly rabbits and domestic grazing animals will have to be excluded from the site. The abandonment of a 100 metre strip of land around Wytham Wood in Oxfordshire has begun to duplicate woodland after a period of 25 years.

Other habitats will have to be managed. Adjacent to Wytham Wood, Gibson *et al.* (1987) have shown that grazing is essential in duplicating calcareous grassland. In

Table 2.1. Habitat creation, re-creation and habitat duplication strategies (After Hopkins, J.J., pers. comm. 1987)

Objective of habitat creation re-creation or duplication	Time perspective	Establishment cost	Management expertise	Short term visual amenity	Potential long-term scientific interest	Between site variation	Optimal location	Comments
Amenity [ie visual, educational, "wildlife experience"]	Short	High	High (Prone to failure)	Moderate/ High	Reduced/ Low	Low	Biologically impoverished landscapes Centres of population	Habitat creation
Nature Conservation (Ratcliffe 1977)	Long	Low	Low	Low?	High	High	Biologically rich landscapes Stressed environments	Habitat duplication or Habitat re-creation

this experiment a 10ha, 21 year old arable field was last cultivated in 1981. Grazing treatment started in 1985 after crops were removed annually to impoverish the nutrient store within what are skeletal soils covering Jurassic corallian limestone. By the end of 1986, 43 vascular plant species had colonised the treatments from a potential source of 75 species restricted to patches of old grassland within 2km of the site. Lack of grazing did not affect the chance of species arriving but their establishment was better in the grazed areas. By the end of 1986 the area grazed both in spring and autumn had reached a state similar to that described by Cornish (1954) for ex-arable chalk grasslands. Although the species composition was not yet comparable to a mature calcareous grassland many of the component species had arrived and were more importantly increasing in contrast to the control areas.

Heathland creation management techniques are well researched but nutrient depletion to levels suitable for heathland re-invasion may be an important factor in the success of this form of habitat duplication (Marrs, 1986, and this volume). Wetland communities of fen and open water readily begin to duplicate themselves provided there is continuity of open water and the water is not polluted or enriched.

Time factors

Freshwaters are very dynamic systems, developing rapidly and generally conforming to the time scales envisaged in successional trends by Tansley (1939). If constrained naturally or otherwise, and managed sympathetically as an open water community, they will increase their number of species in time (Godwin 1923). The time perspective is briefly mentioned in Table 2.1. At the other end of the sere, communities develop more slowly: calcareous grassland within 5 years and woodland within 15–25 years. However, despite these examples some of our most species-rich communities will take decades or centuries to achieve near-perfect duplication, and for these the time factor cannot be compressed by habitat re-creation techniques.

Distance from colonisation source

Habitat duplication as a mechanism for enriching that countryside outside a site has one inherent drawback. Seed sources and animal communities may at best only move a few kilometres from the original site. Some species are known to travel great distances: *Orchis simia* was found in the Netherlands 200km from its nearest source (Willems 1982), but what is important here is how far a significant proportion of the community can travel. There is little information on this aspect of habitat duplication. Gibson *et al* (*loc. cit.*) showed that 57 per cent of a calcareous community had begun to duplicate itself, 5 years after the 'abandonment' of arable land, from seed sources at most 2kms distant. Land barriers of alien habitat can markedly affect re-colonisation rates. For example Godwin (1923) showed that there was a marked impoverishment in pond floras of a similar age where a land barrier of a few hundred metres prevented contact with the River Trent. These island biogeographic principles may apply similarly to other types of habitat.

Table 2.2. A general purpose mixture of grasses for field margin re-creation (After Wells, D. pers. comm. 1988)

	%
Agrostis capillaris	10
Cynosurus cristatus	20
Festuca ovina	15
Festuca rubra	35
(Tussock, not creeping form)	
Phleum pratense	10
Poa pratensis	10

Note: Sown at a rate of 25 kg/ha. Species and percentage composition of mixture can be varied.

Habitat re-creation

Habitat duplication cannot be relied on to enrich the countryside beyond semi-natural areas or those fragments of semi-natural habitats remaining in farmland. At present 'islands' of semi-natural habitat are conserved by the NCC, County Naturalists Trusts, the Royal Society for the Protection of Birds, and similar bodies. Where these 'islands' are within a few kilometres of each other, it is suggested that the corridor of streams and existing hedges is enhanced, in places re-created, by setting aside at least 6 metre margins alongside all field boundaries.

Where 'tracts' of arable or intensive grass ley distance the semi-natural habitats from one another then attempts should be made to re-create islands of appropriate habitat. Again existing or re-created corridor networks would link these 'islands' of re-created habitat with the semi-natural. In addition, the setting aside of whole fields could re-create a 'patchwork quilt' of appropriate habitats. Existing hedgerow field boundaries would be expanded to form, in time, a linear woodland 12 or more metres wide. Alternatively, arable strips could be sown with an appropriate mixture of grasses (see Table 2.2), creating corridors of grass edged by a hedgerow. In time, it is hoped that other species will naturally enter this base community of grasses. What cannot be relied on is the duplication of original communities by plant and animal colonisation or the duplication of plant communities arising from the original seed bank. Intensive arable farming practices using fertiliser and pesticide sprays seem to eliminate all but the arable weeds from the seed bank (Chancellor 1986, Roberts 1986).

The strategy of re-creating large islands of appropriate habitat several hundred hectares in size would, according to biogeographical theories, be beneficial. For example, 'islands' of 'chalk downland' could be re-created on the Wiltshire Downs. Similarly, new woodlands could be re-created under the recent Farm Woodlands Scheme. A planting strategy is being published by the NCC (Soutar, in press). Wetlands could be re-created within the Cambridgeshire and Lincolnshire fens by flooding old washlands at present kept dry, by pumping, for intensive arable farming. Other opportunities could exist for enhancing streams and rivers canalised for drainage purposes. Their important corridors could be re-created by a variety of techniques (Newbold *et al.* 1983, Lewis and Williams 1984).

Coastal wetlands could be re-created by flooding poor arable land (see Hill, this volume) particularly where soils suffer from saline intrusion, old sea defence lines could be refurbished inland from the present sea walls. Water authority engineers look on this as one alternative for a cheaper flood alleviation strategy, bearing in mind farming extensification schemes and the prospect of a rise in sea-level. Habitat re-creation placed in this context could enhance or re-create habitats to the benefit of the countryside and perhaps to the benefit and long-term survival of the fauna and flora within existing semi-natural habitats!

Conclusions

Habitat re-creation is, at best, an imperfect method for duplicating community structure. The Nature Conservancy Council consider that it is more appropriate to conserve as a priority all remaining semi-natural habitats in Britain. Strategies attempting to duplicate, albeit imperfectly these semi-natural habitats, could be achieved by setting aside land around such areas. This is a passive form of habitat re-creation. In some areas of Britain there are few semi-natural habitats and they are separated by distances of more than 25km, while 2km is considered at present the maximum distance for habitat duplication. Habitat re-creation of appropriate former habitats is seen, in these circumstances, as the only way of enriching the countryside. Islands of re-created habitats within a network of conservation field margins and river corridors are envisaged, being supported on a framework of SSSIs and NNRs. Habitat creation suggests something 'new' and is considered appropriate only for amenity purposes. It is seen as being a different objective from those habitats which have been duplicated or re-created for nature conservation reasons.

References

Anon, 1984, *Nature Conservation in Great Britain*. Nature Conservancy Council, Peterborough.

Chancellor, R. J., 1986, 'Decline of arable weed seeds during 20 years in soil under grass and the periodicity of seedling emergence after cultivation.' *Journal of Applied Ecology*, **23**, 631–637.

Cornish, N. W., 1954, 'The origin and structure of the grassland types of the central North Downs.' *Journal of Ecology*, **42**, 359–374.

Gibson, C.W.D., Watt, T.A., Brown, V.K, 1987, 'The use of sheep grazing to recreate species-rich grassland from abandoned arable land.' *Biological Conservation*, **42**, 165–183.

Godwin, H, 1923, 'Dispersal of pond floras.' *Journal of Ecology*, **11**, 160–164.

Lewis, G., Williams, G., 1984, *Rivers and wildlife handbook. A guide to practices which further the conservation of wildlife on rivers*. Royal Society for the Protection of Birds, Sandy.

Marrs, R. H., 1986, 'Techniques for reducing soil fertility for nature conservation. A review in relation to research at Rogers Heath, Suffolk, England.' *Biological Conservation*, **34**, 307–332.

Newbold, C., Purseglove, J., Holmes, N.T.H., 1983, *Nature conservation and river engineering*. Nature Conservancy Council, Peterborough.

Ratcliffe, D.A., 1977, *A Nature Conservation Review, Volume I*. Cambridge University Press, Cambridge.

Roberts, H.A., 1986, 'Seed persistence in soil and seasonal emergence in plant species from different habitats.' *Journal of Applied Ecology*, **23**, 639–656.

Tansley, A.G., 1939, *The British Islands and their vegetation*. Cambridge University Press, Cambridge.

Wells, T.C.E., Bell, S., Frost, A., 1981, *Creating attractive grasslands using native plant species*. Nature Conservancy Council, Shrewsbury.

Willems, J. H., 1982. 'Establishment and development of a population of *Orchis simia* Lamk in the Netherlands, 1972–1981.' *New Phytologist*, **91**, 757–765.

3

Just a few oddball species: restoration practice and ecological theory

William R. Jordan III with *Steven Packard*

Introduction: practice and theory

Ecological restoration is generally regarded as a form of 'applied ecology', which depends ultimately on horticultural common sense, or on insights into the nature of the communities and ecosystems. Thus, the implication is that academic ecology is the real source of information, ideas and insights. And in the conversation between theorist and practitioner, it is to be expected that the ecologist, the theorist, will do most of the talking while the practitioner or restorationist does most of the listening.

On the other hand, the history of science makes it clear that science owes a great deal to 'practice'. An awareness of this permeates the writing of Francis Bacon, that early apologist of modern science. And of course Galileo, a pioneer in this tradition, relied heavily on the insights of craftsmen in his own work. He might almost have been addressing us directly when, in his *Two New Sciences*, he comments on his relationship with the first-rank artisans in the Venetian arsenal: 'Conference with them', he wrote, 'has often helped me in the investigations of certain effects, including not only those which are striking, but also those that are recondite and almost incredible.'

This from one of the founders of the modern scientific tradition. But does the same thing hold true now, nearly 400 years later? Clearly, in sciences such as Galileo's own physics and chemistry it is probably true that scientists working at the frontiers of their discipline have little to learn from practitioners in the shop and the market-place.

This, however, is not so much the case in ecology, which deals with more complex phenomena, which has a much shorter history and which has only recently begun to move beyond description to become experimental and predictive. At this early stage we might expect to find that the restoration and management of communities and ecosystems has a great deal to offer theory, just as the art of artillery had a great deal to offer physics in the time of Galileo. To the extent this is true, we might look for a relationship between theory and practice in which the two are seen as intellectual partners, at least a prelude to possibly more systematic, rigorously controlled experimentation, and even as a paradigm for organising basic research in ecology.

What I want to do here, then, is to turn the discussion of restoration and its relationship with ecology deliberately around and to discuss the value of practice to theory. I want to explore the idea that ecology has been, and at this stage in its history continues to be, to a considerable extent nourished from below, as it were, like a fen. If this is not widely recognised, it may well be that much of the history of recent ecology, including its debt to practical work, is simply left out of the official accounts, and that it is therefore grossly underappreciated – a kind of hidden or unacknowledged source from which ecology draws a considerable share of its sustenance and inspiration.

Fortunately, a number of researchers in this area have recently drawn attention to the value of practice. Two of the clearest statements on this matter have been from distinguished British scientists. Thus John Harper has pointed out the debt that ecology owes to agriculture, precisely because it involves synthesis or the test of performance in the attempt to construct and control ecosystems. Furthermore, this same synthetic approach to research provides a paradigm for the development of ecology as an experimental science (Harper, 1982). In a similar vein Anthony Bradshaw has argued that restoration is nothing less than the 'acid test' of ecological theory, thus suggesting that it belongs not at the margins but at the very centre of ecological thinking (Bradshaw, 1983). And recently four ecological managers, writing under a pseudonym, have suggested that most management is best regarded not as a form of applied ecology at all, but as an experiment or at least a prelude to experimentation – that is, as a test of basic ideas (McNab, 1983).

It is important to be very clear here about one thing. There are at least two kinds of experiments – those designed to find out *whether* something works, and those designed to find out *how* something works. The first kind, which we might call 'technical' experiments, are trying something out – some planting procedure or fertilising technique – to find out whether it works or not. This is a perfectly natural and desirable thing. But it is not what Harper and Bradshaw and 'McNab' are talking about. They are concerned with manipulations of the community or ecosystem that lead toward fundamental insights into *why* various techniques work as they do and, ultimately, into how the community or ecosystem itself works. They are talking, in other words, about the fundamental heuristic value of restoration, and its value as a technique for basic research.

What I would like to do here is to offer some thoughts of my own on this subject, thoughts occasioned for the most part by my work at the University of Wisconsin-Madison Arboretum, where I have been in charge of publications and public education for the past eleven years. The Arboretum provides an ideal occasion for thinking along these lines because it has been doing ecological restoration for more than 50 years and represents what we believe to be the oldest and most ambitious such project ever undertaken in an academic setting.

Fifty years of restoration at Madison Arboretum

At the dedication of the Arboretum in 1934, Aldo Leopold himself expressed the view that the restoration of the ecological communities planned for the Arboretum would serve as a 'first step' towards exhaustive study of the communities. The idea, in other words, was first to create the communities, then to study them.

Of course there were difficulties with this view, as Leopold knew as well as or better than anyone. For one thing, you do not generally start studying a thing by making it, you construct it in order to show that you have come to understand it. And what is true of pocket watches or diesel engines or musical instruments is also true of ecological communities: despite their considerable capacity for self-repair, you really can't expect to make one until you have a pretty clear idea of how it works.

Nevertheless, though Leopold himself acknowledged this difficulty on numerous occasions, there had always been a tendency at the Arboretum to take the process of restoration more or less for granted. By the time I arrived in 1977 the weaknesses of this point of view were evident. The Arboretum then had an extensive collection of restored and partly restored communities. Some were reasonably authentic in terms of floristic composition, but arguably none were really accurate replicas of the natural communities they were meant to represent.

The very practical result of this was that the value of the Arboretum as a facility for basic ecological research was quite limited. Why do your research on a prairie that is small, is infested with exotic plants and has missing species, or artificial distributions of species, when you can get in the car, drive an hour or two and work in a natural prairie that you won't have to apologise for when you come to write up your results?

It was in worrying about these defects in our communities that I began to think that perhaps we were missing the whole point. Many if not all of them were the result of some aspect of the restoration effort that had failed, and therefore constituted a test of the idea on which the effort was based. Furthermore, it seemed likely that a lot of interesting and also 'unnatural' situations must have been set up more or less willy-nilly in the course of the restoration effort. These might be expected to tell us something about the communities that would not be revealed by communities growing naturally and comfortably on their own.

In fact, when I began thinking about this I realised that nearly all the best research that had come out of these communities over the years had been done not in spite of their ecological deficiencies but actually *because* of them. The obvious example here was the classic work on the ecological effects of fire that had been carried out as a direct result of the early efforts to restore tallgrass prairie. When the prairie restoration effort began at the Arboretum in 1935 it was known from historical records that fires had been frequent events on the midwestern prairies for more than a century, although their significance was not clear. As a result, the early prairie restoration efforts at the Arboretum were undertaken without fire, and it was not until several years later, when these efforts were proving unsuccessful, that a series of experimental burns was carried out.

The results of this work, published by John Curtis and Max Partch in 1948, made it clear that fire did indeed play an important role in favouring prairie species over the exotic Kentucky bluegrass. This short paper turned out to be a minor classic, and a significant contribution to the rediscovery of the ecological role of fire that was just then getting under way in the United States. It provided a real insight into the ecology of prairies, and helped the resolution of a debate that went back more than a century by the test of *performance*, synthesis, restoration.

Other projects that had been carried out at the Arboretum suggested a similar pattern of insights achieved as a direct result of an attempt to restore or re-create an ecological community. In setting out to restore a community, you run into trouble and the trouble points to some things you don't know about. So you hunt for ways to fix them, and in the process you come up with new ideas, which you test by introducing them into the system. The result is a series of new ideas constantly being tested not merely by correlation, but by the more stringent test of performance (Bradshaw, 1983).

It was at this point that we coined the term 'restoration (or synthetic) ecology' to refer to restoration efforts undertaken specifically to test an idea about the structure, dynamics or functioning of the community being restored (Aber and Jordan, 1985). Notice that this notion immediately suggests the basis for an intimate, two-way relationship between ecological theory and ecological practice. For one thing, it draws attention to the fact that ecologists routinely use synthetic techniques to study phenomena such as population dynamics, species interactions and nutrient cycling. Furthermore, it recognises that ecologists have also learned a great deal directly or indirectly from actual restoration and management efforts carried out in the field, even if these are often undertaken outside the realm of the formal heuristic procedure that scientists generally account for in reporting their results. The point is that, to the extent these other insights are valid, it might be well to extend the synthetic work of the laboratory into the field and ultimately *to accept synthesis and restoration as a principle or paradigm for carrying out, organising and criticising ecological research.*

This line of reasoning underwent considerable development between about 1977 and 1983. Recently a book on the subject has been published by Cambridge University Press (Jordan, Gilpin and Aber, 1987). I think the results clearly illustrate the value of restoration as a technique for investigating a wide range of ecological phenomena from community-oriented matters such as niche definition, species interactions and minimum viable populations, to ecosystem-oriented questions such as the handling of water and nutrients.

Restoration ecology on the Chicago savannas

For the remainder of this chapter I would like to describe some work recently carried out in the prairie and savanna communities of our area that seems to me a perfect, textbook example of what we have been calling 'restoration ecology'. This work has been carried out by Steve Packard, a manager who works with The Nature Conservancy in its Chicago office, and whose immediate interests and responsibilities, are, therefore, primarily 'practical' rather than theoretical. Nevertheless, in his effort to restore examples of the savanna community that was once abundant in the Midwest, Steve has obtained information that suggests the need for a major revision of current ideas about the composition and dynamics of these communities.

Steve has already described this work in some detail (Packard, 1988a, 1988b). Here, with his help, I will simply offer a short summary of the state of the art of grassland restoration in the American Midwest, while also illustrating the value of doing ecology this way, as it were from the bottom up.

Oak savanna, previously called oak barrens, oak openings or prairie groves, once covered millions of hectares in the American Midwest. They were especially abundant in the zone of transition between the eastern forests and the open prairies, and are often regarded as the characteristic pre-settlement community of this area, which includes southern Wisconsin and much of northern Illinois. Early settlers' accounts abound with enthusiastic descriptions of the congenial, park-like savanna landscapes with their shady groves that must have seemed a relief from the unrelenting openness of the prairies. Thus the French explorer Jean Francois Buisson wrote in 1699, 'The banks of this river (the Des Plaines in northeastern Illinois) are very agreeable; they consist of prairies bounded by small hills and very fine thickets' (Angle, 1968). And two centuries later John Muir, recalling his boyhood as a Scottish immigrant in southern Wisconsin, wrote of his family's 'charming hut in the sunny woods, overlooking a flowery glacial meadow rimmed with water-lilies', in the oak openings of southern Wisconsin (Muir, 1965).

Muir's generation was the last to see the savannas in their pre-settlement condition. Those that were not destroyed outright by ploughing rapidly grew up to brush in the absence of the frequent fires that had shaped and maintained them. In fact, by the 1980s savannas had become so rare, and those that remained were in such poor condition, that there was really little interest in their conservation. Steve Packard got involved in his work with savannas indirectly as a result of ongoing efforts to conserve prairies in the Chicago area both by identifying and preserving existing remnants, and, where possible, through restoration efforts that generally depended on reintroducing fire.

One feature of the prairies he was working on attracted Steve's attention, however, and got him thinking about the old savannas of pre-settlement times. Frequently the prairies were adjacent to old oak groves which dated back to pre-settlement times. They were recognised as degraded remnants of the early savanna, but invariably they had grown up into thickets of brush, typically buckthorn (*Rhamnus cathartica*), and other exotics that were brought in by birds and were able to flourish in the shade of the old oaks.

So why not attempt to restore the savannas as well as the prairies? What Steve now calls the rediscovery of the oak savanna began when he started trying to do this. Since he wanted a savanna, and since there were no healthy savannas around to use as a model, he based his work on the best descriptions that were available. In particular he relied heavily on John Curtis's *Vegetation of Wisconsin* (1959), a comprehensive book based on over a decade of descriptive research, and a standard reference for ecologists and managers concerned with any of the historic ecological communities of the upper Midwest.

Curtis's book offered a fairly detailed description of the oak savanna with extensive species lists, but basically he regarded the savannas as a community transitional between forest and prairie and lacking any species that were really distinctive of the savanna itself. So far as the relationship between prairie and savanna vegetation was concerned, the gist of what he had to say was summed up in two sentences: 'A number of species which are widespread on the prairies actually reach a higher level of importance in the openings, but no species are known which are confined to them. The closest approach would be *Besseya bullii*, whose total range is confined to the area of the Middlewest savannas and which is much more common in the openings than in the prairies' (p. 332).

Following this blueprint, Steve did the obvious thing. He began trying to expand his prairies into the adjacent oak groves, first by encouraging fire from his prairie burns to blast into the grove understorey, then, when he found the fires often would not carry through the brush, by cutting. The next step was to wait to see whether the original vegetation would return from a seed bank that might have been left behind by a vanished prairie vegetation. When this did not occur, Steve did the next obvious thing: he started planting seeds of prairie plants under the oaks.

The results were both surprising and disappointing. A few of the planted prairie species, notably Kalm's brome (*Bromus kalmii*), yellow coneflower (*Ratibida pinnata*) and wild bergamot (*Monarda fistulosa*) grew reasonably well in the more open areas, at least at first. But most species did not. Furthermore, forest species existing on the site showed little inclination to move into the semi-shaded areas under the oaks. In these areas the plantings remained thin, produced little fuel, burned poorly and gradually began filling in with brush and weeds such as Canada and bull thistle, dandelion, briars (*Rubus* spp.) and burdock (*Arctium minus*).

In short, the native plants that Curtis had indicated made up the savanna understorey were refusing to grow there. Apart from the handful of exotics that readily invaded these sites, nothing grew that could be expected to carry the intense fires that were known to have burned through the old savannas and doubtless played a key role in their development. Faithfully following Curtis's blueprint, Steve had encountered not just a problem in restoration and management, but what looked like a defect in the existing description of the savanna.

What was going on here? One possibility was that the prairie plants refused to establish under the oaks because they had to come first in the development of a savanna. In other words, they wouldn't establish there, but, once established on the open prairie, could persist more or less indefinitely as trees moved in and reached maturity, so that in dynamics as well as in composition, the savanna really was a prairie with trees. Though perhaps possible, this seemed improbable, especially considering the longevity of oaks and the sturdy refusal of most prairie species to make any kind of show on Steve's savanna sites.

A second possibility suggested itself, however, when Steve noticed among the burdock, thistle and buckthorn that was invading the cleared areas under his oaks a few natives that were growing well. These included silky wild rye (*Elymus villosus*), cream gentian (*Gentiana flavida*), yellow pimpernel (*Taenidia integerrima*) and a handful of others, including three grasses that were extremely rare in the area.

Though these were all native species, none had been included in the mix of seeds that Steve and the others had been planting under the oaks. All of them had come in on their own. Puzzled, Steve began looking for information about these miscellaneous, 'oddball' natives that were growing where most prairie plants refused to establish. A key source was the compendious *Plants of the Chicago Region* (Floyd Swink and Gerould Wilhelm, 1979). What he found there corroborated his own knowledge of these plants. Some were quite rare in the modern landscape, while others were not and even behaved as weeds which were 'apparently never part of any stable native community'. Many of the 'oddball' species Steve had noticed, were listed as associates of each other, or as having third-party associates in common.

By the summer of 1985 Steve had begun to compile a tentative list of what he was beginning to think of as 'savanna species', and he spent much of the summer prowling golf course roughs, railway rights-of-way and cemetery corners looking for them. He began to find them, and with them other species as well, some so rare that he was unfamiliar with them and had to key them out. He also began collecting seed, and, thinking that it might make sense to encourage the invasion of his oak groves by these presumptive savanna natives, that autumn he began planting them under the oaks.

Although Steve had discovered a list of plants that tended to be found together in shady edges and on old savanna sites, he could not relate this to any kind of comprehensive description of the pre-settlement savannas. That winter, however, while waiting for the results of his savanna planting experiments, he came across a list, compiled by a pioneer doctor named S. B. Mead, and published in a journal called *The Prairie Farmer* in 1846. Remarkably, Mead had not only listed the plants of Hancock County in westcentral Illinois, some 300km west of Chicago, but he also indicated the kinds of community in which they grew. Of the several hundred species listed, 108 were followed by a 'B', for 'barrens', a term often applied by settlers to the oak savannas. Decoded, these turned out to match Steve's list of presumptive savanna species remarkably closely.

And then the savanna began speaking for itself on the sites where Steve and his colleagues had sown their first 'savanna mixes' the autumn before. 'The spring of 1986 arrived on the North Branch savannas with amazing grace', Steve writes. 'Hundreds of thousands of tiny new green cotyledons throve where only black dirt and thistles had been. By the summer of 1987 the space under and near the oaks waved with countless blooming rare and uncommon grasses and flowers . . . An oak grassland community was unfolding, almost entirely from seed we'd held in our hands.'

A challenge for ecology

Now let me conclude with some remarks on the importance of Steve's work in the context of this discussion of restoration ecology. First of all, what he is finding has obvious implications for the conservation of the savanna. For one thing, it has provided the impetus for a fresh search for remnant savannas based on a new notion of what a remnant savanna might actually look like. As recently as a couple of years ago most managers in the area believed the savanna to have vanished for all practical purposes but now it has been possible to recognize – and to begin restoring – a number of degraded savannas. And of course there are now some fresh ideas about how to restore and manage these communities, what sorts of plants to bring in, how to encourage them to grow and how to use fire to maintain them.

To me, work like this, pointing toward the virtual rediscovery of a lost community, promising the literal revival of that community and its replacement in the landscape, is tremendously exciting and a powerful argument for the heuristic value of restoration. Nevertheless, a serious question remains. The new model for the savanna that is emerging from Steve's work is only as good as the information on which it is based. To what extent can we rely on it as a more accurate picture of the old savannas? To what extent can we trust it as a model for efforts to identify, restore and maintain savannas in the future? What relationship does this work have with the science of ecology?

To what extent is it possible not only to reconstruct ecological communities following existing, healthy models, but actually to revise and refine our descriptions of communities that no longer exist using restoration as a way of testing their ecological plausibility?

At the very least this work points toward serious deficiencies in the existing ideas about the savannas. It may not provide the precise, species-by-species, plant-by-plant and process-by-process descriptions that ecologists can provide and have come to expect when working with an existing community. But this work does clearly put the older descriptions to the test of performance, with results that strongly suggest that they may have some serious flaws. It may not be science, finished and complete. But at least it is a prelude to systematic, quantitative research, the kind of experience that provides the basis for the hunches and hypotheses that make real experiments possible and worthwhile.

More than that, it would suggest that Steve's work at least points the way toward a more accurate description of pattern and process on the old savannas. Again, it may not be complete in every detail. The evidence for it may be indirect. But at least it gives us one thing that the old descriptions did not, that is a community that seems to work the way the original savannas must have worked in a landscape strongly influenced by fire.

Perhaps it is fair to say that Steve's work provides the outline for a new model of the savanna community, a revision of Curtis's idea of the savanna as a prairie with trees. But of course this calls for further attention. What predictions does it make that can be tested by observations or by experiments?

One thing to do is to look at the model in detail and to see if the various parts really do fit together with a closeness and precision that suggests mutual adaptation into a functional unit. In fact, Steve has already begun to do this. He has found, for example, that the plants on his savanna list have reproductive strategies and life cycles that are not only quite different from those of prairie species, but which look like adaptations to the savanna community. Many of them, for example, propagate readily by seed, and have strategies for moving short distances in a relatively short time. Steve has pointed out that this is what would be expected of species that had to hold their own in the permanent 'edge' of the rapidly changing savanna community. This may also help explain the fact that many of the species are evergreen, a habit that is rare among prairie plants and distinct from the ephemeral habit by which many forest herbs have adapted to the brief early-season period of light on the forest floor. Might this reflect the need of the savanna plant to hang on to its leaves in order to take advantage of whatever light becomes available on an unpredictable schedule in the dappled shade under the oaks?

Clearly what is needed here is intensive study to explore these and other questions. Some of these may be physiological studies. I myself, as a one-time plant physiologist, would be tempted to study the patterns of photosynthesis in the evergreen plants, comparing them with patterns in prairie and forest plants to find out whether they do show a distinctive pattern that might suggest an adaptation to the economy of the savanna. Even more obvious would be detailed studies of the reproductive strategies and other autecological traits of the plants on Steve's 'savanna' list. Do they really reflect the need for rapid responses, resistance to relatively violent fires and so forth that Steve's work suggests? To answer these questions will take a great deal of work. Do Steve's observations and results merit this sort of attention? And will they attract the attention of ecologists who are in a position to do the necessary work?

So far, unfortunately, the answer to this question is not an encouraging one. The reaction of ecologists to Steve's work has for the most part ranged from indifference to hostility. Only recently have one or two practising, academic ecologists expressed an interest in following up some of the leads suggested by his work. And so we are faced with the question of what will become of his findings. Will they lead to the development of improved management techniques (for there is now little question that this will indeed occur) while being ignored by those in a position to build them into the body of ecological knowledge? Then will it emerge a decade from now in scholarly papers that do not acknowledge the humble origins of the ideas they convey? If so, it seems to me that an injustice will have been done and that ecology will have hindered its own development by its habit of failing to recognise this business of restoration and management as not just the acid test of its ideas, but as the very source of many of them as well.

References

Aber J.D., Jordan, W.R., 1985, 'Restoration ecology: an environmental middle ground', *Bioscience* **35** (7), 399

Angle, P., 1968 (ed.), *Prairie State: Impressions of Illinois 1673–1967 by Travelers and other Observers*, University of Chicago Press, Chicago.

Bradshaw, A. D., 1983, 'The reconstruction of ecosystems', *Journal of Applied Ecology*, **20**, 1–17.

Curtis, J. T., 1959, *The Vegetation of Wisconsin*, University of Wisconsin Press, Madison.

Curtis, J. T., Partch, M., 1948, 'Effect of fire on the competition between bluegrass and certain prairie plants', *The American Midland Naturalist*, **39**(2), 437–43.

Harper, J. L., 1982, 'After Description', in E. I. Newman (ed.), *The Plant Community as a Working Mechanism*, British Ecological Society special publication No. 1, Blackwell, Oxford.

Jordan, W. R. III, Gilpin, M. E., Aber, J. D., 1987, *Restoration Ecology: a Synthetic Approach to Ecological Research*, Cambridge University Press.

McNab, J., 1983, 'Wildlife management as scientific experimentation', *The Wildlife Society Bulletin*, **11**, 397–401.

Muir, J., 1965, *The Story of My Boyhood and Youth*, University of Wisconsin Press, Madison, p. 51.

Packard, S., 1985, 'Rediscovering the tallgrass savanna of Illinois', *Proceedings of the 10th Northern Illinois Prairie Workshop*, Forest Preserve District of Will County, Illinois, 93–98.

Packard, S., 1988b, 'Just a few oddball species', *Restoration & Management Notes*, **6**(1), 13–20.

Swink, F., Wilhelm, J., 1979, *Plants of the Chicago Region*, The Morton Arboretum, Lisle, IL.

SECTION 3
Ecological principles of habitat reconstruction

Habitat reconstruction demands a thorough understanding of the driving ecological factors. Fortunately ecologists are used to working at the community level, but like all scientists they prefer to reduce the number of variables under investigation to a minimum, in order to work more simply. As we have already seen, investigation of the whole habitat requires more than a little intuitive judgement to help the theory along. To begin to understand the ecological functioning of such a complex system, one has to draw on such rules and examples as there are. Ecological theory provides some background, at least in contributing important insights into subjects such as population dynamics, competition and succession.

With the state of our existing knowledge only fragments of these principles can be properly addressed. One obvious point is that we must begin at the site itself, which controls ecosystem development by supplying the necessary physical and chemical environment, including substrate, biomass, nutrients and moisture regime. Another observation is that many areas of semi-natural and species-rich vegetation are also associated with infertile or stressful conditions. This same observation can be studied in reverse, that is by reviewing evidence which links sites of low species diversity with elevated site fertility. Marrs and Gough use this approach to demonstrate the effects of fertiliser additions, successional processes (such as nitrogen-fixation) and pollution inputs, on ecosystem development. This in turn provides a working framework for evaluating management practices which might reduce fertility, and so increase the chances of re-establishing species-rich communities.

Another possible approach is to select rigorously those plant materials which are likely, on a given site, to maintain species-rich communities similar to those found in nature. Hodgson shows how a detailed knowledge of the ecological preferences of plants, defined as their stress-tolerant, competitive and ruderal strategies, can be used to predict ecological processes operating in a variety of habitats. The same logic can be used to devise suitable wildflower seed-mixtures, and even to evaluate the after-effects of habitat transplantation.

Finally, no ecosystem is static and most habitats, especially species-rich ones, require continuous management to maintain them in an interesting and diverse condition. As Bradshaw points out, ecological succession is always waiting to happen, and at any site there is a constant potential towards an increase in biomass and nutrient capital. In habitat reconstruction this is a problem, as the time-scale is relatively short, and complete primary successions to woodland can take place in as little as 100 years. To permanently maintain the community at some intermediate point in the succession may require several different practical solutions. In some way or other these all involve reducing the amount of growth, so as to avoid the competitive exclusion of minor species components, and the prevention of invasion by inappropriate species.

Once again, the ecological principles described in this section reinforce the point that we are dealing with a situation requiring horticultural levels of knowledge and care. Apart from appreciating the principles themselves, we must know exactly which plant materials to use, the site conditions which optimise them, and the cultural conditions and management needed to maintain them in communities.

4

Soil fertility – a potential problem for habitat restoration

R. H. Marrs and *M. W. Gough*

Introduction

Throughout history there have been continual fluctuations in the amount of land being used by man for agriculture, and these fluctuations tend to be greatest on marginal land with the poorest soils. Generally, changes in agricultural production have reflected the prevailing economic conditions (Orwin and Whetham, 1964), but on some infertile soils such as those in the Brecklands of East Anglia, United Kingdom, cultivation and reversion were part of a normal land-use pattern analogous to shifting cultivation in the tropics (Sheail, 1979; Trist, 1979). In the past, where there has been reduction in the intensity of farming, there has been recolonisation by native species, and many such sites now have a very high conservation interest (Wells *et al.*, 1976; Sheail, 1979; Marrs and Proctor, 1979).

In the last 40 years, agricultural economics in Britain have favoured sustained increases in crop production. These increases have been brought about in two main ways, first by an increased cultivation of marginal land (Moore, 1962; Webb and Haskins, 1980; Fuller, 1987), and second by improved crop husbandry which has included advances in plant breeding, improved pest control and much greater use of fertilisers (Gasser, 1982; Elsmere, 1986). However, as we approach the 1990s, current forecasts of agricultural economics suggest that within the European Economic Community (EEC) the pendulum has swung in the other direction (Green, Way, both in this volume).

Compared to earlier reversions of agricultural land there are two main factors which have changed significantly during this most recent period of intensive cultivation, which may reduce the likelihood of natural recolonisation by many native plant species. These are:

1. Seed availability. Previously agricultural land was set in an extensive matrix of native vegetation which acted as a seed source, facilitating natural colonisation between short periods of agricultural activity. Today, however, sources of inoculum from semi-natural communities have been much reduced and attempts to re-create early-successional communities from later ones, relying solely on natural seed inputs, have proved difficult, often giving unpredictable results (Lowday, 1984, 1986). Nevertheless, there are many good examples of semi-

natural communities growing on former agricultural land (Marrs and Proctor, 1979; Wells *et al.*, 1976) and these communities must have developed naturally from seed banks and adjacent sources.

2. Soil fertility. In recent years there have been consistent increases in the amounts of fertilisers used. The resulting residues remaining in the soil may be a problem during attempts to establish and maintain many semi-natural plant communities with high conservation interest, given that existing examples of these communities, for example floristically-rich grasslands, are often associated with infertile soils (Grime, 1973, 1979, 1987; Grubb, 1977; Rorison, 1971).

The problem of seed availability can be circumvented through the use of a suitable seed mixture containing native species (Wells *et al.*, 1981; Wells *et al.*, Brown, both in this volume). In this chapter therefore, we concentrate on the problem of soil fertility by (1) reviewing the evidence that suggests that an elevated soil fertility can be a major problem for the maintenance of high species-diversity and hence conservation interest, (2) presenting data from soil fertility studies where we have compared agricultural soils with those of adjacent successional sequences and (3) discussing potential management scenarios for reducing fertility with special reference to their likely success and the time-scales involved.

The influence of soil fertility on habitat diversity

Many areas of semi-natural vegetation of high conservation interest, such as floristically-rich grasslands and heathlands, are associated with infertile soils. Moreover, most of these communities need to be managed, for example by grazing, cutting or mowing, in order to maintain their conservation interest (Green, 1980). This management helps to maintain diversity through its effects on inter-specific competition and by preventing the colonisation of later-successional species, but in many cases it also acts to maintain soil infertility by removing nutrients from the system regularly. Although we have little information on the establishment and maintenance of high-diversity habitats on fertile old fields in Britain, there are three main areas where some ecological information is available about the effects of increased fertility on species-rich ecosystems – where the high fertility has been brought about through (1) fertiliser addition, (2) successional processes and (3) pollution inputs. The information from these sources will be reviewed briefly, because it provides a framework for the prediction of problems likely to be encountered when attempting to create species-rich habitats on fertile abandoned agricultural land.

The effects of fertiliser addition

Perhaps the best known example of the damaging effects of high fertiliser additions to species-rich grassland is the Park Grass experiment at Rothamsted Experimental Station (Williams, 1978), set up by Lawes and Gilbert in 1856 and continued to the present time, to ascertain the optimum amounts and combinations of fertilisers

Table 4.1 Changes in the number of species in selected treatments of the Park Grass experiment

*Summary of treatments	Fertiliser given	Plot number	Number of species	
N treatment	Other elements given annually		1862	+1973–6
None since 1856	None	3	28	24
None since 1856	P Na Mg	8	29	29
None since 1856	P K Na Mg	7	30	24
96 kg N/ha/yr as NH_4–N	P K Na Mg	9	16	3
144 kg N/ha/yr as NH_4–N	P K Na Mg	11	17	2
96 kg N/ha/yr as NO_3–N	P K Na Mg	14	20	15

* = data summarised from Williams (1978)
+ = last sample between these two dates is presented

Source: Williams, 1978.

needed to obtain maximum hay yield. The untreated plot in this experiment has maintained 24 species throughout the study period, but where ammonium fertilisers in particular have been added, there are now only two or three species remaining (see Table 4.1; Williams, 1978). This decline in species-richness is described in Figure 4.1, for plot 11/d of this experiment, which has been given a treatment of 144 kg N/ha/year as ammonium-nitrogen plus P, K, Na and Mg but no lime. When the cumulative species abundance is plotted versus its rank at different sampling times, it is obvious that there has been a consistent reduction in species-richness through time at this plot (Digby and Kempton, 1987).

It is perhaps of interest to note that these reductions in species-richness brought about by fertilisation are complex. For example, where nitrogen was added as an ammonium salt without lime additions, there has been a decrease in soil pH from 5.3 in untreated plots to <4.0, which might in itself reduce species diversity. Moreover, where the vegetation was given 96 kg N/ha/year as nitrate-nitrogen, there was only a slight reduction in species number, and where P, K, Mg and Na were given without nitrogen the species number was maintained at similar levels to the untreated plots (see Table 4.1).

In addition to these effects on species number, clear treatment-specific differ-

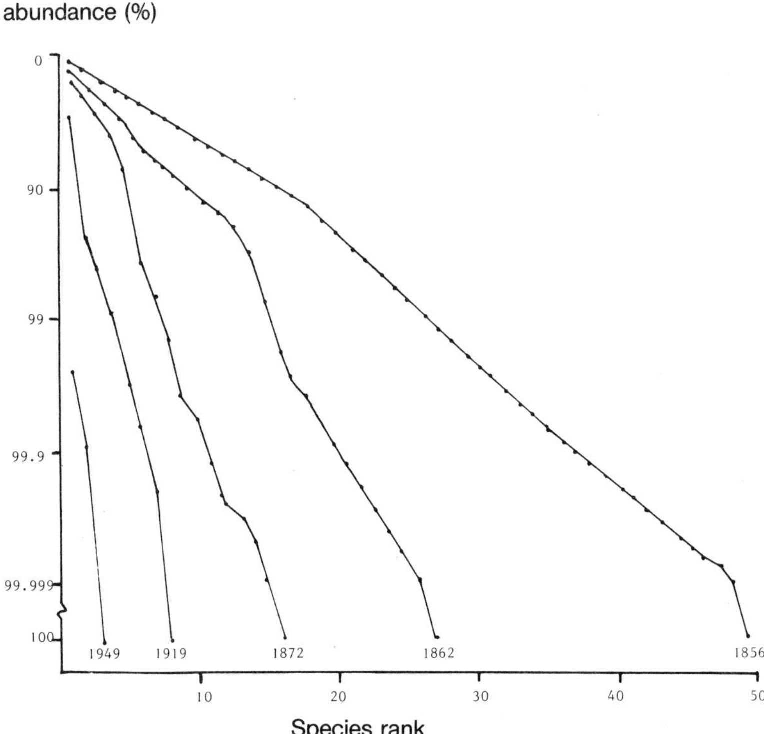

Figure 4.1 A plot of cumulative species frequency against rank for species abundance on Park Grass plot 11/1d from 1856 to 1949, showing a clear reduction in species diversity through time (Digby and Kempton, 1987).

ences were found in the balance of species present, with for example *Festuca ovina* and *Agrostis capillaris* the most common grasses in the untreated plot, *Agrostis capillaris* where P, K, Na and Mg only were given, *Alopecurus pratensis* and *Anthoxanthum odoratum* where ammonium-nitrogen and *Alopecurus pratensis* and *Arrhenatherum elatius* where ammonium-nitrogen and P, K, Na and Mg were given (Digby and Kempton, 1987).

Another interesting study was made of fertiliser addition to sand dune vegetation at Braunton Burrows in north Devon (Willis and Yemm, 1961; Willis, 1963). Laboratory and field applications of nitrogen, phosphorus and potassium increased plant yield, with the greatest response found where nitrogen was added. However, they also found a marked change in species composition with increases in two grasses (*Festuca rubra* and *Poa pratensis* ssp *subcaerulea*). For one species, *Agrostis stolonifera*, growing in a dune slack, a 15-fold increase in height and a 7-fold increase in dry matter was found after two years. Boorman and Fuller (1982) in a similar study confirmed these results, but showed that the high rabbit grazing pressure at their study site helped to counteract the effect of fertiliser addition. Nevertheless, three major conclusions about the effects of fertiliser additions emerged from this study: first the growth of unpalatable ruderals was encouraged, second the abundance of annuals, bryophytes and lichens declined and third the overall diversity decreased.

The effects of successional processes

There is a wealth of observational and experimental evidence to show that high nutrient supplies reduce species-diversity and encourage the growth of more 'competitive' species. Green (1972, 1980) reported a 'seral eutrophication' under gorse (*Ulex europaeus*), a nitrogen-fixing, leguminous shrub, growing on former chalk grassland. When the gorse was cut, but not removed, a much lusher sward developed with invasion by the competitive 'nitrophiles' *Rubus idaeus, Chamaenerion angustifolium* and *Agrostis stolonifera*. Soil analyses indicated an increase of *c.* 770 kg N/ha, which was in the same order of magnitude as the amount of N in the above-ground biomass of a 12-year-old stand of gorse (*c.* 500 kg N/ha). Moreover, an increase of *c.* 20 µg/g in the instantaneously extractable inorganic nitrogen fraction was found under gorse, an almost 50 per cent increase over the grassland soils. Rapid invasion by competitive species has also been observed after the clearance of *Hippophae rhamnoides*, another nitrogen-fixing scrub species, from sand dune sites (Marrs, 1984, 1985), but in this work no measurements of soil chemistry were made.

It might be construed from this line of argument that increased soil fertility under scrub is found only under species which can fix atmospheric nitrogen at high rates. However, this is not entirely correct, because where non-nitrogen-fixing scrub species, such as *Crataegus monogyna*, are cleared from chalk grassland soils, there is the same trend towards a vegetation of competitive species. Grubb and Key (1975) reported an invasion and profuse growth of *Galium aparine* and *Cirsium vulgare* on such soils, and they ascribed this trend to an increased supply of both mineralisable nitrogen from 7 to 34 µg N/g and also of extractable phosphorus from 5 to 11 µg P/g under the scrub. Unfortunately we do not fully understand the exact mechanisms through which this increase in soil fertility occurs, but various factors may be important including: (1) increased foraging for nutrients by deep roots, (2) accumulation of nutrients over a long period in the scrub biomass followed by subsequent rapid release, (3) increased turnover of litter both above and below the ground, (4) increased capture of nutrients from the atmosphere and from biotic sources (bird and small mammal activity) and (5) increased efficiency of mycorrhizas.

The effects of pollution inputs

There has been a continued increase in rainfall inputs of nutrients, particularly in nitrate-nitrogen over the last 100 years (Brimblecombe and Pitman, 1980; Brimblecombe and Stedman, 1982), largely because of man's activities. In Europe, in particular, increasing deposition of nitrogen, sometimes locally enriched by intensive agricultural activities, has been suggested as the main driving variable influencing adverse vegetation change in chalk grasslands (During and Willems, 1986; Bobbink and Willems, 1987) and in heathlands (Diemont and Heil, 1984; Heil and Diemont, 1983). Small experimental additions of nitrogen fertiliser (28 kg N/ha) appear to support this view at least for heathlands (Heil and Diemont, 1983). This level of fertilisation is below measured estimates of bulk precipitation on these sites (*c*. 40 kg N/ha/year). Indeed, if all of this precipitation input was captured, theoretically it would have the same sort of effect as the smallest nitrogen treatment on the Park Grass Experiment (48 kg N/ha/year).

Potential problems for habitat creation on fertile soils

In attempts to create diverse species-rich habitats, high fertility could have two major effects on the vegetation, irrespective of whether the land was abandoned and left to natural colonisation processes, or artificially sown:

1. High fertility would tend to favour high yielding competitive species to the detriment of slower growing species. Where seeds of many species were sown, there would be a change over time to a sward with few species, so that the higher costs of sowing the richer seed mixture would be wasted. Furthermore, ecosystem development would tend to prevent invasion by either bryophytes or lichens.
2. High fertility would favour late-successional species over early-successional species. This could take two forms, either a rapid invasion of late-successional species, helped by the high fertility (a man-modified relay floristics succession: Clements, 1916), or a shorter period of dominance by early-successional species (an accelerated initial floristics composition succession: Egler, 1954).

In any of these scenarios ecosystem development on abandoned agricultural land will depend on both the residual fertility of the soil, the soil requirements of the desired semi-natural communities and the continued application of maintenance management treatments.

Soil fertility: a comparison of soils from agricultural land and from semi-natural communities

Introduction

In an attempt to provide information on soil fertility that could be used to guide habitat restoration policies for both agricultural land and late-successional stages, we have recently compared the fertility of soils under current agricultural use with soils supporting examples of semi-natural vegetation on a range of parent materials (Gough and Marrs, in press). The aims of this work were twofold; first to determine how fertile typical agricultural land was in relation to semi-natural successional sequences, and second to determine which of the main plant nutrients – nitrogen, phosphorus and potassium – contributed most to observed differences in fertility.

The answers to these two questions are crucial in developing habitat restoration policies for the creation of diverse habitats on fertile soils, because clearly we need to know if there is a problem, and to have some estimate of the magnitude of differences between the fertile soil and the fertility requirements of the required community, in order to assess whether the project is feasible and to be able to predict the likely time-scales involved. Moreover, if successional sequences can be ranked with respect to their fertility (for example as has been suggested already for chalk grassland soils and invading scrub – Grubb and Key, 1975) on a scale that encompasses or comes close to agricultural soils, then it may be sensible to select the nearest successional stage on the fertility scale as the management objective (for example choosing a scrub habitat rather than a species-rich grassland). Moreover, it is important to identify the nutrient element responsible for the fertility. In many cases fertility will be a function of at least the two main elements, nitrogen and phosphorus (as in the case of the scrub soils on chalk, Grubb and Key, 1975), but if this is not a general phenomenon, and there is uncoupling of nitrogen and phosphorus influences at some sites or stages, different problems may occur. For example, we have already seen that nitrogen had a much greater effect on species number in the Park Grass Plots (see Table 4.1) than phosphorus.

Methods and results

Six sites were studied, two from each of three parent materials, clay, sand and limestone. At each of these sites, soils were collected from four areas: (1) a semi-natural grassland, (2) a scrub habitat, (3) a woodland and (4) an adjacent area of agricultural land. Clearly this sampling strategy is biased in favour of marginal agricultural land, because it is adjacent to unimproved areas, but it should approximate to types of land likely to be suitable for habitat restoration. Soil fertility was assessed using standard chemical analysis procedures (Allen *et al.*, 1974), but augmented using comparative growth bioassays, where growth and nutrient uptake by three test species were used as indices of soil fertility.

Two main soil factors were identified as important determinants of soil fertility – extractable phosphorus and mineralisable nitrogen. Soil extractable phosphorus was greater in the agricultural soils than grassland soils on all parent materials, but higher values were also found on scrub soils on clay and woodland soils on limestone (see Table 4.2). Using the fertility classification for phosphorus of Allen *et al.* (1974), where soils were classified as low, intermediate and high fertility based on extractable phosphorus concentrations of <0.2, 0.2 – 2.0, and >2.0 mg P/100g, then all grassland soils fell in the intermediate range, none being >1.3 mg P/100g. Most agricultural soils could be classified as fertile, being near 2.0 (Coombe Hill/Northchurch Common 1.9 mg P/100g) or greater. The agricultural soil from Sherwood Forest was lower, but the reasons for this will be discussed later. These soil chemical results were confirmed by plant growth studies designed to highlight differences in phosphorus supply (Gough and Marrs, in press).

In contrast to the results for extractable phosphorus, the nitrogen mineralisation rate of the agricultural soils was generally either lower than, or in the same order of

Table 4.2 A comparison of (a) extractable phosphorus and (b) mineralisable nitrogen concentrations in soils on agricultural land and from semi-natural communities on three parent materials

	Parent material	Clay		Sand		Limestone	
	Site	Hatfield Forest	Coombe Hill/ Northchurch Common	Clumber Park	Sherwood Forest	Lathkill Dale	Monks Dale
a. Extractable phosphorus (mg/100 g)							
Agricultural land		6.8	1.9	29.1	1.2	3.1	1.6
Grassland		0.9	0.7	0.5	1.3	0.6	0.3
Scrub		2.4	2.0	0.4	1.2	0.3	0.4
Woodland		0.3	1.2	0.8	0.5	4.1	1.2
L.S.D. ($n = 8, p<0.05$)		1.9	1.1	0.9	0.8	1.2	0.3
b. Mineralisable nitrogen (mg/100 g/14d)							
Agricultural land		8.3	4.0	2.4	6.6	14.3	13.2
Grassland		14.2	6.3	2.8	3.4	9.7	16.2
Scrub		15.4	6.6	4.8	2.7	14.9	14.9
Woodland		2.2	3.4	1.8	3.0	19.3	16.0
L.S.D. ($n = 8, p<0.05$)	7.0	3.2	2.0	2.3	4.8	5.5	

magnitude as, the grassland soils (see Table 4.2), with the exception of the Sherwood Forest site (discussed later). However, nitrogen mineralisation rates were greater at two scrub sites, one on limestone (Lathkill Dale) and one on sand (Clumber Park), but rates for woodland were generally lower than grasslands, except on the limestone, where rates were either similar to (Monks Dale), or greater than (Lathkill Dale), grassland soils (see Table 4.2).

Discussion

Three main points have emerged from this study that have a bearing on habitat restoration: (1) the major factor determining residual soil fertility in agricultural soils compared to species-rich grasslands is the phosphorus supply, (2) the initial supply of nitrogen on agricultural soils appears to be unimportant and (3) there was a great deal of variation in soil fertility between the various successional sequences studied.

The extractable phosphorus concentration of the agricultural soils was variable, and was generally greater than the grassland soils, ranging from 2.7 to 58 times those of the grassland, with one exception: the Sherwood Forest site on the sandy parent material, where the value was less than the grassland soil. This site was, however, unusual in that it was an abandoned field that had received similar fertiliser dressings to the Clumber Park site until *c.* 1975 when it was abandoned and allowed to revert naturally to a grassland. At this site, the results suggest that soil extractable phosphorus can decline from agricultural to grassland levels within twelve years. It is also interesting to note that Sherwood Forest should be the only agricultural soil to have a greater nitrogen mineralisation rate than grasslands. We have no explanation for this but can hypothesise that its higher pH (6.0±0.3 compared to 3.5±0.1 in semi-natural grassland) may in itself allow faster rates of nitrogen mineralisation and nitrification. However, grassland development on this site has also involved an increase in organic matter, perhaps initially accelerated by the high phosphorus supply. Together the higher organic nitrogen capital and high pH may combine to increase the potential nitrogen supply as the new ecosystem develops. This increasing nitrogen supply through time may cause future deleterious change in species composition, and the relationship between declining phosphorus supply and increasing nitrogen supply during this period is clearly worthy of further study.

The data on soil fertility from the different successional stages emphasise the variability between stages on different parent materials. On two scrub soils phosphorus supply was greater than in grassland soils, on another two scrub soils nitrogen supply was greater and on the remainder there was no difference (see Table 4.3). These results contrast with those of Grubb and Key (1975), who showed that on chalk soils the supply of both nitrogen and phosphorus was increased under scrub. Moreover, woodland soils were found to be more fertile compared to grassland soils only on the limestone, where phosphorus supply was greater at one site, and both nitrogen and phosphorus greater at the other (see Table 4.3). These results suggest that under some late-successional communities, but not all, there is an increase in soil fertility over early-successional communities, and in different

Table 4.3 A summary of sites where phosphorus (P) and nitrogen (N) supplies were greater in scrub and woodland soils relative to unimproved species-rich grassland

Site	Successional stage	
	Scrub	Woodland
Hatfield Forest	P	–
Coombe Hill/Northchurch Common	P	–
Clumber Park	N	–
Sherwood Forest	–	–
Lathkill Dale	N	PN
Monks Dale	–	P

communities on different parent materials the fertility increase may be because of nitrogen and/or phosphorus. As yet we do not know the reasons for these differences, but observations show that where the supply of both elements are increased over grassland levels (Grubb and Key, 1975), there is a rapid invasion of fast growing weed species. The extent to which this invasion would occur if the fertility increase was related to only one element remains to be investigated.

Techniques to reduce soil fertility

Very little research has been directed towards the deliberate impoverishment of soils for re-creating semi-natural communities, although some information is available from some studies of old field succession, and agricultural studies designed to assess the economic value of fertiliser residues. Because we have so little information to go on, we can but speculate on the general principles involved, and make educated guesses at optimal methodologies and time-scales of treatment required.

Briefly there are two main methods that can be used to impoverish soil fertility, by managing the systems so that (1) nutrient exports are greater than imports (with an emphasis on reducing the total nutrient capital), and/or (2) nutrients are sequestered in unavailable pools, where onward circulation is blocked or reduced to a low level (with an emphasis on reducing available pools). From a management point of view, the methods can be classified into three strategies: by (1) managing the site to promote natural processes that will impoverish nutrient supply, (2) indirect removal using some form of continuous cropping and (3) direct soil nutrient pool removal.

Natural processes

Included under this heading are management practices which will promote natural soil degradation processes such as leaching and incorporation of soil organic matter which has a low decomposition rate. We envisage three strategies under this heading,

(1) allowing natural early-successional development, (2) sowing native species which are able to impoverish the soil and (3) continuous fallowing.

Natural successional development of a relatively species-rich grassland on the Sherwood Forest arable sandy soil was shown in the previous section to have resulted in a decline in extractable phosphorus to grassland levels within twelve years. This result is very similar to the value of ten years required for American sandy old fields to lose phosphorus, calcium and potassium (Odum, Pinder and Christiansen, 1984). However, the potential increase in nitrogen mineralisation during the initial ten years in our work is worrying, and requires further study.

Similarly sowing *Bromus erectus* on the same fertile scrub soils overlying chalk studied by Grubb and Key (1975) has reduced soil fertility (measured both chemically and by plant bioassay) to adjacent grassland levels within twelve years (Grubb and Pakeman, unpublished). Unfortunately, we have very little information on the mechanisms involved in producing this observed effect, and further work on this topic, including testing a range of native species, is currently in progress.

During continuous fallowing the nutrient producing processes in the soil continue, with little uptake by vegetation (see Figure 4.2). In the absence of uptake, there is a potential for nutrients to be leached down the profile, or even for erosion loss through surface run-off. This process is directly analogous to the herbicide treatment in the Hubbard Brook Ecosystem Study (Bormann and Likens, 1978), where the effect of delaying the recovery of plant uptake after cutting was a longer reorganisation phase, with increased nutrient loss over a longer period. The practical use of this approach will be limited by the adverse appearance of the site, as continuous bare ground would have limited conservation appeal.

Indirect removal by continuous cropping

There are two main problems with continuous cropping; first the cropping must result in a net removal of nutrients and second its effect becomes less successful as the nutrient pools decline, and if the management objective has not been achieved then it is economically difficult to justify further treatment ('law of diminishing returns').

Cropping cereals without fertiliser addition has been used by the Nature Conservancy Council to reduce soil fertility at one sandy arable site in Breckland (Marrs, 1986), and this approach has been adopted as the initial part of the strategy to restore native semi-natural grasslands and heathlands to arable soils in the new Environmentally Sensitive Area of Breckland (MAFF, 1988). In the test area Marrs (1986) found that offtake as grain was greater than estimated inputs (× excess, written in parentheses) for phosphorus (× 8) and potassium (× 3), but not for nitrogen. When the straw was also removed there was then a net drain of all three major elements – nitrogen (× 2), phosphorus (× 16) and potassium (× 13). Long-term cereal cropping studies have, however, shown that it can take many years to deplete soil reserves, especially for phosphorus and nitrogen. Fertiliser residues derived from either organic additions or inorganic fertilisers may take a very long time to be depleted in this manner, with estimates in some cases of >70 years (Johnston and Poulton, 1977). Where continuous grass cropping has been done, similar time-scales are involved (Johnston and Penny, 1972).

Grazing is an alternative form of cropping, but with this technique there may be very little net loss of nutrients. In the study at Llyn Llydaw (Perkins, 1978), offtake was greater than inputs for phosphorus (× 1.2) and potassium (× 4), but for no other element studied. It may be possible to increase removal efficiency if daylight grazing were enforced, with removal to holding pens at night. This type of management is practised successfully in Holland (Gimingham and De Smidt, 1983) and in Germany (Thompson, 1979) as part of ongoing conservation management. The grazing can be carefully controlled, and the manure is collected and sold, thus providing additional income. This type of management might prove to be attractive for managing nature reserves and restored areas elsewhere, because with some imaginative planning and the use of native or unusual breeds there is great scope for publicity and educational purposes.

Direct removal of soil nutrient pools

Techniques such as topsoil stripping to expose less fertile subsoils appear at first sight to be rather drastic, but it is only a slightly more severe action than conservation practices already carried out in some areas (e.g. peat digging, sod-cutting or 'plaggen' (Netherlands) and turbary). However, for agricultural soils surface stripping may be one of the best alternatives. The soils are already cultivated, and thus already have some form of crop rotation, and topsoil is a readily saleable commodity. In a previous paper (Marrs, 1986), it was suggested that one small old field site (12 ha) could generate an income of between £72,000 and £120,000 from topsoil sales, and profit from such sales could pay for the purchase of additional native seeds.

An alternative approach could be to deep-plough sites, so that the fertile surface layers are diluted through the profile. This approach is effectively utilising natural processes, such as leaching, to remove nutrients from the deeper layers, and makes the assumption that these losses occur more rapidly than uptake by foraging deep roots.

Conclusions

Soil fertility reduction may be a prerequisite to the establishment of some habitats with a nature conservation interest on abandoned fertile arable land. The magnitude of the reduction required depends on both the residual fertility, the elements involved and the type of vegetation to be established. Irrespective of how the problem of soil fertility is tackled, and on many sites it may not be a problem, the site must be managed to maintain semi-natural communities. Grazing or cutting are often used for this purpose, and these treatments are also often used to counterbalance deleterious effects of fertility (Rorison, 1971). It is also perhaps possible to use other management techniques in their place, for example the incorporation of toxic waste materials (lead mine/colliery waste) in the surface soil, where the toxicity stress (*sensu* Grime, 1979) may act as a counterbalance to fertility and help to maintain species-diversity.

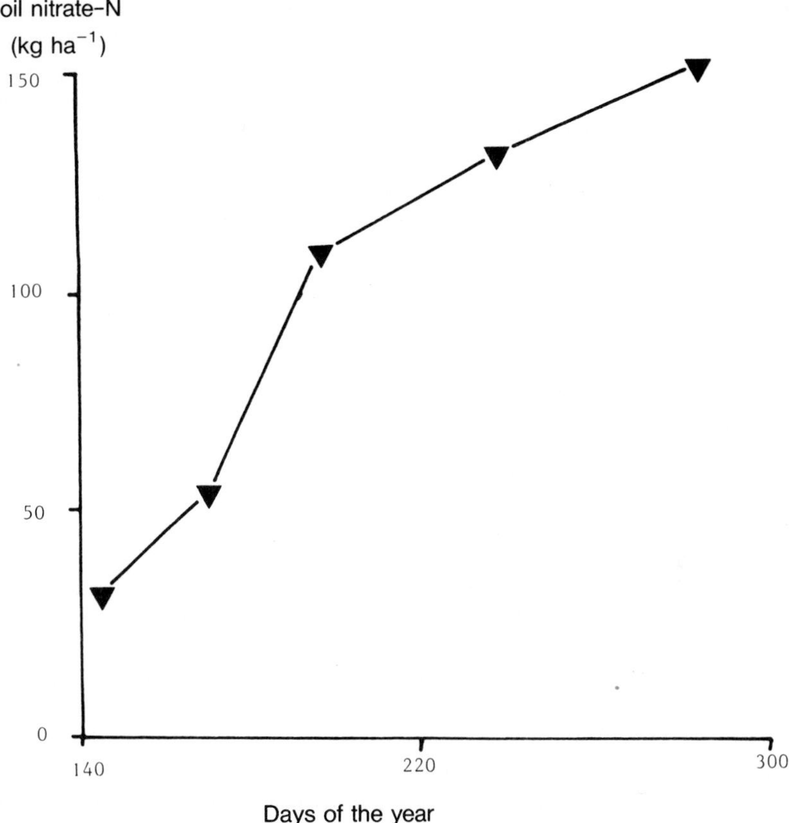

Figure 4.2 Seasonal accumulation of nitrate-nitrogen in the soil during a fallow year. If there was a high rainfall during this period there is a potential for leaching loss of nitrate, because this anion is rapidly leached. (Data from Lamb, Peterson and Fenster, 1985).

Acknowledgements

We thank our colleague Dr Peter Grubb of the Botany School, University of Cambridge, for help and advice during this work, and for permission to quote his unpublished data. This work was funded by the NERC special topic scheme on Agriculture and Conservation.

References

Allen, S. E., Grimshaw, H. M., Parkinson, J.A.B., Quarmby, C., 1974, *Chemical Analysis of Ecological Materials*, Blackwells Scientific Publications, Oxford.

Bobbink, R., Williams, J.H., 1987, 'Increasing dominance of *Brachypodium pinnatum* (L.) Beauv. in chalk grasslands: a threat to a species-rich ecosystem', *Biological Conservation*, **40**: 301–14.

Boorman, L. A., Fuller, R. M., 1984, 'Effects of added nutrients on dune swards grazed by rabbits', *Journal of Ecology*, **70**, 345–55.

Bormann, F. H., Likens, G. E., 1978, *Pattern and Process in a Forested Ecosystem*, Springer-Verlag, New York.

Brimblecombe, P., Pitman, J., 1980, 'Long-term deposit at Rothamsted, southern England', *Tellus*, **32**, 261–7.

Brimblecombe, P., Stedman, D. H., 1982, 'Historical evidence for a dramatic increase in the nitrate component of acid rain', *Nature*, **298**, 460–1.

Clements, F. E., 1916, *Plant Succession: an Analysis of the Development of Vegetation*, Carnegie Institute, Washington.

Diemont, W. H., Heil, G., 1984, 'Some long-term observations on cyclical and seral processes in Dutch heathlands', *Biological Conservation*, **30**, 283–90.

Digby, P.G.N., Kempton, R. A., 1987, *Multivariate Analaysis of Ecological Communities*, Chapman & Hall, London.

During, H. J., Willems, J. H., 1986, 'The improverishment of the bryophyte and lichen flora of the Dutch chalk heaths in the thirty years 1953–1983', *Biological Conservation*, **36**, 143–58.

Egler, F. E., 1954, 'Vegetation science concepts, I. Initial floristics composition, a factor in old field development', *Vegetatio*, **14**, 412–17.

Elsmere, J. I., 1986, 'Use of fertilizers in England and Wales, 1985', *Report Rothamsted Experimental Station, 1985*, 245–51.

Fuller, R. M., 1987, 'The changing extent and conservation interest of lowland grasslands in England and Wales; a review of grassland surveys 1930–84', *Biological Conservation*, **40**, 281–300.

Gasser, J.K.R., 1982, 'Agricultural productivity and the nitrogen cycle', *Philosophical Transactions of the Royal Society of London*, **296B**, 303–14.

Gimingham, C. H., De Smidt, J. T., 1983, 'Heaths as natural and semi-natural vegetation', in M.J.A. Werger and I Ikusima (eds.), *Man's Impact on Vegetation*, Junk, The Hague, pp. 185–99.

Gough, M. W., Marrs, R. H., in press, 'A comparison of soil fertility between semi-natural and agricultural plant communities: Implications for the creation of species-rich grassland on abandoned agricultural land,' *Biological Conservation*.

Green, B. H., 1972, 'The relevance of seral eutrophication and plant competition to the management of successional communities', *Biological Conservation*, 4, 378–84.

Green, B. H., 1980, 'Management of extensive amenity areas by mowing', in I. H. Rorison and R. Hunt (eds.), *Amenity Grasslands: an Ecological Perspective*, John Wiley, Chichester, pp. 151–61.

Grime, J. P., 1973, 'Competitive exclusion in herbaceous vegetation', *Nature*, **242**, 344–7.

Grime, J. P., 1979, *Plant Strategies and Vegetation Processes*. John Wiley, Chichester.

Grime, J. P., 1987, 'Dominant and subordinate components of plant communities: implications for succession, stability and diversity, in A. J. Gray, M. J. Crawley and P. J. Edwards (eds.), *Colonization, Succession and Stability*, Blackwell Scientific Publications, Oxford, pp. 413–28.

Grubb, P. J., 1977, 'The maintenance of species-richness in plant communities: the importance of the regeneration niche', *Biological Reviews*, **52**, 107–45.

Grubb, P. J., Key, B. A., 1975, 'Clearance of scrub and re-establishment of chalk grassland on the Devil's Dyke', *Nature in Cambridgeshire*, **18**, 18–22.

Heil, G., Diemont, W. H., 1983, 'Raised nutrient levels change heathland into grassland',

Vegetatio, **53**, 113–20.

Johnston, A. E., Penny, A., 1972, 'The Adgell Experiment 1848–1970', *Report Rothamsted Experimental Station 1971, Part 2*, pp. 39–68.

Johnston, A. E., Poulton, P. R., 1977, 'Yields on the Exhaustion Land and changes in the NPK content of the soils due to cropping and manuring 1852–1975', *Report Rothamsted Experimental Station 1976, Part 2*, pp. 53–86.

Lamb, J. A., Peterson, G. A., Fenster, C. A., 1985, 'Fallow nitrate accumulation in a wheat-fallow rotation as affected by a tillage system', *Soil Science Society of America Journal*, **49**, 1441–6.

Lowday, J. E., 1984, 'The restoration of heathland vegetation after control of dense bracken by asulam', *Aspects of Applied Biology*, **5**, 283–90.

Lowday, J. E., 1986, 'Restoration of *Calluna* heathland following bracken clearance', in R. T. Smith and J. A. Taylor (eds.), *Bracken: Ecology, Land Use and Control Technology*, Parthenon Press, Carnforth, pp. 233–8.

MAFF, 1988, *Breckland Environmentally Sensitive Area: Guidelines for Farmers*, BL/ESA/4, MAFF, London.

Marrs, R. H., 1984, 'The use of herbicides for nature conservation', *Aspects of Applied Biology*, **5**, 151–60.

Marrs, R. H., 1985, 'Scrub control', in P. Doody (ed.), *Sand Dunes and their Management*, Focus on Nature conservation No. 13, Nature Conservancy Council, Peterborough, pp. 243–51.

Marrs, R. H., 1986, 'Techniques for reducing soil fertility for nature conservation purposes: a review in relation to research at Roper's Heath Suffolk, England', *Biological Conservation*, **34**, 307–32.

Marrs, R. H., Proctor, J., 1979, 'Vegetation and soil studies of the enclosed heathlands of the Lizard Peninsula, Cornwall', *Vegetatio*, **41**, 121–28.

Moore, N. W., 1962, 'The heaths of Dorset and their conservation', *Journal of Ecology*, **50**, 369–91.

Odum E. P., Pinder, J. E., Christiansen, T. A., 1984, 'Nutrient losses from sandy soils during old field succession', *American Midland Naturalist*, **111**, 148–54.

Orwin, C. S., Whetham, E. H., 1964, *History of British Agriculture 1846–1914*, Longmans, London.

Perkins, D. F., 1978, 'The distribution and transfer of energy and nutrients in the *Agrostis-Festuca* grassland ecosystem', in O. W. Heal and D. F. Perkins (eds), *Production Ecology of British Moors and Montane Grasslands*, Springer-Verlag, Berlin, pp. 375–95.

Rorison, I. H., 1971, 'The use of nutrients in the control of the floristic composition of grassland, in E. Duffey and A. S. Watt (eds.), *The Scientific Management of Animal and Plant Communities for Conservation*, Blackwell Scientific Publications, Oxford, pp. 65–78.

Sheail, J., 1979, 'Documentary evidence of the changes in the use, management and appreciation of the grass-heaths of the Breckland', *Journal of Biogeography*, **6**, 277–92.

Thompson, J. A., 1979, 'Impression of the Luneburg Heath and nature reserve', *Nature and National Parks*, **17**(64), 20–1.

Trist, P.J.O., 1979, 'An introduction to Breckland', in P.J.O. Trist (ed.), *An Ecological Flora of Breckland*, EP Publishers, East Ardsley, pp. 9–14.

Webb, N. R., Haskins, L. E., 1980, 'An ecological survey of heathlands in the Poole Basin, Dorset, England, in 1978', *Biological Conservation*, **17**, 281–96.

Wells, T.C.E., Bell, S. A., Frost, A. J., 1981, *Creating Attractive Grasslands Using Native Plant Species*, Nature Conservancy Council, Shrewsbury.

Wells, T.C.E., Sheail, J., Ball, D. F., Ward, L. K., 1976, 'Ecological studies on the Porton Ranges: relationships between vegetation, soils and land-use history', *Journal of Ecology*, **64**, 589–626.

Williams, E. D., 1978, *Botanical Composition of the Park Grass Plots at Rothamsted*, Rothamsted Experimental Station, Harpenden.

Willis, A. J., 1963, 'Braunton Burrows: the effects on the vegetation of the addition of mineral nutrients to the dune soils', *Journal of Ecology*, **51**, 353–74.

Willis, A. J., Yemm, E. W., 1961, 'Braunton Burrows: mineral nutrient status of the dune soils', *Journal of Ecology*, **49**, 377–90.

5

Selecting and managing plant materials used in habitat construction

J. G. Hodgson

Introduction

Arguably there are three major challenges to those engaged in habitat reconstruction. The first is to revegetate areas chronically hostile to plant growth (e.g. coal mine spoil, slate tips and waste rich in heavy metals) resulting from the extractive industry. The second, which relates only to sites more conducive to plant growth, is to create and subsequently maintain attractive species-rich vegetation for amenity purposes. The third is to transfer ecosystems from one site to another with sufficient care and after-care to ensure that their floristic and faunistic interest is retained.

Unfortunately, the impact of ecological theory on these and related problems has not been as great as it ought to have been. In part this is because our understanding of the way communities are formed and subsequently function is both fragmentary and incomplete. However, the success of, for example, Bradshaw and his co-workers in revegetating toxic mine waste (Bradshaw, Humphreys and Johnson, 1978), and that of Wells (1979) in creating species-rich grassland from seed, indicates just how much can be achieved if our limited ecological knowledge is used wisely. However, not all those involved in the manipulation of vegetation are as ecologically well informed, or as successful.

An appreciation of ecological theory is perhaps most important for those attempting to create or transplant species-rich vegetation where there is frequently a bewilderingly large range of species and/or potential management options available. Accordingly this paper will concentrate on species-rich plant assemblages. Initially some key ecological attributes and processes involved in the creation and maintenance of species-rich vegetation will be briefly described. Subsequently it will be shown how, using autecological information from Grime, Hodgson and Hunt (1988) supplemented by data for rarer species (Hodgson *et al.*, unpublished), the effects of these ecological processes on the establishment and survival of species in natural and sown vegetation can be identified or at least predicted. The dataset will also be used to suggest what in theory might constitute a good general wild flower seed mixture. Finally, some of the problems in transporting vegetation and reconstituting it at a new site will be discussed.

Ecological theory

The following ecological attributes and processes are amongst those which critically influence the species composition of vegetation in general and the creation and maintenance of species-rich vegetation in particular.

*Established strategy (*sensu *Grime, 1979)*

Grime (1979) identifies two factors which limit the accumulation of biomass, 'stress', which constrains the rate and extent of growth, and 'disturbance', which results directly in the destruction of biomass. Stressed habitats (e.g. bare rock surfaces, unproductive calcareous pasture) are exploited primarily by plants described as 'stress-tolerators'. This grouping, exemplified by many lichens, sheep's fescue (*Festuca ovina*) and rock-rose (*Helianthemum nummularium*), consists of slow-growing, long-lived evergreen species which are both capable of surviving long periods under conditions not conducive to growth and are also relatively unresponsive to any amelioration of the environment. These species, because of their slow growth rate, are particularly sensitive to damage. By contrast, in frequently disturbed fertile habitats (e.g. arable fields) 'ruderals' prevail. These plants (e.g. chickweed, (*Stellaria media*)) are rapid growing and short lived. They produce flowers and seeds at an early stage of development. Where there is an abundance of resources and the intensity of disturbance is low (i.e. where conditions for plant growth are close to optimal), a third group of species, 'competitors', are found. These large, fast-growing species (e.g. stinging nettle, *Urtica dioica*) tend to monopolise the available resources leading to the competitive exclusion of most other potential components of the vegetation. In addition to these three primary strategies a number of intermediate strategies may be recognised and in Figure 5.1 these are positioned on a triangular diagram in the manner illustrated in Appendix 1 of Grime (1986b).

An equally important prediction of strategy theory is that there are no species which can exploit sites which combine low productivity and high levels of disturbance. Thus no flowering plants are able to survive in the heavily trampled, unproductive recreational areas so characteristic of many parts of upland Britain and erosion of footpaths can only be checked by (a) rerouting footpaths, (b) providing footpaths with artificial surfaces or (c) the addition of fertilisers to encourage higher productivity and the establishment of faster growing, more wear-resistant species.

Readers should be aware that strategy theory, which is described more fully in Grime (1979), is controversial and is likely to remain so at least in the short term. In the longer term, the Integrated Screening Programme devised at the Unit of Comparative Plant Ecology, Sheffield University, to characterise the essential ecology, physiology, biochemistry, morphology and anatomy of key species of ecological (and economic) importance should provide an appropriate dataset for a rigorous testing of the theory. However, despite these reservations, plant strategy theory does in practice appear to provide sensible answers to a variety of ecological problems. In particular, it has proved exceptionally useful in studies of the effects of

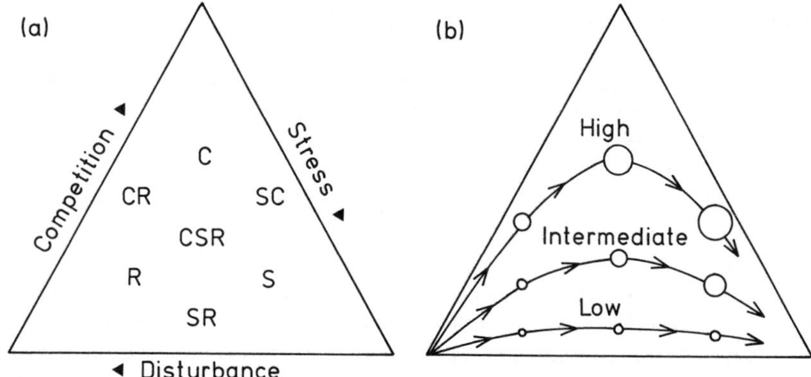

Figure 5.1 Plant strategies *sensu* Grime (1979) and their relationship to succession.
Note: In (a) the three primary strategies (competitor (C), ruderal (R) and stress-tolerator (S)) together with four intermediate strategies (competitive ruderal (CR), stress-tolerant ruderal (SR), stress-tolerant competitor (SC) and C–S–R strategist (CSR)) are positioned in a triangular diagram in the manner illustrated in Grime, 1986a (p. 183). In (b) the path of succession is illustrated at low, intermediate and high potential productivity. The size of biomass at each stage is indicated by circles. Diagram redrawn from Grime, 1979 (p. 151).

land-use on commonness and rarity (Hodgson, 1986). Here it will be used to interpret ecological processes and to make recommendations for improving the species-richness of sown and transplanted grassland.

Succession

If an arable field is left fallow, its species composition will change. This process, succession, is described in detail by Bradshaw (in this volume). The initial vegetation in which annuals and other short-lived species predominate is, with time, typically replaced, firstly by perennial, non-woody species and secondly by shrubs and trees. Changes in plant strategy (*sensu* Grime, 1979) associated with the succession from arable land and two other successional pathways are illustrated in Figure 5.1b.

The production of species-rich vegetation often requires management to arrest succession since (a) the successional process also leads to an increase in plant biomass and (b) there is a relationship between potential species density and annual maximum standing crop plus litter (Grime, 1979). Relatively few species are able to exploit situations where the standing crop is low as a consequence of high stress or disturbance. Equally, because of competitive exclusion by dominants, species densities are low where values for maximum standing crop are high. However, at intermediate values of standing crop (perhaps within the range 350–750 g m^{-2}) species-rich vegetation may occur. This coincides with a situation where the effects of stress and disturbance on the one hand, and of competitive exclusion on the other, are not too severe. It is thus clear that to achieve species-richness in practical terms the site must either be relatively unproductive to start with or management (e.g. cutting and the removal of litter) must be carried out at frequent intervals.

Phenology

The British flora appears to a large extent compartmentalised into species which grow best during the cool conditions of spring (and often autumn as well) and those exploiting the warmer summer months (see Wells, 1971; Grime and Mowforth, 1982). One of the most spectacular and familiar examples of this dichotomy is associated with bluebell woods. In spring, during the 'light phase', before the development of the tree canopy, bluebells (*Hyacinthoides non-scripta*) carpet the ground. By summer, when the leaf canopy of the trees is fully expanded, all that remains of the bluebells are the dead seed heads and bulbs buried deep in the soil.

Grime and Mowforth (1982) have shown that many species which grow in early spring have a high nuclear DNA content and large cells (*H. non-scripta*, DNA per nucleus 42.4 pg), while those exploiting summer conditions have a low value and small cells (rose-bay willowherb, *(Chamaenerion angustifolium*) 0.7 pg). This relationship may be explained by the greater sensitivity of cell division as opposed to cell expansion to low temperatures. Under summer conditions, because the formation of large cells takes a disproportionately long time, the rapid production of small cells (with low DNA amounts) appears to be the optimal solution. For species with large cells (and high DNA amounts) it has been suggested that rapid early growth results from the expansion of cells formed but not expanded during the previous summer or autumn. However, an insufficient number of these large cells are generated to sustain rapid summer growth.

Problems of dispersal

It is tempting to assume that vegetation, particularly that associated with older well established plant communities, contains only the species which are best adapted to the habitat. Two examples are given to illustrate that this is not necessarily so:

1. The common rock-rose (*Helianthemum nummularium*) is not recorded from Ireland, presumably because the seed is poorly dispersed, and the rare hoary rock-rose (*H. canum*), which elsewhere in Britain is restricted to xeric grassland, exploits habitats which appear more suited to *H. nummularium* in Western Ireland (Grime *et al*., 1988).
2. The low mobility of many woodland herbs is well known and, for example, in bluebell (*Hyacinthoides non-scripta*) and ramsons (*Allium ursinum*), most seed is shed within 0.4 and 2.5 m of the parent plant respectively (Grime *et al*., 1988). This limited dispersal may account for the tendency of many woodland herbs to be restricted to ancient woodland (see Peterken, 1974).

Thus the chance of a species colonising an area will be influenced by, among other things, its suitability for the prevailing habitat conditions, its mobility and its distance from the site to be colonised. However, although we can classify dispersules according to their morphological and anatomical adaptations for dispersal, we do not understand fully the processes of dispersal and colonisation which distinguish between rapid and slow colonists. For this reason we may sometimes be guilty of trying to create vegetation in situations where natural colonisation might lead to a more interesting flora (see Smart, in this volume).

Seed biology

In landscape reconstruction, an insufficient awareness of aspects of seed biology can lead to invasion of weed species, the failure of preferred sown species, or both. Many unwanted weeds, e.g. dock (*Rumex crispus* and *R. obtusifolius*) produce a large and long-persistent seed bank in the soil (seed bank type IV *sensu* Thompson and Grime, 1979). Inappropriate management (e.g. creation of areas of bared soil, excessive application of fertilisers) may encourage the establishment and spread of such species from seed dormant in the soil. Species in which a smaller proportion of seeds is incorporated into a persistent seed bank (type III) may be similarly troublesome. Where the unwanted species have seeds which persist in the soil for less than one year (type I (germination occurs soon after shedding) and type II (germination delayed until spring)), fewer problems are posed.

A further problem relates to speed of establishment. Most seedlings need bare ground or vegetation gaps for establishment, and early germination, to pre-empt growing space, is often crucial to successful establishment under conditions where there is intense competition between seedlings (Fenner, 1985). Species in which germination is slow may have difficulty in establishing from sown seed mixtures.

The season in which germination occurs must also be considered; this can be illustrated by comparing two common British species, ribwort plantain (*Plantago lanceolata*) and burnet saxifrage (*Pimpinella saxifraga*). *Plantago lanceolata* germinates over a wide range of conditions and establishment from either spring or autumn-sown seed would be expected. By contrast, seeds of *Pimpinella saxifraga* have a chilling requirement and germinate only in spring. For this species, spring-sown seed is likely to fail.

There are two further differences between the two species: (a) unlike *Plantago*, *Pimpinella* lacks a persistent seed bank and (b) *Pimpinella* establishes successfully both in vegetation gaps and in a short grassy sward, while *Plantago* mainly colonises gaps (Grime *et al.*, 1988).

These two species, which frequently occur together in calcareous grassland, illustrate the variation in regenerative strategies which commonly occurs within a single plant community. The understanding of this complexity remains a challenge to plant ecologists and provision for this regenerative complexity presents a challenge to those re-creating species-rich swards.

The colonisation process in practice

The extent to which species lists from the literature can be used in conjunction with Grime *et al.* (1988) and Hodgson *et al.* (unpublished) to predict ecological processes will now be examined. Firstly, as a test of the method, the colonisation of quarries will be considered using the species list in Table 6 of Davis (1982). (This study also gives some clues as to when intervention in the colonisation process is desirable.)

Secondly, the question 'Why can some wild flowers be grown from seed more easily than others?' will be considered using Tables 3 and 4 of Wells (1987). Thirdly, an attempt will be made to understand some of the problems in transferring vegetation between sites using Table 1 of Worthington and Helliwell (1987).

Natural colonisation in quarries

On theoretical grounds, successional processes in derelict chalk and limestone quarries would be expected to follow a different but predictable course on rock surfaces and soil heaps and to involve the replacement of species of bare soil and open rocky habitats by those of calcareous grassland. On soil heaps plants with a ruderal strategy *sensu* Grime (1979) will give way to C–S–R strategists and later to stress-tolerators, as nutrients become sequestered within the increasing plant biomass (see Figure 5.1). On rocks the expected sequence of succession would follow that associated with sites of low productivity in Figure 5.2. An increase in species-richness with time, as the colonists gradually reach the site, is also predicted. Data comparing the ecological characteristics of early and later colonising species of chalk and limestone quarries (see Tables 5.1 and 5.2A and Figure 5.2) are consistent with these expectations. With regard to the regenerative phase of the life history, it may be predicted that many common early colonists have unusually small seeds (Salisbury, 1942; Hodgson and Mackey, 1986) and data in Table 5.2A(3) confirm this expectation. Similarly the high proportion of wind-dispersed early colonists and the absence of a relationship between colonising potential and type of seed bank (see Table 5.2) is in accord with data on the colonists of spoil tips given in Grime (1986a).

It should, however, also be noted that an alternative succession could theoretically result directly from the establishment of other colonisers, including some trees and ferns, on raw quarry spoil. Certainly trees are often among the early colonists but whether they survive to reproductive maturity is less certain. In Central

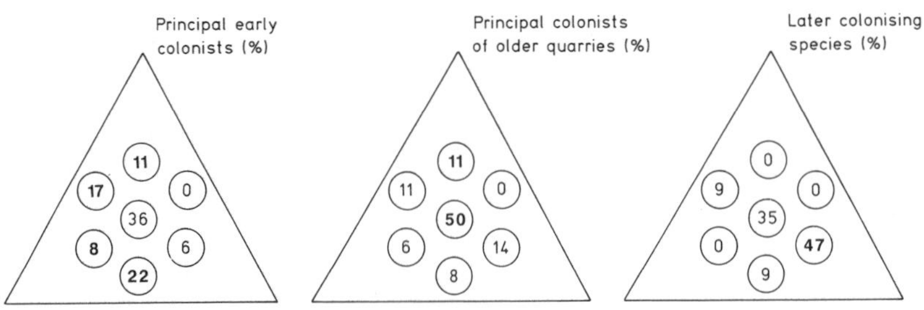

Figure 5.2 A comparison of the plant strategies *sensu* Grime (1979) of early and late colonising species of quarries.

Note: Sources of data as in Table 5.1 and abbreviations of plant strategies here and in the remaining figures as in Figure 5.1. Values are percentages. The highest value for each strategy is given in bold.

Table 5.1 A comparison of the commonest habitats of early and late colonizing species of quarries

Data relate to the fifty-seven commonest herbaceous and small woody species in chalk and limestone quarries in England listed in Table 6 of Davis (1982, p. 13), who sampled twelve quarries in each of four age classes (<15, 15–35, 35–55, >55 years). Three groupings were separated from within this data set: principal early colonists (present in >66% of young (<15 year-old) quarries), principal colonists of older quarries (present in >66% of older (>15 year-old) quarries), late colonising species (common species of quarries but present in <30% of young and <56% of older quarries). Four of the fifty-seven species fell into none of these groupings.

Commonest habitat	Principal early colonists (%)	Principal colonists of older quarries (%)	Late colonising species (%)
Quarry spoil	28	22	12
Other rocky habitats	17	17	0
Meadows and 'improved' pasture	33	33	12
Wasteland	11	6	18
Calcareous pasture	6	11	53
Other habitats	6	12	6
No. of spp.	18	18	17
Kruskall – Wallis one-way ANOVA $x^2 = 13.19$, $p<0.05$			

Source: Ecological data were mostly derived from Grime *et al.* (1988) and for a few less common species unpublished survey data of Hodgson *et al.* were utilised.

Note: Sums of percentages do not always total 100 due to conventional rounding.

England the male fern (*Dryopteris filix-mas*) is most commonly found on rock ledges and walls, but the crevices it occupies do not seem of a sufficient size to maintain an adult plant and the species sporulates mainly in woodland (Grime *et al.*, 1988). A similar failure to attain reproductive maturity may sometimes obtain for trees growing on quarry spoil.

Another feature inconsistent with the scenario of a simple succession from short-lived species to the long-lived perennials of calcareous grassland is the high proportion in Table 5.1 of species of agricultural grassland (meadows and 'improved' pastures) present amongst both the principal early colonists and principal colonists of older quarries. Moreover, there is a highly significant correlation between number of quarries colonised and general abundance in the landscape both for herbs (including grasses – see Table 5.2A(2) and trees and shrubs (see Table 5.2B(1)). Also, species with the capacity to spread by clonal growth as well as by seed become established in a larger number of quarries (see Table 5.2A(6)). These features suggest that the dispersal and establishment phases pose considerable barriers in quarry colonisation. It appears that, in addition to the ecological 'suitability' of the species concerned, the quantity of seed present in the landscape and its proximity to the quarry strongly influence the probability of successful colonisation. Consequently, much of the

Table 5.2 An ecological comparison of the flora of recently abandoned and older quarries

In A2 and A3 species were arbitrarily classified as of low, medium and high colonising ability if recorded from <33, 34–66 and ⩾67% in the case of young, and ⩽45, 46–66, ⩾67% in the case of older quarries. In B1 fast and slow colonisers are recorded from <50% and >50% of new and <50% and >50% of old quarries.

As in Hodgson and Mackey (1986), mean seed weight and standard deviation were obtained after log-transformation of data. For ease of interpretation these values have been antilogged.

Both here and in the remaining tables and figures, the Kendall correlation coefficient is abbreviated to *K* and *z* is calculated by the Mann-Whitney U test. Data indicated with an asterisk excludes monocarpic species (*n*=45).

Requirements for seed germination were also tested but differences were not statistically significant.

Species attribute	Years since dereliction	
	<15 years	>15 years
A. Herbaceous species		
n=57		
1. Mean spp/m² with which typically associated (%)		
10.1–14.0	15	12
14.1–18.0	27	20
18.1–22.0	49	51
>22.0	9	17
	K=+0.13, NS	
2. Commonness (% occurrence ± standard deviation in vegetation survey in C. England)		
spp. with low colonising ability	2.6±3.2	3.6±4.9
spp. with medium colonising ability	6.6±6.1	3.8±4.3
spp. with high colonising ability	6.5±5.8	8.2±5.9
	K=+0.29, *P*<0.001	*K*=+0.34, *P*<0.001
3. Seed weight (mg) (lower limit of SD – mean – upper limit of SD)		
spp. with low colonising ability	0.2 – 0.8 – 2.8	0.1 – 0.5 – 3.5
spp. with medium colonising ability	0.1 – 0.5 – 2.3	0.2 – 0.5 – 1.8
spp. with high colonising ability	0.1 – 0.3 – 1.4	0.1 – 0.4 – 1.6
	K=–0.17, *P*<0.05	*K*=–0.04, NS
4. Dispersal type (% quarries colonised ± SD)		
wind dispersed	45±25	58±18
other	59±24	60±19
	z=–1.68, *P*<0.10	*z*=–0.30, NS

5. Seed bank		
(% quarries colonised ± SD)		
transient	48±28	60±15
persistent		
usually small (type III)	55±22	61±21
often large (type IV)	50±27	54±11
	K=+0.06, NS	K=−0.07, NS
6. Extent of (attached) vegetative patch*		
(% quarries colonised ± SD)		
<100 mm	42±24	54±19
101–250 mm	45±28	60±14
>250 mm	58±26	64±18
	K=+0.20, P<0.05	K=+0.19, P=0.05
B. Trees and shrubs		
n=9		
1. Commonness of seedlings and young saplings		
(% occurrence ± SD in vegetation survey in C. England)		
spp. with low colonising ability	2.1±1.9	2.2±1.9
spp. with high colonising ability	5.8±4.2	5.7±4.2
	K=+0.47, P<0.05	K=+0.43, P<0.10

Source: Grime *et al.*, 1988, Hodgson *et al.*, unpublished.

colonising flora may not be particularly well adapted to the quarry habitat and there may be many 'underexploited niches' occupied by mobile, rapidly colonising species. In many areas quarries are an important habitat for orchids (Orchidaceae), and sometimes also clubmosses (Lycopodiaceae) and round-leaved winter-green (*Pyrola rotundifolia*). These species are all heavily dependent upon mycorrlingal infection for seedling, or sporeling, establishment, and have an extended juvenile phase. Their survival may be related both (a) to the persistence in quarries of open unproductive habitats and (b) to the apparently low level of adaptation of, and presumably low level of competition from, the quarry flora. It would be difficult to reproduce such conditions artificially.

Establishment from sown wild flower mixtures

Attempts to create species-rich vegetation from wild flower mixtures are often relatively unsuccessful. In an effort to find out why this should be so, a comparison of some of the ecological characteristics of successful and unsuccessful sown species has been carried out. The results are included in Tables 5.3–5.5.

Some species attributes (species-richness of habitat, seed weight, capacity for lateral vegetative spread) are similar for species that establish readily and those that do not (see Tables 5.3(3), 5.4A(1), 5.4B(2)) and provide no clues as to reasons for

Table 5.3 A comparison of some habitat characteristics of successful and unsuccessful sown species

High, moderate, low and no success refer to 100%, 50–99%, 1–49%, 0% establishment from sown seed in the 'real world' situations (e.g. housing estates and county parks) in which wild flower mixtures are utilised.

'Waste land' here includes spoil, rocky and other unmanaged dryland habitats.

		Success			
		High	Moderate	Low	None
1. Commonest habitat (%)					
Meadow	$K=+0.42$, $P<0.001$	44	30	14	2
'Improved' pasture	$K=+0.13$, NS	9	10	0	2
Road verge	$K=+0.08$, NS	13	10	7	7
Waste land	$K=+0.05$, NS	30	30	50	37
Calcareous pasture	$K=-0.24$, $P<0.01$	0	20	29	28
Other habitats (arable, mire, woodland)	$K=-0.26$, $P<0.01$	4	0	0	23
No. of species		23	10	14	43
2. Abundance in a local flora (% occurrence in survey of C England ± standard deviation)		5.8 ±7.0	3.4 ±3.1	2.0 ±3.4	1.4 ±2.9
			$K=+0.31$, $P<0.001$		
3. Mean spp/m^2 with which typically associated (%)					
10.1–14.0		9	0	7	23
14.1–18.0		57	10	50	28
18.1–22.0		30	60	21	33
<22.0		4	30	21	16
				$K=-0.01$, NS	

Source: Wells, 1987, p. 66; statistical comparisons as in Table 5.2.

establishment or failure. However, there are also many differences between successful and unsuccessful colonists. Successful establishment appears to be associated with species of agricultural grassland (meadows and 'improved' pasture) and those with C–S–R, competitive ruderal or competitive strategies (see Tables 5.3(1) and 5.5). Least successful are stress-tolerators and species of calcareous pasture and habitats such as arable land, mire and woodland. Species with a chilling requirement for germination also seem unusually prone to failure (see Table 5.4A(2)), perhaps as a result of seed being sown in spring rather than in autumn. This may be due either to failure to germinate or, more probably, to an inability to establish within the established sward of fast-growing species (see above). Equally, late-flowering species, which bloom from July onwards (typically after the time when management procedures will commence), appear disadvantaged (see Table 5.4B(1)). There is also

Table 5.4 A comparison of some regenerative attributes of successful and unsuccessful sown species

		Success			
		High (%)	Moderate (%)	Low (%)	None (%)
A. Features relating to initial establishment from sown seed					
1. Seed weight (mg)					
<0.2		4	20	7	34
0.21–0.50		26	0	29	17
0.51–1.00		35	20	28	10
1.01–2.00		13	30	0	20
>2.0		22	30	36	20
				K=–0.01, NS	
2. Germination requirements					
a. Non-dormant		48	30	43	37
				K=–0.05, NS	
b. Dormant*					
Dry storage	K=+0.06, NS	20	43	17	29
Scarification †	K=+0.22, P<0.05	70	43	50	19
Chilling	K=+0.26, P<0.01	10	14	33	52
B. Features influencing capacity to regenerate (and thus persist) *in situ*					
1. Flowering by June		94	100	50	56
			K=+0.31, P<0.001		
2. Capacity for lateral vegetative spread					
a. Usually monocarpic		17	30	29	28
				K=–0.08, NS	
b. Size of (attached) vegetative patch (mm)**					
<100		47	57	60	52
100–250		32	29	20	29
>250		21	14	20	19
				K=–0.03, NS	

Source: Wells, 1987, p. 66; statistical comparisons as in Table 5.2.

Note: In A2b and B2b the calculations exclude non-dormant (*) and monocarpic species (**) respectively. †Mostly legumes

Table 5.5 The percentage distribution with respect to plant strategy *sensu* Grime (1979) of successful and unsuccessful sown species

	Success			
	High	Moderate	Low	None
C-S-R	59	50	36	31
CR	13	10	4	8
C	7	0	4	2
R	9	5	14	13
SC	0	0	4	4
S	7	20	25	32
SR	7	15	14	12

Source: Wells, 1987, p. 66.

a positive correlation between commonness and successful establishment (see Table 5.3(2)). Dr Campbell and Professor Grime have unpublished experimental evidence that common species are tolerant of a wider range of productivity and disturbance than less common ones (see Figure 5.3). While the origin of this tolerance, which operates *within* populations, is uncertain, the relationship does emphasise the potential advantages of sowing common species.

These differences between successful and unsuccessful species suggest that:

1. sites utilised for sowing wild flower mixtures are often relatively productive, encouraging competitive exclusion of the sown species by robust, fast-growing grasses and tall clonal herbs;
2. species sown are often ecologically inappropriate for other reasons (e.g. where species of calcareous pasture are sown on non-calcareous soils); and
3. the creation of a species-rich sward should in the 'real world' be a two-stage process.

Firstly a number of common species which are both ecologically appropriate and easy to establish should be sown (autumn sowing would, theoretically at least, allow a greater range of species to be utilised). Subsequently, at least one year later, additional species, which may be more difficult or expensive to establish, should be introduced as seed in especially created vegetation gaps, or as established plants or rooted cuttings. It will be necessary to monitor the progress of newly introduced species and perhaps, at least temporarily, to introduce new management to ensure their establishment.

However, the high incidence of failure to establish from seed indicates that the use of wild flower mixtures is not always an optimal solution. As illustrated by Smart (in this volume), sites of moderate productivity situated adjacent to corridors for dispersal (e.g. railway sidings) may naturally develop visually attractive and scientifically interesting vegetation. Occasional disturbance by bulldozer or other

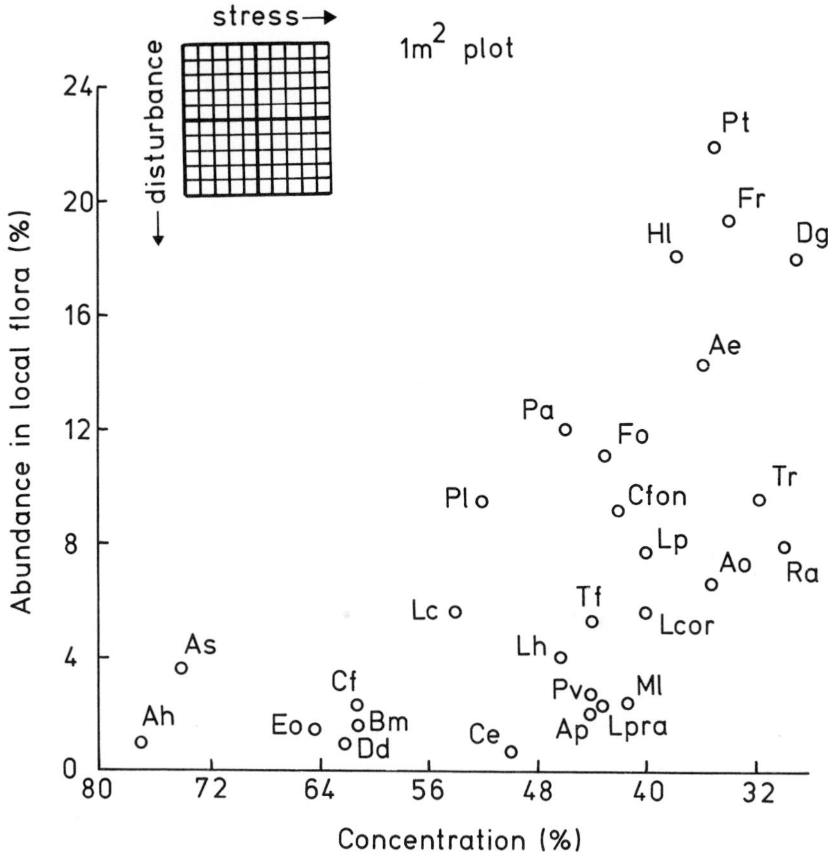

Figure 5.3 Relationship between abundance in a local flora and distribution with respect to stress and disturbance (Campbell and Grime, unpublished data).

Note: The data relate to a field experiment in which species were established in 1m square plots by seed under conditions of moderate fertility. After the seedlings had established a linear gradient of nutrient availability along one axis of the plots was produced and maintained by use of nutrient solution of different strengths. Along the other axis a gradient of disturbance was created by varying the frequency with which clipping and simulated trampling were carried out. This resulted in the production of a grid of 100 × 10 cm^2 sub-units differing from each other in severity of nutrient stress and mechanical disturbance. Treatments were maintained for 18 months. The number of sub-units in which each species persisted, and their distribution, were then measured for each plot.

Widely distributed species have similar values in each of the four quarters of the plot and a concentration of *c.* 25%. Species narrowly distributed with respect to disturbance and stress will have high values for % concentration.

Abundance in the local flora is abstracted from Grime *et al.* (1988).

Species names are abbreviated as follows: Ae, *Arrhenatherum elatius*; Ah, *Arabis hirsuta*; Ao, *Anthoxanthum odoratum*; Ap, *Avenula pratensis*; As, *Anthriscus sylvestris*; Bm, *Briza media*; Ce, *Centaurium erythraea*; Cf, *Carex flacca*; Cfon, *Cerastium fontanum*; Dd, *Danthonia decumbens*; Dg, *Dactylis glomerata*; Eo, *Euphrasia officinalis*; Fo, *Festuca ovina*; Fr, *Festuca rubra*; Hl, *Holcus lanatus*; Lc, *Luzula campestris*; Lcor, *Lotus corniculatus*; Lh, *Leontodon hispidus*; Lp, *Lolium perenne*; Lpra, *Lathyrus pratensis*; Ml, *Medicago lupulina*; Pa, *Poa annua*; Pl, *Plantago lanceolata*; Pt, *Poa trivialis*; Pv, *Prunella vulgaris*; Ra, *Rumex acetosa*; Tf, *Trisetum flavescens*; Tr, *Trifolium repens*.

crude management tool to recommence the succession when the vegetation becomes more rank and less visually attractive may be carried out to perpetuate the ecosystem. Similar management techniques have been recommended for quarries by Jefferson and Usher (1987).

Devising a suitable wild flower mixture

From the results given in Tables 5.3–5.5, it appears that a species-rich meadow associated with moderately fertile non-calcareous soils of neutral pH would be an appropriate ecological system to create in many of the 'real world' situations, where seed mixtures are currently utilised.

Species for inclusion in this mixture can be selected using criteria which relate to the ecological suitability of species for particular locations and to their ease of estab-

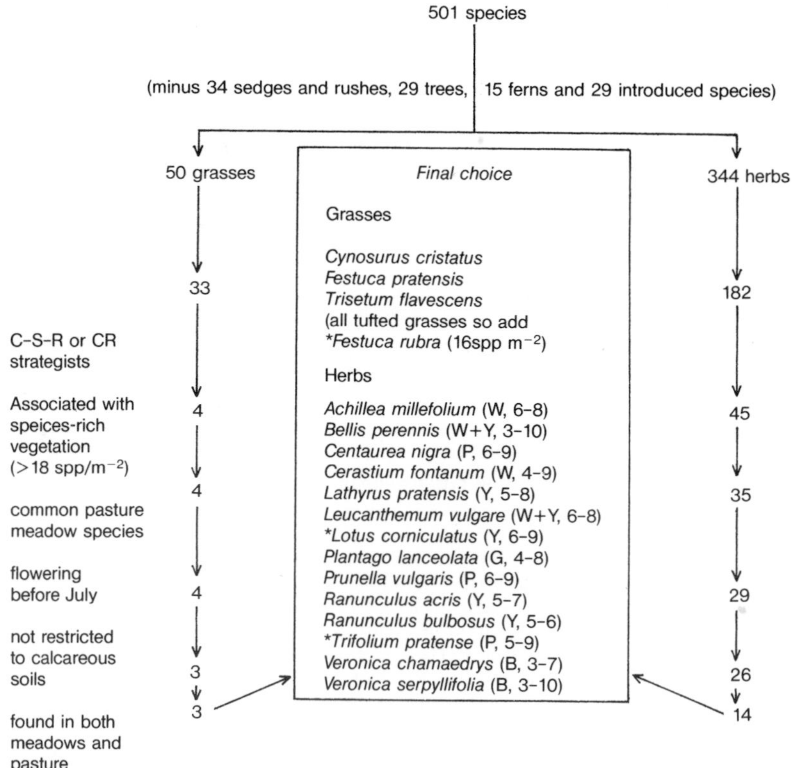

Figure 5.4 How a general purpose seed mixture for meadows may be selected using ecological criteria and the dataset for 501 species of Grime *et al.* (1988).

Note: Letters and numbers after species names refer to flower colour (B = blue, G = greenish, P = purple, W = white, Y = yellow) and months when in flower respectively.

*species often supplied as competitive cultivers. Only the less vigorous varieties occurring in old, unimproved grassland should be used.

Table 5.6 The degree of correspondence between the floristic composition of the meadow seed mixture proposed in Figure 3.6 and that for the *Lathyrus pratensis* sub-community of the *Centaureo-Cynosuretum cristati* meadow community recognised in the National Vegetation Classification (co-ordinated by Dr Rodwell and due to be published by the Nature Conservancy Council in 1990).

Constancy in NVC meadow community	Total no. of spp. in meadow community	No. of spp. in proposed mixture	List of species in proposed mixture
81–100%	4	4	*Cynosurus cristatus, Festuca rubra, Lotus corniculatus, Plantago lanceolata*
61–80%	9	3	*Centaurea nigra, Ranunculus acris, Trifolium pratense*
41–60%	10	7	*Achillea millefolium, Bellis perennis, Cerastium fontanum, Lathyrus pratensis, Leucanthemum vulgare, Prunella vulgaris, Ranunculus bulbosus*
21–40%	16	3	*Festuca pratensis, Trisetum flavescens, Veronica chamaedrys*
1–20%	36	0	–
0%	–	1	*Veronica serpyllifolia*
Mean spp. per 2 × 2 m square	22		

lishment and manipulation. In Figure 5.4 a preliminary list of appropriate species for use in the creation of species-rich meadows is given, together with reasons for the choice. For the purposes of this exercise, it has been assumed that the sward is to be cut in June and subsequently mown (simulating the mixed cutting and grazing regimes characteristic of many hay meadows).

We do not understand fully how ecosystems, particularly complex ones, function. Therefore, it is advantageous to check whether the species selected would normally coexist. The Nature Conservancy Council's 'National Vegetation Survey', co-ordinated by Dr Rodwell and due for external publication in 1990, allows such a check and it can be seen from Table 5.6 that the list of species suggested for sowing shows a high correspondence to the herb-rich *Lathyrus pratensis* subcommunity of the *Centaureo-Cynosuretum cristati* community. This community, characteristically grazed from August to April and mown annually in the late spring, is 'the typical grassland of grazed hay-meadows treated in the traditional fashion on circumneutral brown soils throughout the lowlands of Britain. It is becoming increasingly rare as a result of agricultural improvement.' Species-rich 'old' meadows fall within this community. Thus, after the reintroduction of species, the proposed seed mixture offers the prospect of a practicable and desirable end-product.

Modifications to the list may be necessary however. Some of the smaller proposed species, particularly I suspect *Ranunculus bulbosus*, may not establish easily. In

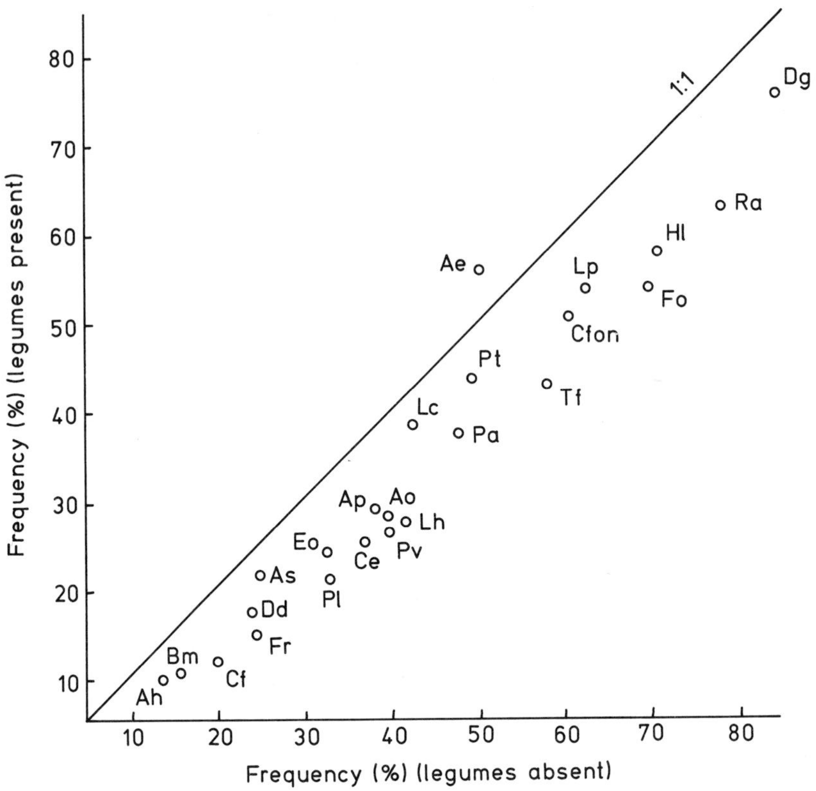

Figure 5.5 Comparison of the range of environments in which species survive in the presence and absence of legumes (Campbell and Grime, unpublished data).

Note: Frequency refers to the percentage of combinations of stress and disturbance in which a species survived during the experiment described in Figure 5.3.

addition, and controversially, one might also argue that legumes should not be sown. In a field experiment (see Figure 5.5) it was shown that above-ground biomass of turf was higher and species-diversity lower over a wide range of productivity and disturbance treatments when legumes were present. Nevertheless, this example illustrates how a preliminary species mixture can be chosen for a specific combination of soil conditions and management regimes and how the end-product can be shown to be ecologically appropriate.

Habitat transplantation

Objectives and practical constraints

Habitat transplantation should ideally involve (1) the transference of whole ecosystems from one site to another with minimal damage to the constituent species (in practice only the transference of flora and microfauna is feasible), (2) the re-establishment of flora and fauna in an appropriate reception site and (3) the provision of suitable after-care to minimise the loss of scientific interest and/or amenity value. Prior to transference a thorough survey of flora and fauna should be undertaken and information should be collected on the management of the original site (e.g. intensity and duration of grazing and type of stock used), together with other salient features (such as annual variation in the water-table and rooting depth of the vegetation).

Unfortunately the events both prior and subsequent to habitat transference tend to militate against this idealised approach. Developers are, for example, understandably reluctant to delay work until after the optimal time for a floristic and faunistic survey, to transfer material any greater distance or with any greater care than is absolutely necessary and to be involved in expensive after-care and monitoring. Thus, in practice the effectiveness of habitat transplantation may often be constrained by (1) an inadequate appreciation of the species composition of the site and of the ecological processes regulating it, (2) problems in finding appropriate reception sites and transporting material safely and (3) limitations in the amount of monitoring and after-care available after the transference of vegetation has been completed.

These practical aspects, which are discussed fully elsewhere (e.g. Penny Anderson, in this volume), are outlined here simply to illustrate the complex array of factors which, *in toto*, determine the success or otherwise of the transplantation process and confound attempts to predict the long-term outcome of attempts at transplantation. Nevertheless, ecological theory can, in a limited way, be used both to assess the success of the procedures adopted, and to suggest improvements. This will now be illustrated using the dataset of Worthington and Helliwell (1987).

Grassland and marshland transfer (see Helliwell, in this volume)

In this experiment vegetation and soil from moist grassland and mire were transferred to a receptor site. Methods similar to those used by a gravel company in the reinstatement of land for agriculture were adopted. Thus the vegetation was treated relatively roughly and the aim of the experiment was to create vegetation that was similar rather than identical in floristic composition to the original.

In general terms the transference was a success (see Helliwell, in this volume). There were few changes either in species composition or in number of species per m^2 after transference and there was no indication of any gradual replacement of species characteristic of species-rich vegetation by those of more species-poor communities (see Table 5.7(1)). Equally, although the rarest species present, *Oenanthe fistulosa*, which within the London area is restricted to just seven tetrads (Burton, 1983) has been lost, there is no evidence that species which have decreased are any rarer than those which have increased (see Table 5.7(2)). However, there are perhaps lessons to be learned by comparing the ecological attributes of species which appear to have increased with those which have decreased as a result of the procedures adopted.

Problems associated with transference of soils and vegetation

Species may survive transplantation to a suitable reception site, either because they have been transferred extremely carefully or because they recover quickly from the

Table 5.7 A comparison of the ecological attributes of species which have decreased and those which have increased following their transference to Tomrod, Staines Moor.

Species were subdivided into wetland or dryland groupings. Subsequently those whose frequency six years after transference was at least 150% of that in the original vegetation were classified as 'increased' and those where a >50% decline was recorded as 'decreased'.

Data for attributes 4 to 7 refer only to polycarpic perennials and *Ranunculus bulbosus*, a summer dormant (and relatively slow colonist), whose foliage, it is assumed, was not fully expanded in the October when the original vegetation survey was carried out, and shrubs were also omitted.

	Dryland			Wetland		
	Decreased	Inter-mediate	Increased	Decreased	Inter-mediate	Increased
1. Mean spp/m² with which typically associated (%)						
<10.1	0	0	0	36	13	25
10.1–14.0	29	11	15	29	13	38
14.1–18.0	36	22	55	29	38	25
18.1–22.0	29	56	25	7	25	13
>22.0	7	11	5	0	13	0
		K=O, NS			K=+0.10, NS	
2. Abundance in London area (% tetrads ± SD)	86±17	93±6	79±25	22±14	42±21	25±12
3. Monocarpic species (%)	18	18	38	22	0	11
		K=+0.20, P=0.05			K=–0.18, NS	
4. Seed bank						
Transient (types I–II)	29	38	53	17	0	0
Persistent, usually small (type III)	29	63	41	8	33	33
Often large (type IV)	43	0	6	75	67	67
	(n=14)	(8)	(17)	(12)	(3)	(6)
		K=–0.39, P<0.05			K=–0.03, NS	

5. Size of (attached) vegetative patch (mm)						
⩽100	43	44	35	21	13	0
101–250	7	44	25	21	0	25
251–1,000	29	0	20	29	50	50
>1,000	21	11	20	29	38	25
		K=+0.02, NS			K=+0.12, NS	
6. Canopy height*						
⩽100 mm	29	22	5	7	0	13
101–299	50	56	25	21	25	13
300–599	7	22	40	50	38	13
⩾600	14	0	30	21	38	63
		K=+0.38, P<0.01			K=+0.22, P<0.1	
7. Nuclear DNA (pg cell^{-1})	5.4±5.7 (12)	6.9±5.4 (9)	10.8±8.6 (15)	12.7±11.4 (9)	8.3±4.3 (3)	5.3±6.0 (3)
		K=+0.26, P<0.05			K=–0.33, NS	

Note: *Data for dryland monocarpic species (K=+0.29, P<0.1), dryland grasses (+0.51, P<0.05) and dryland herbs (+0.33, P<0.05) are also statistically significant.

Source: Floristic data are abstracted from Worthington and Helliwell, 1987, pp. 306–8.

disturbance associated with transference and replanting. In this experiment plant material has quite deliberately been handled in a rough manner. This being the case, we might predict that species which normally exploit disturbed habitats would be more prominent in the reconstituted vegetation. The increase in abundance of monocarpic grassland species (annuals and biennials) (see Table 5.7(3)) is consistent with these expectations. The soil seed bank may also be expected to have contributed to the recolonisation of bare earth exposed as a result of the transference. However, rather surprisingly, apart from the increase in docks (*Rumex* spp.), which have been pulled up by hand, there is no evidence of a general tendency for species with a persistent seed bank to have increased and for those with a transient seed bank to have decreased, rather the reverse (see Table 5.7(4)). Equally, while creeping thistle (*Cirsium arvense*), which regenerates from root fragments to form extensive patches (Grime *et al.*, 1988), has been invasive, there is no general relationship between capacity for vegetative spread (maximum patch size) and success following transference (see Table 5.7(5)). Moreover, there are no major differences between the range of strategies in the transferred vegetation and that of the original turf (see Figure 5.6). Thus, like Worthington and Helliwell, I conclude that the present composition of the transferred grassland vegetation is primarily the product of the successful regrowth of a majority of the transferred grassland species. (It should be emphasised however that the species involved in the transference are mainly associated with relatively productive habitats and include many C–S–R strategists and competitive ruderals (see Figure 5.6). Stress-tolerators, which tend to be less resilient and to have a slower growth rate, might not have responded so favourably.)

Trends within the wetland datasét are rather different. Firstly, monocarpic species tend to have decreased (see Table 5.7(3)). Secondly, a majority of species, both increasing and decreasing, have a persistent seed bank (see Table 5.7(4)). The extent of regeneration from buried seed (presumed to be great for rushes (*Juncus* spp.)) is therefore difficult to assess. Thirdly, the range of plant strategies is rather different from that present in the original turf (Figure 5.6). The proportion of competitors (C) and of stress-tolerant competitors (SC) has increased at the expense of other strategy types. These two increasing strategies (C and SC) are characteristic of sites with low disturbance. Therefore, since the processes of vegetation transference itself involved a high level of disturbance (Worthington and Helliwell, 1987), it is assumed that the trends identified for wetland species relate, not to the revegetation process, but to the effects subsequent of management.

Problems relating to after-care of the transferred vegetation

The data in Figure 5.6 suggest that grazing pressure is much lower in the wetland areas used as a receptor site. This explanation would also account for the decreased abundance of wetland monocarpic species (annuals and biennials) which typically exploit disturbed conditions. Reduced grazing leads to an increase in the height of the vegetation and an increase in taller species at the expense of shorter ones. Therefore, to assess whether the intensity of grazing pressure is too low, the maximum canopy heights of increasing and decreasing species are compared. The high incidence of 'tallness' amongst increasing species, particularly those of grassland (see

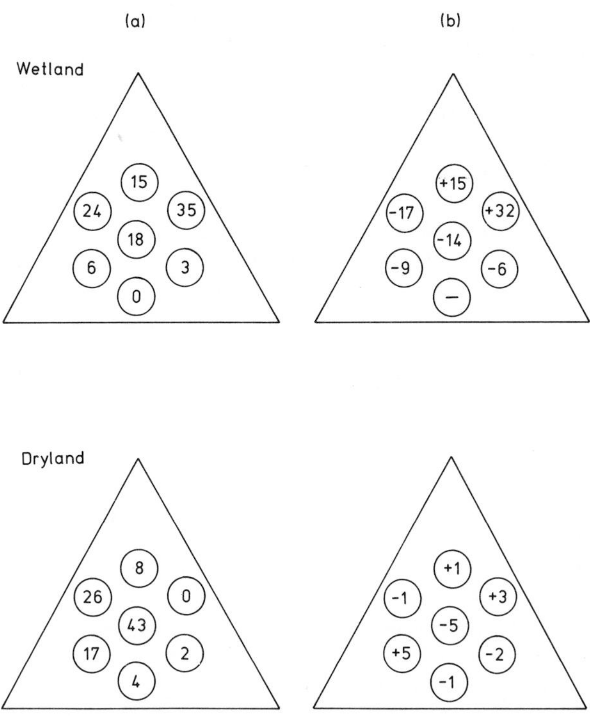

Figure 5.6 Plant strategies *sensu* Grime (1979) of species recorded in the transplantation experiment at Tomrod.

Note: Increased and decreased species are defined in Table 5.7.

Table 5.7(5)), supports the suggestion that the site is, or has in the recent past been undergrazed.

The other critical aspect of management relates to its timing. As already noted above, species differ in their capacity to exploit the cool (spring and autumn) and warm (summer) seasons of growth. An alteration in the period of management is likely to alter the balance between cool-season species (with high DNA amounts) and warm-season species (with low amounts). If the period of management is brought forward to spring, the summer growth of species with low DNA content may be encouraged, while a delay of management until summer might promote species with rapid spring growth.

In the case of grassland plant species with high DNA amounts (cool-season plants) tend to have increased at the expense of those with low amounts (warm-season plants) (see Table 5.7(7)). At present the site is grazed during May to July, again in September, with additional grazing in October or early March (Worthington and Helliwell, 1987). The results given in Table 5.7(7) imply that grazing in the original, parental grassland area commenced earlier (or at least was more intense during spring, or even that there was formerly more winter flooding, which would also restrict species with early spring phenologies).

In contrast to grassland plants, increased wetland species tend to have a low DNA amount (associated with summer growth) (see Table 5.7(7)). This result, taken in conjunction with the increased abundance of competitors and stress-tolerant competitors (which exploit undisturbed habitats), suggests a relatively low impact of grazing and other forms of summer management on the floristic composition of the wetland area. Two possible explanations can be suggested. Firstly, because of the low stocking rates and late commencement of grazing, there may be sufficient herbage in the grassland area to allow the stock to avoid the wet area with its abundance of unpalatable rushes. Alternatively, the water-table in summer may now be higher than at the original site, making the wetland vegetation less accessible to grazing stock. Unfortunately it is not possible to discriminate between these two alternatives on the basis of the existing dataset. In either circumstance some modification to the existing management regime is recommended while the floristic diversity of the site still remains relatively high.

Conclusions

The available evidence suggests that attempts to establish or to transplant species-rich vegetation in the 'real world' have operated on a trial-and-error basis (see Jordan, in this volume). It is argued that results would be more successful if practitioners appreciated more fully the ecological properties of the species that they are attempting to manipulate and the ecological processes influencing plant distribution. This would lead to more appropriate species being chosen and suitable after-care being implemented. It is to be hoped that autecological data in Grime *et al.* (1988) will go some way to making such information more accessible. It is also suggested that, because our knowledge of the way ecosystems function is inadequate, an appreciation of which species normally occur together (i.e. phytosociology) is very valuable. Finally, it should be emphasised that this general approach, based on an appreciation of species ecology, is relevant not just to problems related to species-richness but can be used in all situations in which vegetation is being reconstructed or modified.

Acknowledgements

This work draws heavily on ecological theories developed by Professor J. P. Grime and on data jointly collected with Professor Grime and Dr R. Hunt (Grime *et al.*, 1988). I would also like to thank Dr B. Campbell and Professor Grime for permission to reproduce unpublished experimental results, Dr T.C.E. Wells for access to much valuable data and the Nature Conservancy Council for permission to utilise information from the National Vegetation Survey prior to its publication.

References

Bradshaw, A. D., Humphreys, M. O., Johnson, M. S., 1978, 'The value of heavy metal tolerance in the revegetation of metalliferous mine waste', in G. T. Goodman and M. J. Chadwick (eds.), *Environmental Management of Mineral Waste*, Sijhoff and Nordhoff, Netherlands, pp. 311–34.

Burton, R. M., 1983, *Flora of the London Area*, London Natural History Society, Spottiswoode Ballantyne Ltd., Colchester and London.

Davis, B.N.K., 1982, 'Regional variation in quarries', in B.N.K. Davis (ed.), *Ecology of Quarries*, Institute of Terrestrial Ecology, Cambridge, pp. 12–19.

Fenner, M., 1985, *Seed Ecology*, Chapman & Hall, London.

Grime, J. P., 1979, *Plant Strategies and Vegetation Processes*, John Wiley, Chichester.

Grime, J. P., 1986a. 'The circumstances and characteristics of spoil colonisation within a local flora', in M. W. Holdgate (ed.), Proceedings of the Royal Society discussion meeting on quantitative aspects of the ecology of biological invasions, *Philosophical Transactions of the Royal Society*, **B314**, 637–54.

Grime, J. P., 1986b, 'Manipulation of plant species and communities', in A. D. Bradshaw, E. Thorpe and D. A. Goode (eds.), *Ecology and Design in Landscape*, BES Symposium **24**, Blackwell Scientific Publications, Oxford, pp. 175–94.

Grime, J. P., Hodgson, J. G., Hunt, R., 1988, *Comparative Plant Ecology: a Functional Approach to Common British Plants*, Unwin Hyman, London.

Grime, J.P., Mowforth, M.A., 1982, 'Variation in genome size – an ecological interpretation', *Nature*, **299**, 151–153.

Hodgson, J. G., 1986, 'Commonness and rarity in plants with special reference to the Sheffield Flora. Part II. The relative importance of climate, soils and land use', *Biological Conservation*, **36**, 253–74.

Hodgson, J. G., Mackey, J.M.L., 1986, 'The ecological specialization of Dicotyledonous families within a local flora: some factors constraining optimization of seed size and their possible evolutionary significance', *New Phytologist*, **104**, 479–515.

Jefferson, R. G., Usher, M. B., 1987, 'The seed bank in soils of disused chalk quarries in the Yorkshire Wolds, England: implications for conservation and management', *Biological Conservation*, **42**, 287–302.

Peterken, G. F., 1974, 'A method for assessing woodland flora for conservation using indicator species', *Biological Conservation*, **6**, 239–45.

Salisbury, E. J., 1942, *The Reproductive Capacity of Plants*, Bell, London.

Thompson, K., Grime, J. P., 1979, 'Seasonal variation in the seed banks of herbaceous species in ten contrasting habitats', *Journal of Ecology*, **67**, 893–921.

Wells, T.C.E., 1971, 'A comparison of the effects of sheep grazing and mechanical cutting on the structure and botanical composition of chalk grassland', E. Duffey and A. S. Watt (eds.), *The scientific management of animal and plant communities*, BES Symposium 11, Blackwell, Oxford, pp. 497–515.

Wells, T.C.E., 1979, 'Habitat creation with respect to grassland', S. E. Wright and G. Buckley (eds.), in *Ecology and Design in Amenity Land Management*, Wye College and Recreational Ecology Research Group, Wye, Kent, pp. 128–45.

Wells, T.C.E., 1987, 'The establishment of floral grasslands', *Acta Horticulturae*, **195**, 59–69.

6

Management problems arising from successional processes

A. D. Bradshaw

Introduction

Any plant or animal community, whatever species compose it, has specific requirements, in terms of the physical, chemical and biotic factors on which it depends and which influence it. The community is, of course, just part of a complex interacting system, an ecosystem, which may be very finely balanced indeed. The existence and stability of this ecosystem depend on many factors and their interaction. Change in any factor is likely to cause change in the plant and/or animal community within the ecosystem.

In any particular situation, however, one factor may have a dominating, controlling, influence. Common examples are grazing and burning, both of which prevent tree growth. In these circumstances, if the controlling factor is removed, then spectacular changes will occur in the community, not only towards greater biomass but also towards larger, more vigorous species, a process well known as succession, or, properly, secondary succession since it is from an already developed state.

In Britain, as in many countries, most communities, except woodland and high mountain, are affected by man and grazing animals, and are in what can be termed an arrested stage of succession, in which the balance of factors can be very fine. A good example is calcareous grassland. Its precise composition varies from place to place in relation not only to intensity of grazing, but also to soil factors such as soil depth and phosphorus (Gittins, 1969).

All this is well known to ecologists. However, in the context of habitat reconstruction, there are some important matters to consider.

Ecosystem requirements

When community, or really ecosystem, reconstruction is being undertaken, it is obvious from many contributions in this volume that all the requirements of the ecosystem must be met. In most situations the soil, particularly, has been damaged or destroyed. Successful restoration then depends on the treatment of all the soil problems – physical, nutrient and toxicity factors – as well as the biological problems set by the species themselves (Schaller and Sutton, 1978; Bradshaw and Chadwick, 1980).

What is crucial is to ensure that these problems are properly treated. If they are not then reconstruction will fail. For some, such as soil acidity, the appropriate amount of neutralising material, lime, must be given at the outset (Costigan *et al.*, 1981). For others, developing deficiencies can be made up later. In our experience the most important problem to overcome, and the problem where failure to provide successful treatment is most common, is lack of nitrogen (Bloomfield *et al.*, 1982). This arises because a great deal of nitrogen must be accumulated in the soil to provide an appropriate capital from which adequate supplies of mineral N can be released by mineralisation annually (Bradshaw, 1983). The more advanced and productive the community, the greater is the amount of nitrogen required, which may present problems (Marrs *et al.*, 1983).

In the long term this nitrogen capital is accumulated by natural processes, by atmospheric deposition and more particularly by biological fixation commonly associated with legumes. The latter can accumulate over 100 kg N $ha^{-1} yr^{-1}$ even on difficult substrates (Dancer *et al.*, 1977). Invasion by legumes or other N-fixing species is a notable part of most successions (Marrs *et al.*, 1983). Because the nitrogen is accumulated in plant materials with a low C/N ratio, decomposition and release of mineral N is particularly rapid (Jefferies, *et al.*, 1981; Lanning and Williams, 1979). Legumes are therefore an important element in degraded land restoration.

Successional processes as a problem

Succession is, indeed, always waiting to happen, even on most extreme materials, whether glacial moraines, lava flows, derelict quarries or disused railway lines. This process of primary succession can achieve remarkable results very quickly – complete forest cover in less than 100 years in most cases. It is due both to allogenic factors originating outside the plant and animal community, and to autogenic factors arising from within, which progressively improve the habitat and therefore allow the plant and animal community to develop a progressively greater complexity, biomass and activity.

In this the part played by autogenic factors, and particularly the nitrogen accumulation provided by N fixing species, must not be underrated. It can be the driving force in the development of the ecosystem, for example on glacial moraines (Crocker and Major, 1955), iron-ore spoil banks (Leisman, 1957) and china clay wastes (Roberts *et al.*, 1981). The potential effect of N-fixers has already been alluded to in relation to clovers. A demonstration of the actual effect of an N-fixer on china clay waste, *Alnus* sp., is given in Figure 6.1. The nitrogen accumulated by the alder allows the sycamore to grow at normal rates; without it the sycamore is completely moribund.

The developing vegetation will have other effects. Its ability to scavenge nutrients such as phosphorus and potassium from lower layers and accumulate them at the surface as a result of litter fall is very important. The vegetation will also take up, retain and then contribute to the soil, nutrients such as nitrogen and potassium being deposited from the atmosphere. The result is considerable improvement in the fertility of the surface soil (Knabe, 1973).

In parallel with this is a progressive invasion of new species into the developing community. Part of this is due to the stochastic processes of dispersal which are

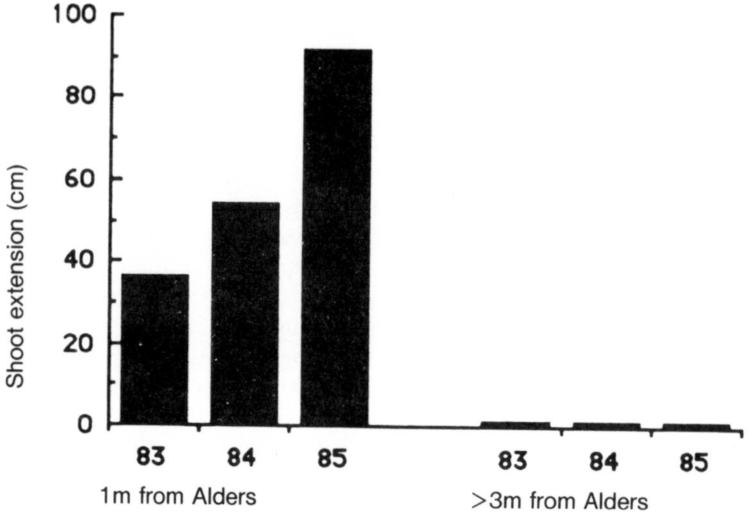

Figure 6.1 The influence of nitrogen accumulation on growth in successional situations: the effect, over three years, of proximity to alder (*Alnus glutinosa* and *incana*) on the growth of sycamore (*Acer pseudoplatanus*) established on china clay mining waste for eight years.

unlikely to deliver immediately all the species that can grow on a site. But it is partly as a result of the improvements occurring in the habitats due to the effects of the existing plants, a process which can be called facilitation.

The exact balance between these processes is still a matter for debate (see for example Connell and Slatyer, 1977; Miles, 1978). But no matter what processes are occurring, succession can be very important in creating new communities of great value such as Hampstead Heath (see Figure 6.2). However, it can equally occur where the habitat has been reconstructed and a particular community put in place, and can be a potential cause of trouble.

There are then two possibilities. If the habitat has been reconstructed to a level of fertility, etc., which is more than that required by the community being established on it, there may be fairly rapid change in the community, analogous to what occurs in secondary succession, until a new community develops which is in equilibrium with the habitat.

If, on the other hand, the habitat has been reconstructed to a fertility level with which the community is in equilibrium, there will not be any rapid or immediate change. But the sort of progressive improvement in fertility which we associate with primary succession almost certainly can, and will, occur (see Marrs and Gough, in this volume). As a result the community will again change, but at a slower rate than in the previous case.

It would be easy to put the former problem on one side because it really arises from poor reconstruction technique, which ought not to occur, and concentrate only on the latter, which is an inevitable characteristic of all communities until they

Figure 6.2 The beneficial effects of natural succession: a photograph of the gravel pits situated at the top of Hampstead Heath, London, in 1870, and exactly the same view today – no artificial planting has been carried out.

have reached their final, relatively stable, climax condition. However, in practice it may be very difficult to achieve the latter. So whatever is the origin of the successional change, we have to find ways by which it can be prevented if we want to achieve accurate reconstruction of habitats and communities.

The problems in practice

Since our aims are practical – how to achieve successful reconstruction – we must now translate all this theory into simple practice. Just what have we to deal with? In effect there are only two problems: (1) growth and (2) invasion by other species.

Growth

Growth is both an essential attribute of all communities and potentially one of their most destabilising processes. From it can arise large accumulations of plant material, particularly if recycling cannot keep pace. And those species which grow faster, taller, more aggressively, will eliminate the other weaker species by competition. If this growth, therefore, is not removed then not only does the morphology of the community change radically, such as a short grass turf changing to a rank overblown sward, but the species composition will change, perhaps totally. The best evidence for what can happen was provided many years ago by grassland agronomists (e.g. Jones, 1933). The richness of species to be found in old grazed pastures is witness to the crucial importance of growth removal.

In reconstructed habitats the problem of growth and competition from vigorous species will obviously be most marked where the reconstruction has in a sense been too successful and the fertility has been raised to too high a level. Many people are interested in the re-creation of herb-rich swards. If these are established on soil of high fertility, perhaps brought in specially to achieve success, the result may well be failure, because of the excessive growth and competition of those species, particularly grasses, which have the ability to respond to the high fertility. It is good therefore to see in many wild seed catalogues produced by commercial companies the repeated recommendation that low fertility situations are best.

Of course what is required is not just an appropriate low fertility but also a management system by which growth is removed to an appropriate level and in a manner appropriate to the species involved. One of the most floriferous semi-natural habitats, the lowland meadow, has at the North Meadow at Cricklade, Wiltshire, been subject for centuries to a very precise and consistent regime of growth removal, the annual hay cut in July, followed by a carefully regulated grazing regime from August to February.

Any discussion on growth management could occupy a whole book, which is not possible here. There is, however, one other aspect besides the general control of growth that deserves mention – this is the relationship between herbaceous vegetation and trees, particularly in those situations where trees are being planted into grass dominated swards, as transplants, whips or even standards. In these conditions the growth of herbaceous vegetation can provide quite enormous competition against the trees, not just reducing the growth of the trees but even killing them. A recent publication of the Forestry Commission makes the situation very clear (Davies, 1987). The situation can be exacerbated if the trees are fertilised because this actually increases the competitive effect of the grass more than it improves the resistance of the tree (Gilbertson *et al.*, 1987).

Invasion by other species

The inevitability of invasion by species not previously present in the community has already been commented upon. In reconstructed habitats, whether this will be a problem will depend particularly on the level to which the reconstruction has been taken in terms of soil fertility. In any situation where the habitat has been substantially improved there is obviously the possibility that species which are adapted to the higher fertility will invade, particularly if no arresting factor is applied. The consequences of this are now apparent on every motorway verge, where, since the cessation of cutting, shrubs and trees are rapidly invading wherever immigrants are available. Over the next 30 years the grassland verges of motorways, which were all carefully topsoiled, will turn into woodland.

There are many situations where nutrient addition is important to ensure effective restoration of damaged communities (for many examples see Bradshaw and Chadwick, 1980). One good example is the restoration of eroded sand dunes, where sewage sludge has been shown to be a very effective tool to ensure the re-establishment of marram grass (*Ammophila arenaria*). However, excessive use of sewage sludge can be counter-productive because it can raise the levels of N and P to those where species alien to sand dunes particularly nettles (*Urtica dioica*) will invade and become permanently established. Many will interpret this as a good reason for not using such treatments; but it is not, because restoration without nutrients often leads to failure. It is merely an argument for achieving the right level of habitat improvement in reconstruction.

One particular invasion which may be most serious because of its autogenic effects, is invasion by nitrogen-fixers. If therefore it is the early stages of succession which are required, particularly communities dominated by heather (*Calluna vulgaris*), then it is imperative to prevent invasion by gorse and broom (*Ulex europaeus* and *Cytisus scoparius*), and on china clay waste, for instance, the very successful alien tree lupin (*Lupinus arboreus*). In the reconstruction of woodland they would, by contrast, all be invaluable.

Practical solutions to successional problems

Faced with these two major problems, what is the restoration ecologist to recommend? We have to look for simple, readily applicable, and inexpensive solutions. There are in fact several methods, nearly all self-evident and easy to apply.

Introduce the correct species at the outset

This may seem very obvious. Where only a few species are involved it may not be difficult. But in complex mixtures, species may be unwittingly included which are aggressive or have crucial facilitation effects. A particular example would be the inclusion of ryegrass (*Lolium perenne*) and a vigorous legume such as red clover (*Trifolium pratense*) in a wild flower mix for neutral soils. Many people would assume that this would never happen, but it does, commonly. If these species are not included, others almost as troublesome, such as the cultivar S59 of red fescue (*Festuca rubra*), and apparently (but not actually) more innocent legume species such as black medick (*Medicago lupulina*), are included instead.

Ensure absence of aggressive weed species

Although the original seeds mix may be chosen with great care, the site materials may contain aggressive, inappropriate species. This is almost inevitable when topsoil is used, unless great care is taken, because the supply of topsoil is very uncontrolled (Bloomfield *et al*., 1981). The species which can appear are not just annual or biennial broad leaved weeds, some of which, such as broad-leaved dock (*Rumex obtusifolius*) can be very aggressive, but also very difficult perennials such as the grasses *Dactylis glomerata* and *Elymus repens*. The simple solution is to avoid using topsoil as the starting point unless it has been specifically obtained from the ecosystem to be reconstructed.

Rely on lack of immigrants

In some situations the species which could invade are not present in the vicinity sufficiently commonly to cause trouble. This is very possible in the case of tree and shrub species, where the community being reconstructed is in an extremely open, perhaps arable situation. There is no doubt that the slowness of tree and shrub colonisation of our motorway verges following the end of mowing has been largely due to this; where suitable seed sources are available such colonisation can be very rapid. Once a few individuals attain seed-bearing age the whole process of colonisation can become immensely faster; this is the history of the invasion of much chalk grassland by hawthorn (*Crataegus monogyna*). So it is by no means a perfect method.

Rely on competition to exclude unwanted species

One aspect of the growth of the wanted species is that they can, for a time at least, exclude the unwanted. Grass growth, as we have seen, can restrict tree and shrub establishment. There is no doubt that this can, in some places, have very helpful effects, e.g. in slowing down the succession of grassland to woodland where grazing is absent. But it certainly does not prevent it. The regeneration niche plays a crucial part in the life history of many species (Grubb, 1977) and is contantly being produced by the activities of animals and the death of existing plants.

Ensure soil fertility is properly restricted

As was argued earlier, this can be a very powerful method. Obviously the reconstruction of an acid peaty soil in the restoration of heathland will preclude the invasion by species of more neutral habitats. Similarly the reconstruction of a highly calcareous soil in the restoration of a calcareous grassland will ensure the exclusion of the more nutrient-demanding species of neutral soils. An excellent example of this is the extraordinary flora of the alkali waste heaps left by the Le Blanc industry in north-west England. Despite the fact that they have been completely surrounded by dense vegetation typical of normal neutral soils, for one hundred years, they have maintained a very unique calcicole flora (Bradshaw, 1983). Indeed, it is a very restricted flora because many of the calcicole species that could invade are too far away to be able to get there.

If the fertility of these types of habitat is improved, for instance by the addition of NPK fertiliser, then several of the more demanding species will establish. At the same time there is a great transformation of the composition of the community, as the more vigorous species drive out the less vigorous (Ash *et al.*, 1989). In any reconstruction operation involving these sorts of habitats it is therefore absolutely crucial to keep control over fertility levels. Sufficient depth of soil from the original ecosystem must be taken, as for example was done with limestone grassland at Thrislington (Park, in this volume). At the Liverpool International Garden Festival, in a specially reconstructed heathland garden, *Rumex obtusifolius* grew splendidly amongst the turves of *Calluna vulgaris* because the depth of the soil transferred was too little. In a major transfer of the unique calcicolous community occurring on the old lime beds at Witton, Cheshire (now a recognised site of nature conservation importance), to allow the extension of a domestic refuse disposal site, a minimum of 1.5 metres of the lime waste is being transferred to ensure that species cannot get at the nutrients in the refuse beneath.

It is possible that even this depth of material will not be enough. But it will at least maintain most of the essential features of the original habitat. In this respect it is good that the treatment of motorway cutting faces with topsoil, so deplored by Nan Fairbrother (1972) because of its destruction of potentially unique and extreme habitats (see also Penny Anderson, in this volume), is no longer being applied so universally. The outcome at various sites on the M2 and M5 motorways is already very interesting. Certainly the normal range of species of neutral grassland is not invading.

Rely on other site factors

There will, of course, be other site factors such as drainage and soil texture which can exert important controlling effects, and must not be forgotten. But it may be difficult to match exactly what operated in the original site.

Cutting or grazing

Either of these obviously restricts growth and therefore competition. Their effects on overall growth and on individual species are both obvious and documented, and well discussed in relation to management of natural and semi-natural plant communities (Wells, 1980). What is disappointing is that despite the very careful work carried out in relation to important habitats, particularly roadsides (Parr and Way, 1988), there has been only restricted adoption of properly designed cutting programmes, despite their economic and wildlife benefits.

Although cutting will not always have quite the same effects as grazing because it is non-selective, it is nevertheless an effective substitute (Wells, 1971). If the cuttings are removed this does allow some removal of nutrients and lowering of fertility, as discussed by Marrs and Gough (in this volume).

Burning

Cutting and grazing can be expensive and not necessarily easy to carry out. In some situations burning is an important alternative (Green, 1986). It will not have the same detailed effects as cutting and grazing, but it is a powerful method for preventing succession, and appropriate for particular plant communities such as heathland. It does of course lead to substantial losses of nitrogen and sulphur, although rather little in relation to amounts already in store.

Removal of individual species

If the previous treatments cannot be applied, or at least may be applied insufficiently to prevent successional changes, this technique of last resort can be adopted. The removal of species in their entirety by hand or machine is certainly a very positive, if rather extreme and laborious method. Cutting of individual species is an alternative. Unfortunately regrowth of cut stumps, especially of trees and shrubs, may make the second situation worse than the first.

Herbicides provide a very positive alternative. But if they are to be used extensively rather than selectively, then great care will be needed in their choice and application (Haggar, 1980). It may be much better to use them on a more selective, spot treatment basis where uncritical application is less likely to cause trouble. But they will still be expensive and troublesome to use.

Conclusions

The characteristics of all communities are the outcome of interaction of community and habitat factors within the delicate balance of ecosystems. If the habitat factors are changed then it is almost inevitable that the community will change. Consequently, if an accurate reconstruction of any ecosystem is required, it is necessary to put everything back exactly as it was. This includes all the attributes of the habitat, including any controlling factors which appear to be arresting succession, because succession is a feature of nearly all ecosystems.

If we do not do this, we can bring into play particular alternative techniques. Species removal by hand, for instance, may be a substitute for proper fertility control or grazing. But in few cases will the application of a new controlling factor be a proper substitute for the operation of the original factor, or set of factors, since in most cases the qualities of a particular community are determined by several factors acting in consort.

In planning for reconstruction we should aim to put back all factors as they were in order to achieve a 'genuine' restoration. If this cannot be done then we must be realistic and truthful and admit that we cannot exactly replace what was there before. In many cases this may not matter because, from a wildlife point of view, we can certainly reproduce something not far removed from what was there before, of similar value and interest.

If, within the constraints of what resources are available, we cannot manage to put back anything similar to what was there before, it should still be possible to produce something else which will not only add to the diversity of habitats in this country but also be beautiful and interesting. Anyone who does not believe this should look at the woodland in the old sand pit on Hampstead Heath or the complex of communities in the Millersdale Quarry Nature Reserve. There is more to reconstruction than narrow restoration.

References

Ash, H. J., Gemmell, R. P., Bradshaw, A. D., 1989, 'Colonisation of industrial waste heaps and the introduction of native plant species', (in preparation).

Bloomfield, H. E., Handley, J. F., Bradshaw, A. D., 1981, 'Top soil quality', *Landscape Design*, **135**, 32–4.

Bloomfield, H. E., Handley, J. F., Bradshaw, A. D., 1982, 'Nutrient deficiencies and the aftercare of reclaimed derelict land', *Journal of Applied Ecology*, **19**, 151–8.

Bradshaw, A. D., 1983, 'The reconstruction of ecosystems', *Journal of Applied Ecology*, **20**, 1–27.

Bradshaw, A. D., Chadwick, M. J., 1980, *The Restoration of Land*, Blackwell Scientific Publications, Oxford.

Connell, J. H., Slatyer, R. O., 1977, 'Mechanisms of succession in natural communities and their role in community stability and organisation', *American Naturalist*, **111**, 119–44.

Costigan, P. A., Bradshaw, A. D., Gemmell, R. P., 1981, 'Reclamation of acidic colliery spoil. I. Acid production potential', *Journal of Applied Ecology*, **18**, 865–78.

Crocker, R. L., Major, J., 1955, 'Soil development in relation to vegetation and surface age at Glacier Bay, Alaska', *Journal of Ecology*, **43**, 427–48.

Dancer, W. S., Handley, J. F., Bradshaw, A. D., 1977, 'Nitrogen accumulation in kaolin mining wastes in Cornwall. II Forage legumes', *Plant and Soil*, **48**, 303–14.

Davies, R. J., 1987, *Trees and Weeds*, H.M.S.O., London.

Fairbrother, N., 1972, *New Lives, New Landscapes*, Penguin Books, Harmondsworth.

Gilbertson, P., Kendle, A. D., Bradshaw, A. D., 1987, 'Root growth and the problems of trees in urban and industrial areas', in D. Patch (ed.), *Advances in Practical Arboriculture*, H.M.S.O., London, pp. 59–66.

Gittings, R., 1969, 'The application of ordination techniques', in I. H. Rorison (ed.), *Ecological Aspects of Mineral Nutrition of Plants*, Blackwell, Oxford, pp. 37–66.

Green, B. H., 1986, 'Controlling systems for amenity', A. D. Bradshaw, D. A. Goode and E. Thorp (eds.), *Ecology and Design in Landscape*, Blackwell, Oxford, pp. 195–210.

Grubb, P. J., 1977, 'The maintenance of species-richness in plant communities: the importance of the regeneration niche', *Biological Reviews*, **52**, 107–45.

Haggar, R. J., 1980, 'Weed control and vegetation management by herbicides', in I. H. Rorison and R. Hunt (eds.), *Amenity Grassland: An Ecological Perspective*, John Wiley, Chichester, pp. 163–73.

Jefferies, R. A., Willson, K., Bradshaw, A. D., 1981, 'The potential of legumes as a nitrogen source for the reclamation of derelict land', *Plant and Soil*, **59**, 173–7.

Jones, M. J., 1933, 'Grassland management and its influence on the sward', *Journal of the Royal Agricultural Society*, **94**, 21–41.

Knabe, W., 1973, 'Investigations of soils and tree growth in five deep-mine refuse piles in the hard coal region of the Ruhr', R. J. Hutnik and G. Davis (eds.), in *Ecology and Reclamation of Devasted Land*, Gordon & Breach, New York, Vol. 1, pp. 307–24.

Lanning, S., Williams, S. T., 1979, 'Nitrogen in revegetated china clay sand waste. I Decomposition of plant material', *Environmental Pollution*, **20**, 147–59.

Leisman, G. A., 1957, 'A vegetation and soil chronosequence on the Mesabi iron range spoil banks, Minnesota', *Ecological Monographs*, **27**, 221–45.

Marrs, R. H., Roberts, R. D., Skeffington, R. A., Bradshaw, A. D., 1983, 'Nitrogen and the development of ecosystems', J. A. Lee, S. McNeil and I. H. Rorison (eds), in *Nitrogen as an Ecological Factor*, Blackwell, Oxford, pp. 113–36.

Miles, J., 1978, *Vegetation Dynamics*, Chapman & Hall, London.

Parr, T. W., Way, J. M., 1988, 'Management of roadside vegetation: the long term effects of cutting', *Journal of Applied Ecology*, **25**, 1075–88.

Roberts, R. D., Marrs, R. H., Skeffington, R. A., Bradshaw, A. D., 1981, 'Ecosystem development on naturally colonised china clay wastes. I Vegetation changes and overall accumulation of organic matter and nutrients', *Journal of Ecology*, **69**, 153–61.

Schaller, F. W., Sutton, P. (eds.), 1978, *Reclamation of Drastically Disturbed Lands*, American Society of Agronomy, Madison.

Wells, T.C.E., 1971, 'A comparison of the effects of sheep grazing and mechanical cutting on the structure and botanical composition of chalk grassland', in E. Duffey and A. S. Watt (eds.), *The Scientific Management of Animal and Plant Communities for Conservation*, Blackwell, Oxford, pp. 497–515.

Wells, T.C.E., 1980, 'Management options for lowland grassland', in I. H. Rorison and R. Hunt (eds.), *Amenity Grassland; An Ecological Perspective*, John Wiley, Chichester, pp. 175–95.

SECTION 4

Opportunities for habitat reconstruction

Opportunities for habitat reconstruction continually crop up as land use patterns evolve. Each successive upheaval in the development of a landscape produces new niches to be exploited; and so it pays for the restorationist not only to look for opportunities within the present land pattern, but as Bunce and Jenkins point out, to note where shifts and developments have taken, and are taking place. Historically, the forest clearances created by primitive agriculture created the conditions for semi-natural communities, many of which (such as meadows and heathlands) are themselves legitimate models for today's habitat reconstruction exercise. But it is the more recent shifts of land use, some ephemeral and apparently unpromising, that have really provided the spare land for the deliberate cultivation of artificial habitats.

The most important of these was the Industrial Revolution and the wide range of new habitats which it created through urbanisation – dense housing areas, factories, mineral extraction sites and their ancillary transport systems of railways, canals and roads. Vast areas of derelict and industrial wasteland were generated, and with them some of the most bizarre examples of (accidental) habitat creation. Colonies of orchids developing on pulverised fuel ash, calcicole floras colonising old lime beds in calcifuge surroundings, and the discovery of metal-tolerant grasses growing on non-ferrous mine wastes not only made exciting ecological studies: they also pointed the way towards imaginative reclamation, and drew attention to the feasibility of habitat construction itself.

Urban wastelands also provide innumerable situations and substrata on which to practise habitat reconstruction. While the urban conservation movement has drawn attention to the fragments of semi-natural habitat caught up in and around the city, there are other communities more at home in this environment. Jane Smart addresses the potential of 'wasteland' or pioneer communities developing on wharves, railway sidings, gardens, abandoned allotments and similar areas. 'Soils' consisting of brick rubble, glass, tarmac, gravel, clay or subsoil are inherently variable and infertile, and are often suitable substrates upon which attractive and interesting vegetation can be sustained by relatively simple management. Such areas soon become a focus for educational interest, for community involvement, and for wildlife.

Too often there has been a tendency to rationalise urban and mineral extraction sites with topsoil coverings and to adopt dull landscaping solutions when, as has been shown in the previous section, this does not best serve the ecological goals of habitat reconstruction. A more difficult opportunity is therefore the formal, urban park, recreational playing field and school grounds. In the special case of school grounds, Funnell shows how primarily educational interests can drive habitat creation work, and can even succeed in forcing the issue with grounds maintenance staff who are used to doing things differently. If carefully rationalised, less formal landscaping solutions can in these situations be cost-effective.

The countryside itself presents further opportunities for habitat reconstruction, particularly as the semi-natural areas within it have been so reduced and degraded by the pressures of modern agriculture, commercial forestry and urbanisation. Outside nature reserves and protected areas the countryside simply does not look after itself, and ways to integrate semi-natural communities into farming and forestry practice must be positively sought.

A rationale for identifying farmland surplus to agricultural requirements is described by Green and Burnham, who highlight areas of 'mismatch' of demanding crops with poor soils, high conservation potential and environmental vulnerability. Such areas are clearly candidates for habitat reconstruction and enhancement, but more information is needed on how they function ecologically. For example, Bunce and Jenkins petition for a much more detailed analysis of the pattern of farm landscapes, particularly with respect to the ecological dynamics of fragmented areas, in order to evaluate the relationship between connectivity and the persistence of plant and animal populations in edge habitats.

Much the same is true of commercial forestry plantations, where the dynamics of import, maintenance and export of plant and animal species are stressed by Anderson. Although there are obvious limitations for semi-natural communities under closely-spaced tree canopies, diversity can be encouraged by treatments and planting arrangements which are designed to break up the pattern of the canopy at both the forest and compartment scales. Here again, edges and interfaces are important as places in which to sustain and manage the target communities.

The opportunities for habitat reconstruction may be there, but they will only be realised if sympathetic land use policies are in place and operate effectively. In the urban environment, growing public interest in nature and the environment will make this easier to achieve on land already in public ownership. In the countryside, although policies and grant structures relating to forestry and agriculture have never been better disposed towards conservation management, there is no guarantee that habitat opportunities will be actively promoted by individual land owners. As Way points out, what is actually done will depend largely on its income-generating potential.

7

Land potential for habitat reconstruction in Britain

R.G.H. Bunce and *N. R. Jenkins*

Introduction

The majority of papers in the present volume are concerned with the practicalities of habitat reconstruction. This chapter is intended as an overview of current acreages of suitable urban land and countryside in Britain within which such activities could occur. Areas of potential change which depend on future agricultural policies are also considered, including the possible pathways by which the native botanical capital could colonise less intensively managed farmland.

Three broad stages may be separated in the development of the British landscape. Initially, new habitats were created following forest clearances, and many are now much prized in conservation terms, e.g. the heathlands and ancient anthromorphic grasslands which over the centuries have developed a balance of native species. These must now be maintained in their seral stages in order to hold the equilibrium of conditions which gave rise to them originally. The second major development came with the great engineering works of the nineteenth century. These created new types of habitat, e.g. railway, canal and mineral extraction sites, together with massive urban connurbations. Subsequently, more areas have become available due to the decline of traditional heavy industries in inner cities, and as a result of major public works such as motorways. The growth of the environmental movement has added further impetus to the awareness of the potential for creating new habitats on otherwise highly disturbed sites. However, although an observer travelling by road or rail around the country can see many vacant or underused areas, there have been relatively few efforts to quantify the actual area of land concerned because of the disparate nature of data sources (Bunce and Heal, 1984). The available figures are reviewed here and compared with independent estimates from work carried out by the Institute of Terrestrial Ecology (ITE).

To complete the cycle, a third phase is now likely, due to the pressures on agriculture caused by food surpluses. This trend is still new and is difficult to quantify at present. Green (in this volume) presents figures as to the possible extent of land which might become available for habitat re-creation, and the implications are briefly discussed.

Urban land

The inherent difficulties of defining urban areas are summarised by Coppock and Gebbett (1978). Fordham (1974) used systematic point samples to estimate the area of urban land in Britain and derived a figure of 17,885 km^2 which, allowing for the expansion at 149 km^2 per year, gave 20,517 km^2 in 1978 (the baseline of our ITE survey). Best (1976), using a method derived from local authority records, quoted a figure of 18,750 km^2. The figure derived by ITE in 1978 is 29,974 km^2 from a stratified sample of 256 × 1km^2 sample squares (Bunce and Heal, 1984), (see Table 7.1). For all these estimates, the errors and measurements are comparable. Table 7.1 also gives the figures for the rural area with further categories being available (Bunce *et al.*, 1984).

Derelict land is even more fragmented than urban land, as emphasised by Coppock and Gebett (1978) and Barr (1969). Dennington (1979) summarised the difficulties of using the Department of Environment (DOE) figures for different counties and metropolitan areas. More recently in 1982, the DOE have co-ordinated data for England giving a figure of 45,683 ha which is comparable to the 96, 200 ha produced by ITE for the whole of Britain. A separate survey of derelict land was carried out by the DOE in 1988 and the estimates will be made available in the near future. The advantages of using the 1978 ITE survey referred to above is that these areas can be further broken down into categories of the type of derelict land concerned. Initial figures indicate a further breakdown into natural waste 19,200 ha; derelict vegetated land 13,000 ha; and derelict unvegetated land 70,000 ha. The latter area is comparable with the figure of 72,000 ha quoted by a co-ordinated study published by the DOE (1986).

In relation to the present topic, these broad figures need to be separated into areas which have potential for habitat re-creation. It is necessary to produce mutually exclusive categories, in order that the total extent of each type is fully available and that resources can be allocated efficiently. Thus, in a 1984 ITE survey, detailed mapping was carried out in towns so that the data could provide a more comprehensive breakdown than is currently available. For example, within the urban areas a sub-sample showed that almost 50 per cent of the residential land was not covered by buildings, and was therefore potentially available for wildlife enhancement. The figures produced by Bunce *et al.* (1984) also show that roads outside urban areas cover a considerable area, i.e. some 340,000 ha from the initial study. Included in this area are verges, roundabouts, motorway fringes and other similar areas which can also provide new habitats.

Further work is needed to establish the full range of land available for new wildlife habitat. Railway land, estimated at 54,400 ha by Bunce *et al.* (1984), presents a considerable area for habitat re-creation, as emphasised by the work of Sargent (1984) who described its current wildlife status as well as its potential. The character and importance of roads is described by Way (1973). The area of quarries, although small, is potentially important, as Davis (1982) has shown, for the establishment of species which are frequently rare in the surrounding countryside. Although the figures represent relatively clear-cut categories, there is potential for further refinement, as is emphasised by the estimate of 230,500 ha for recreation areas (Bunce *et al.*, 1984). These are generally intensively managed with fertiliser and gang mowers,

Table 7.1 Allocation of the land surface of Britain to various land-uses

Land use Area in hectares % of total	Land use	Area in hectares	% of total	Land use	Area in hectares	% of total
				Wheat	1,061,270	4.6
All crops	Cereals	3,426,148	14.9	Barley	2,117,240	9.1
4,428,207 ha				Other cereals	247,638	1.1
19.60%	Other crops	709,013	3.4	Ploughed/fallow	107,398	0.7
				Sugar beet	146,959	0.6
	Horticulture	293,046	1.3	Animal fodder	190,955	0.9
				Mixed crops	263,701	1.2
				Short-term leys	2,357,204	10.4
All grass	Leys	3,483,270	15.2	Other leys	1,098,820	4.8
6,385,070 ha						
27.80%				Generally reseeded	1,447,997	6.4
	Permanent grass	2,901,800	12.7	Older grassland	1,453,806	6.3
All wood				Copses	781,433	0.3
2,207,300	Broad-leaved wood	561,010	2.4	Shelter belts	34,552	0.2
9.60%	Conifer	1,404,737	6.1	Scrub	241,556	1.1
	Scrub	241,556	1.1	Woodland	1,853,050	8.1
				Rough grass	505,270	2.2
	Rough grassland	2,177,424	9.5	Mixed rough grass	513,960	1.4
				Bracken	360,530	1.6
Semi-natural				Rushes	374,358	1.6
5,514,553 ha				Mountain grass	621,306	2.7
24.00%						
				Purple moor grass	761,907	3.3
	Moorland	3,337,129	14.5	Cotton grass	669,994	2.9
				Heather	1,260,201	5.5
				General moorland	645,027	2.8
Unavailable	Aquatic	726,394	3.2			
4,334,471 ha				Buildings etc	2,277,890	9.9
18.90%	Human	2,992,440	13	Communications	719,550	3.1
				Inland rock	169,403	0.7
	Unavailable	615,637	2.6	Maritime	446,234	1.9

Note: This is based on estimates derived from field survey in eight sample squares in 1978 in each of 32 land classes. Conversion into a Great Britain basis is made by estimating the proportion of the kilometre squares belonging to each land class.

Source: Bunce *et al.*, 1984

but could become available for the development of new habitats (see Baines, Funnell, in this volume). The work by Good and Munroe (1981) on trees in urban areas emphasises the importance of such land that is present within towns. There are therefore many categories for which data are not available, and for which exciting new studies could be carried out.

Rural land

Turning now to the rural landscape (see Table 7.1), there are also areas which are frequently overlooked. Small areas in the corners of fields, and especially those associated with linear features, are also important in the landscape (see Tables 7.2 and 7.3), the latter representing refuges whereby species can be maintained in otherwise open landscapes. Changes in the amounts of these linear features are important (Barr *et al.*, 1984) and have been considerable in the period 1947–1985 (see Table 7.2). Simple estimates of width can convert these lengths of hedgerows, etc., to areas, giving an indication of the area of ground involved and their importance in habitat terms.

The hedgerow figure is an underestimate in that most boundaries between fields (even those without hedgerows) still have 'residual' species which may be important for insects. There are also headlands and streamsides within agricultural landscapes (see Table 7.4). To compare the effects of these features on species diversity, samples were taken from 8 random 1 × 1km squares in all of the 32 main land classes in Britain (Bunce *et al.*, 1981). Within each 1km^2 sample, the number of vascular plant species found in two 1 × 10m quadrats placed at random along streams, roadside verges and hedgerows were compared with those present in five random 200m^2 quadrats. Table 7.4 shows that the figures for hedges, streams and quadrats are quite comparable in terms of their common species, but there are more unique species associated with streams than in either of the other categories. There are also differences between lowlands and uplands in that hedgerows in the uplands (i.e. in high land class numbers) are relatively species-poor. The implication is that

Table 7.2 Linear features for England and Wales, 1947–85

	1947	1969	1980	1985
Hedgerows	796	703	653	621
Fences	185	193	199	210
Walls	117	114	111	108
Banks	151	140	132	128
Open ditches	122	116	111	112
Woodland fringe	241	241	243	243

Note: Total length of features in 1,000 km

Source: Hunting Surveys and Consultants Ltd, 1986

Table 7.3 Lengths of linear landscape features in Great Britain

	Hedges	Fences	Walls	Ditches	Streams
England	519.7	737.4	84.3	244.5	80.8
Wales	870.8	118.9	309.9	24.6	20.8
Scotland	666.7	336.1	111.2	62.4	103.3
Total	2057.2	1192.3	505.4	331.5	204.9

Note: 1. Lengths in km × 1,000
2. Based on estimates derived from field survey in 12 sample squares in each of 32 land classes recorded in 1984. Conversion to Great Britain basis is made by estimating the length of the features in the kilometre squares and the proportion of the kilometre squares belonging to each land class.

there is considerably greater potential for species conservation in the management of linear features within agricultural landscapes than would be expected from the areas concerned. This is due to the intense management which now takes place in much of the open countryside. A strategy for habitat renewal at a national level would need to consider this, as well as the ecological possibilities of encouraging 'unique' species into the surrounding agricultural areas.

A major conclusion from this type of study is that the potential for habitat creation in the rural countryside is still surprisingly high. The loss of species has perhaps been over-emphasised because residual populations associated with linear features have not been adequately recognised. A typical example is the expansion of cowslips onto motorway verges. The total botanical capital may still be quite high in many of these areas and the actual number of species left behind could well be higher than has been frequently reported.

The agricultural potential for change

Agricultural changes are particularly important at the present time, with the emphasis in many areas on the surpluses which have accrued in food production, particularly in cereal areas.

Milk quotas have already been imposed by the European Commission and it is suggested that in many quarters cereal production will be similarly controlled. The widely discussed 'set-aside' policy and possible extensification programmes for surplus land also point to considerable potential for habitat creation within agricultural landscapes. However, it must always be borne in mind that these policies are meant to restrict production, and are not primarily intended as environmental improvements. If, however, conservationists prepare appropriate strategies then it is possible that they will be able to influence policies.

These problems have stimulated a range of different studies summarised by Green (in this volume). A wide range of figures are presented, covering the range 1–6 million hectares, depending on which assumptions have been used. It is there-

Table 7.4 Species unique to two 10m × 1m quadrats compared with those from five $200m^2$ quadrats, for the 32 British land classes.

Land Class	Verges Total	Verges Unique	Hedges Total	Hedges Unique	Streams Total	Streams Unique	Quadrats Total	Quadrats Unique
3	84	11	102	17	123	26	179	42
12	100	18	92	17	39	23	229	35
11	89	11	84	11	62	23	110	61
25	78	8	67	9	67	45	106	47
2	102	11	118	8	127	8	211	120
4	119	13	130	20	125	22	205	39
9	131	14	119	11	158	39	238	54
14	107	7	77	16	110	41	145	51
26	92	10	90	11	117	38	148	72
10	85	8	91	11	110	28	150	49
1	68	8	62	18	70	32	123	67
15	72	12	67	8	68	24	96	84
6	84	11	73	20	117	21	173	75
27	65	7	64	9	113	43	132	77
6	97	10	103	15	141	22	221	73
16	88	12	71	8	143	53	147	53
13	108	10	102	8	154	36	164	68
8	79	15	47	14	136	25	193	50
7	54	10	119	16	120	35	155	87
17	89	13	61	12	168	40	181	35
28	65	8	58	2	139	44	161	53
20	49	11	61	3	116	44	170	58
19	54	5	0	0	90	31	111	63
18	54	14	47	3	158	23	162	57
31	87	4	0	0	115	6	129	84
22	103	7	0	0	164	22	212	71
32	111	10	0	0	142	15	210	46
29	103	9	0	0	197	28	212	54
21	98	6	0	0	183	22	208	43
24	66	11	0	0	172	32	150	37
23	0	0	0	0	87	31	184	52
30	81	12	0	0	96	49	138	33

Note: Species unique to two 10m × 1m quadrats placed at random along linear features, where present, in each of 32 land classes (Bunce *et al.*, 1981), in comparison with those from five $200m^2$ quadrats placed at random within the squares.

fore necessary to consider a variety of scenarios in order to assess the likely direction of change. These scenarios are not predictive in a statistical sense, but are rather projections following from a set of assumptions. An example is the Reading model (Harvey *et al.*, 1986), which is constructed in three stages:

1. The macro-economic model (developed at Newcastle University) for expressing the relationships of the gross national product of British agriculture.
2. The data base on land use in England and Wales and its link with habitat composition derived from ITE surveys.
3. A linear programme linking the two, so that changes in economics can be translated into implications at a land use, and eventually at a habitat, level.

The Reading model showed that there was a range of responses depending upon the assumptions used in the construction of the model. Changes in the agricultural enterprises are linked to ecological change, so that, for example, a change from cereals to less intensive grassland has ecological implications, not only in terms of invertebrate populations but also in the variety of plant species growing in the fields. The model identifies certain regions, such as the Midlands, as being susceptible to changes either in sheep numbers or cereal acreages. Other studies, for example by Mathias *et al.* (1986), within the Reading model also point to the Midlands as being an area likely to change in future. Sheep enterprises in the uplands remain stable because the linear programme identifies this enterprise as the optimum for this type of land. Further work is required to examine the likely changes covered by such enterprise shifts into countryside implications, e.g. by linking changes in the cereal acreage to the removal of hedgerows. The initial Reading model is now being redesigned to incorporate more sophisticated inputs and to quantify the countryside impacts. Such impact assessments must be quantitative, and must also be capable of ready interpretation at a policy level. A proper understanding of the pattern of the countryside can lead in turn to subtle manipulation and redesigning of existing landscapes. Sufficient is known about the ecology of some species to show how comparable areas displayed either as large blocks or scattered areas might be more beneficial for wildlife management. One example is given by Harms and Knaapan (1988) who showed that groups of small woodlands covering a similar area to a large block are more beneficial for red squirrels, which could colonise them provided the groups were a certain minimum distance apart. Further work has to be done to show how these patterns in the landscape are important to other species.

An approach recently being developed in ITE employs the grouping of botanical species into classes which indicate the management status of a particular area. Examples of such groups derived from multivariate analysis for cultivated grasslands are given in Table 7.5, and their relationships are shown in Figure 7.1. Thus species groups present in a particular area can indicate the potential for manipulating that land. An indication of status relative to nutrient gradients can also be determined. The work by Marrs (in this volume) shows how nutrient levels might be managed to facilitate the introduction of other species into the landscape. Current assessments of land status are necessary in order to provide the baseline from which more detailed reconstructions of habitat can be made.

Landscape patterns are important both in the terms of hedgerows and field patterns, but also in the way that species are present within these fields. The relationship of the different elements to each other within the landscape is critical in estimating colonisation rates and the availability of propagules. For example, in upland landscapes hedgerows are absent and therefore play no role in the way certain species would move or recolonise. It is necessary to emphasise that in some

Table 7.5 Eight species groups from managed agricultural grasslands

Group	Description	Species names
Arable weeds:	growing either within crops or or in short-term grassland	*Aphanes* spp. *Capsella bursa-pastoris* *Chenopodium album* *Matricaria matricarioides* *Myosotis* spp. *Sonchus asper* *Spergularia arvensis* *Veronica arvensis*
Cultivated herbage:	although wild these species are widely planted as an agricultural crop	*Dactylis glomerata* *Lolium perenne* *Lolium multiflorum* *Phleum pratense* *Poa trivialis* *Trifolium repens*
Initial grassland colonisers:	species that may grow within crops where persistent seed from the soil has often enabled them to become established in gaps	*Elymus repens* *Cirsium arvense* *Cirsium vulgare* *Convolvulus arvensis* *Galium aparine* *Heracleum sphondylium* *Plantago major* *Poa annua* *Polygonum aviculare* *Polygonum persicaria* *Potentilla reptans* *Potentilla anserina* *Rumex crispus* *Rumex obtusifolius* *Stellaria media* *Urtica dioica*
Early grassland colonisers:	species that occur widely in the early stages of establishment of grasslands	*Agrostis stolonifera* *Bellis perennis* *Cardamine hirsuta* *Cerastium fontanum* *Ranunculus repens* *Taraxacum* agg. *Veronica serpyllifolia*

cases species are completely removed from landscapes, whereas in others there is only progressive modification. Currently ITE are carrying out a preliminary survey of buried and windblown seed in order to assess the potential for colonisation of habitats if extensification should take place on a large scale. Together with the sample data collected in 1978, exhaustive species lists from the marginal habitats are also being collected in order to establish the total botanical capital for an area. From this

Table 7.5 continued

Group	Description	Species names
Secondary grassland colonisers:	species which generally only become established when the sward has been sown for some time but which can become established earlier	*Brachythecium rutabulum* *Cynosurus cristatus* *Eurynchium* spp. *Festuca rubra* *Holcus lanatus* *Plantago lanceolata* *Poa pratensis* *Trifolium pratense* *Veronica chamaedrys*
Intermediate neutral grassland:	species which generally belong to older grasslands and are indicators of stability although they can compete with some added fertiliser	*Achillea millefolium* *Prunella vulgaris* *Ranunculus acris* *Ranunculus bulbosus* *Rumex conglomeratus* *Senecio jacobaea*
Neutral meadows and hayfields:	species of older meadows that have generally not received heavy fertiliser	*Alopecurus pratensis* *Arrhenatherum elatius* *Bromus mollis* *Cardamine pratensis* *Centaurea nigra* *Leucanthemum vulgare* *Equisetum arvense* *Filipendula ulmaria* *Rumex acetosa*
Impeded drainage:	species of badly drained grasslands or wet hollows that can be rapidly affected by drainage	*Cirsium palustre* *Deschampsia cespitosa* *Juncus articulatus* *Juncus effusus* *Mnium undulatum* *Stellaria alsine*

Note: From 200m^2 quadrats derived from the full set of 256 sample 1km squares from the 32 land classes (Bunce *et al.*, 1981). The groups were derived by multivariate analysis with outlying members reallocated according to ecological knowledge of management relationships.

it can be established whether the plants are actually available to recolonise declining agricultural areas, or whether new species and new plants have to be introduced.

Conclusions for future work

The whole subject of the analysis of pattern in landscape, such as the networks of hedgerows or woodland distributions, is important in landscape and has been

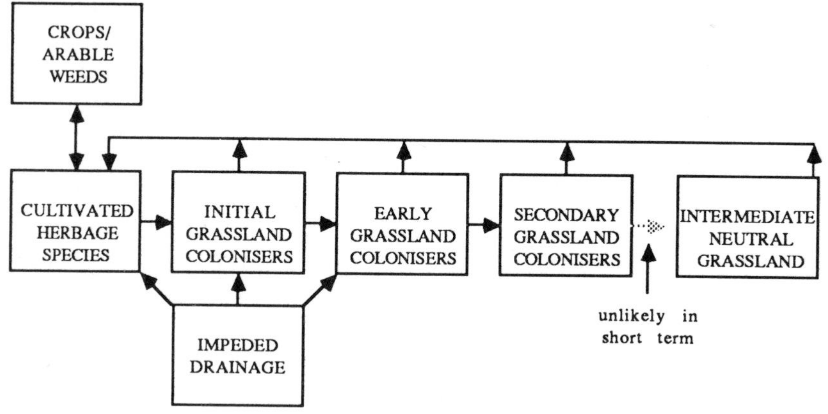

Figure 7.1 Relationships between the groups of species shown in Table 7.5.
Note: The arrows indicate the directions of change which are likely to take place.

relatively little studied in Britain. The International Association of Landscape Ecology symposium in Münster in 1987 (Schreiber, 1988) emphasised the important work which has been done on the Continent on island biogeography and connectivity. Within agricultural landscapes, work currently in progress at ITE has emphasised the importance of linear features in the maintenance of botanical capital and, by implication, herbivorous species. Further work is required to determine whether possible extension of species readily takes place and whether the role of small areas can be greater than they appear. For example, the Game Conservancy and others are looking at the details of distribution insect species within narrow strips of vegetation separating crops (see Boatman *et al.*, in this volume). Field headlands provide just one example of the importance of islands of less intensively managed vegetation connected into a network within the agricultural area. More detail is also required on species dispersal between habitats in order to find out whether they have the potential to reach and colonise available areas.

Ultimately it is necessary to assess owner attitudes and the importance of the socio-economic factors linked to ecological status. The many examples of habitat reconstruction presented in this volume emphasise that it is the commitment and the will of individuals which leads to the creation of new habitats (see Way, in this volume). Although the technical potential for habitat reconstruction might be generally well understood, the decision to undertake that activity is the one which eventually determines landscape pattern.

References

Barr, J., 1969, *Derelict Britain*, Penguin, London.

Barr, C. J., Benefield, C. B., Bunce, R.G.H., Ridsdale, H. A., Whittaker, M., 1986, *Landscape Changes in Britain*, ITE, Abbots Ripton, Huntingdon.

Best, R. H., 1976, 'The extent and growth of urban land', *The Planner*, 1, 8–11.

Bunce, R.G.H., Barr, C. J., Whittaker, H. A., 1981, 'Land classes in Great Britain: preliminary descriptions for users of the Merlewood method of land classification', Merlewood research and development paper No 86, Institute of Terrestrial Ecology, Grange-over-Sands.

Bunce, R.G.H., Heal, O. W., 1984, 'Landscape evaluation and the impact of changing land-use on the rural environment: the problem and an approach', in R. D. Roberts and T. M. Roberts (eds.), *Planning and Ecology*, Chapman & Hall, London, pp. 164–88.

Bunce, R.G.H., Tranter, R. B., Thompson, A.M.M., Mitchell, C. P., Barr, C. J.., 1984, 'Models for predicting changes in rural land use in Great Britain', in D. Jenkins (ed.), *Agriculture and the Environment*, ITE, Cambridge, pp. 37–44.

Coppock, J. T., Gebbett, L. F., 1978, *Land Use and Town and Country Planning*, Pergamon Press Ltd., on behalf of the Royal Statistical Society and Social Science Research Council.

Davis, B.N.K., 1982, *Ecology of Quarries: the Importance of Natural Vegetation*, ITE, Cambridge.

Dennington, V. N., 1979, *Derelict and Degraded Land in England*, Energy Technology Support Unit, Harwell.

Department of the Environment (Planning, Regional and Countryside Directorate), 1986, *Survey of Derelict and Despoiled Land in England*, Department of the Environment.

Fordham, R. C., 1974, *Measurement of Urban Land-use*, Dept. of Land Economy, Cambridge University Press.

Good, J.E.G., Munro, R.C., 1981, 'Trees in town and country', in F. T. Last and A. S. Gardiner (eds.), *Forest and Woodland Ecology*, ITE, Abbots Ripton, pp. 16–19.

Harms, W. B., Knaapan, J., 1988, 'Landscape planning and ecological infrastructure: the Ranstad study', in K. F. Schreiber (ed.), *Connectivity in Landscape Ecology*, proceedings of the 2nd International Seminar of the International Association for Landscape Ecology at Münster, Schoningh, Padeborn, pp. 163–9.

Harvey, D. R., Barr, C. J., Bell, M., Bunce, R.G.H., Edwards, D., Errington, A. J., Jollans, J. L., McLintock, J. H., Thomson, A.M.M., Tranter, R. B., 1986, 'Countryside implications for England and Wales of possible changes in the Common Agricultural Policy', report to the Department of the Environment and Development Commission, Centre for Agricultural Strategy, University of Reading.

Hunting Surveys and Consultants Ltd., 1986, *Monitoring Landscape Change*, Hunting, Borehamwood.

Mathias, C. H., Brown, D.A.H., Measures, A. R., 1986, *Changes in Land Use in England, Scotland and Wales 1985 to 1990 and 2000*, Laurence Gould, Warwick.

Sargent, C., 1984, *Britain's Railway Vegetation*, ITE, Cambridge.

Schreiber, K. F., (ed.), 1988, *Connectivity in Landscape Ecology*, proceedings of the 2nd International Seminar of the International Association for Landscape Ecology in Münster, Schoningh, Paderborn.

Way, J. M., 1973, 'Road verges on rural roads: management and other factors', Monks Wood Experimental Station occasional reports No. 1, ITE, Abbots Ripton, Huntingdon.

8

Environmental opportunities offered by surplus agricultural production

B. H. Green and *C. P. Burnham*

Introduction

British landscapes and their wildlife have been very largely created and maintained by the practice of agriculture. Following the Neolithic revolution forest clearance was early and very nearly complete. It is estimated that by Norman times only some 15 per cent of the land, which had been almost completely forested, remained under trees (Rackham, 1980). Thereafter agricultural impacts on the environment have fluctuated according to the economic dictates of the times. In periods of agricultural prosperity marginal land has come into production and in depression it has reverted back again to moor, marsh and forest.

The Parliamentary Enclosures, which took place mainly between 1750 and 1850, represented a second major surge of change in agriculture. At this time the reclamation of the unenclosed common grazings into hedged fields, and the accompanying drainage of wetlands, was an important part of the agrarian revolution which so early in Britain transformed open field peasant farming to capitalist agriculture. In 1696 an early land-use survey showed uncultivated rough grazings covering more than 25 per cent of England and Wales; by 1901 their extent had been more than halved (Best, 1981).

After a period of high farming British agriculture went into recession around 1870. Much of the land which had been under the plough reverted again to grass. Three-quarters of the Porton Ranges downland in Hampshire was, for example, under the plough between 1856 and 1885 (Wells *et al.*, 1976). There was a period of agricultural recovery during the First World War, but apart from this the agricultural depression persisted until the next great war.

The 'dig for victory' campaign in that war to increase home food production in the blockade initiated a third major agricultural revolution which has had repercussions on the countryside as profound as those earlier. Huge state support for farmers, coupled with technological innovations, have transformed production; Britain's self-sufficiency in temperate foodstuffs has increased from 30 per cent before the war to nearly 80 per cent today. But this has been at substantial environmental cost.

In England and Wales the overall loss of semi-natural vegetation between 1947 and 1980 has been 25 per cent (Countryside Commission, 1986).

We are now on the threshold of another agricultural revolution. There are huge European food surpluses, and major changes in the Common Agricultural Policy are being proposed and implemented. What opportunities does this offer for the restoration of the lost habitats and the creation of new environments?

Land surplus to food production needs

A recent study at Wye College (Burnham, Green and Potter, 1987) has attempted to determine how much land is surplus to agricultural needs and where this land might best be taken out of production, or production made less intensive, in order to maximise environmental benefits.

To assess how much land is potentially surplus to food production needs by the year 2000, agricultural land budget work carried out at Wye College in the early 1970s (Edwards and Wibberley, 1971) has been updated (Edwards, 1986). Recent historical changes in the demand for food and rates of increase in agricultural production have been extrapolated within a broad range of values and assumptions made about different levels of United Kingdom self-sufficiency in temperate foodstuffs. According to whether conservative or liberal estimates of these parameters are chosen, between 1 million and 6 million hectares (one-third of the UK agricultural area) seem to be potentially surplus by the year 2000. A reasonable estimate in the middle of this range would put the potentially surplus land at 3 to 4 million ha (approximately 15–25 per cent of the UK agricultural area). These estimates are comparable to those arrived at by a variety of means by other investigators (see Table 8.1). In addition to these figures it is estimated that present food surpluses represent some 0.5–1.0 million ha of land (Buckwell, 1986).

Some of this land will be needed for other uses; but the land needed for the main uses of urban expansion and forestry together by 2000 is likely to be well short of 1.0 million ha. It seems therefore that there are awesomely large areas of land which could, in theory, be withdrawn from agriculture or put to less intensive agricultural use.

The environmental effects of set-aside

Land might be taken out of production or farmed less intensively in a variety of ways (see Table 8.2). Numerous proposals have recently been made for doing so by the European Commission and British government (EEC, 1985; MAFF, 1987). They range from farm forestry, fallowing and set-aside of field margins, to the early retirement of older farmers and establishment of Environmentally Sensitive Areas (ESAs) where farmers will be paid to farm in more environmentally benign ways.

Although some environmental benefits are intended, and will undoubtedly accrue from all these measures, this will not be one of the two principal aims of the schemes. These will be the reduction of food surpluses and the maintenance of farm livelihoods. The maintenance of farm livelihoods is one of the main objectives of the Common Agricultural Policy. Small farmers in France and Germany are numerous and politically powerful. Their governments have thus resisted any production control

Table 8.1 Comparison of estimates of land surplus to agriculture, modified to the common date of 2000

Study	Definition employed	Area	Year	Range of estimates m ha	Equivalent for year 2000 m ha
Land Budget, 1986 (Wye)	Farm land available for other uses	UK	2000	0.9–6.1	2.8–4.3
Land Budget, 1971 (Wye)	Farm land available for other uses	UK	2000	4–6	4–6
Mansholt, 1986	Cultivated land to go out of production	EC	1980	(5–6% of total)	1.4
North, 1987	Land surplus to needs assuming economically efficient production	UK	2015	6.3	3.15
Crops balance, 1987 (Wye)	Land surplus to requirements	UK	2000	1.8–3.6	1.8–3.6
Harvey *et al.*, 1985	Surplus areas of major crops	GB	2000	2.4–2.9	2.6
Reading, 1986	Area equivalent of required reduction in intensity	EandW	5yrs	0.2–2.2	1.3–1.9
NFU, 1986/87	Land available for other uses	UK	pa	0.15	2.25

Source: adapted from Bell (1987)

proposals, such as stringent price cuts, which would threaten their survival. The survival of small farmers is also generally regarded as desirable by conservationists in this country, who see them as vital both to the survival of rural communities and for the maintenance of traditional types of unintensive management of the land. The more farmers there are, however, the smaller the slice of the declining overall amount of money they can expect to earn from agriculture. Their existence in the future will thus inevitably depend even more upon income support from the government. This money will probably increasingly be paid for the attainment of environmental objectives rather than food production.

In the Environmentally Sensitive Areas in England and Wales farmers are being paid up to £200 per hectare for farming with restrictions on their use of agrochemicals, dates of hay cutting and requirements to maintain and manage hedgerows, walls and other environmental amenities. Over 70 per cent of eligible land within them has been registered and their success in meeting their relatively modest environmental objectives seems assured. Potter (1988) has shown that many contain a predominance of small farms needing this income supplementation and that some moderate yield savings seem likely.

The new MAFF farm forestry proposals pay farmers up to £190/ha/annum for

Table 8.2 Strategies for land diversion in relation to current farming schemes

	Alternative use			
Set-aside pattern	Wilderness	Forestry	Heritage Farms	Alternative Crops
Fallowing	Weedy fallows temporary ecological benefit (SA)	–	Maintaining cover for game (SA)	Cover crops on erosion-prone land (SA/ESA)
Headlands	Natural regeneration to scrub and trees (SA)	Shelter-belts, hedgerow widening (FWS)	Headland management to maximise species-diversity (SA/ESA)	–
Part or whole farm set-aside	Reversion in uplands to rough pasture, moor, shrubby and grassy heaths (ERS)	Afforestation, farm-forestry FWS	Extensive grazing of re-created wetlands and meadowland (ESA)	Grouse, deer farming (ED)

Notes: SA Set-Aside (Farm Extensification Scheme)
FWS Farm Woodland Scheme
ESA Environmentally Sensitive Area
ERS Early Retirement Scheme? (Pre-pension scheme)
ED Enterprise Diversification

up to 40 years for the establishment of woods on farmland taken out of production. The set-aside scheme pays up to £200/ha/annum for up to five years for the fallowing of land or removal of field margins from production. Both offer obvious opportunities for linking existing environmental features such as ponds and woods, buffering hedges and watercourses from the effects of cultivation and increasing the absolute amount of cover on farmland. It remains to be seen how successful they will be at the levels of compensation offered.

All schemes will not however necessarily fully exploit potential environmental gains. Their take-up will depend primarily on the economic circumstances and predilections of individual farmers, which will rarely coincide with environmental needs (Potter and Gasson, 1987). If the potential environmental benefits are to be maximised, environmental objectives must probably be a more specific aim, as is the case in the current Conservation Reserve Program in the United States. To enter this scheme, which aims to idle some 18 million ha by 1990, a farmer must have land which has erosion rates greater than specified levels (Ervin, 1987).

Targetting land for diversion from intensive agriculture

The environmental vulnerability of land to erosion, flooding or pollution of aquifers with fertiliser-derived nitrate is one set of criteria which might be used to help target land for set-aside in order to maximise environmental benefits. In the Wye study it has been combined with two other sets of criteria, namely 'mismatch' between agricultural land quality and cropping patterns and conservation potential to restore lost ecosystems, in an attempt to map areas where land taken out of agriculture would confer most environmental advantages (see Figure 8.1). For each of these three sets of criteria relevant attributes were mapped in England and Wales in the 10 km squares of the national grid. The information was digitised from existing mapped sources, in particular soil association maps; or collated from existing datasets, notably that held by the ITE at Bangor and on the National Disk of the BBC Domesday package. A simple map overlay computer methodology was used to identify squares where the attributes deemed to maximise benefit from set-aside were aggregated. The areas identified may be compared with more subjectively defined Environmentally Sensitive Areas (see Table 8.3). This methodology is now being refined to incorporate more criteria and increase its spatial resolution.

Countryside management

Whatever methods are used to control agricultural production it seems that large areas of land are going to be farmed less intensively, or specifically for conservation objectives. Some farmland, such as hay meadows or downland, needs to be farmed in a particular way to maintain its ecological and amenity value. The principle of supporting environmentally-friendly farming systems (EFFS) by diverting funds from production subsidies now seems firmly established in ESAs and likely to be more widely embraced in the future; perhaps National Parks will be the next priority.

But not all land requires its present farm management. The South Downs ESA already includes provision for paying farmers to revert from arable to grass. It is possible that we shall also see large-scale reversion of grass to moor, pasture to scrub and woodland and land flood back to marsh and fen. Field margins and new copses will certainly be established on a small scale and new urban, industrial and recreational developments at more generous densities of land-use will probably incorporate parks, golf courses, lakes, forests and other amenities. This urban and suburban green space, incorporated in developments as 'planning gain' to mollify powerful protectionist interests (NIMBYs – 'not in my backyard'), may come to fulfil many of the amenity functions of the present agricultural countryside.

Curiously in Europe this conversion of farmland to non-productive uses, especially to wilderness, is almost invariably regarded as undesirable 'development' or 'abandonment', even by conservationists. Yet the most visited National Park in the USA – Shenandoah – was created in 1936 by just such abandonment of farmland. It now bears magnificent stands of second growth hardwood forest. It was just one environmental gain from the United States Federal Land Utilisation programme which between 1935 and 1946 bought 4.5 million ha of worn-out cropland and restored it to parks, forests and grazing land (Daniels, 1988). One only has to visit

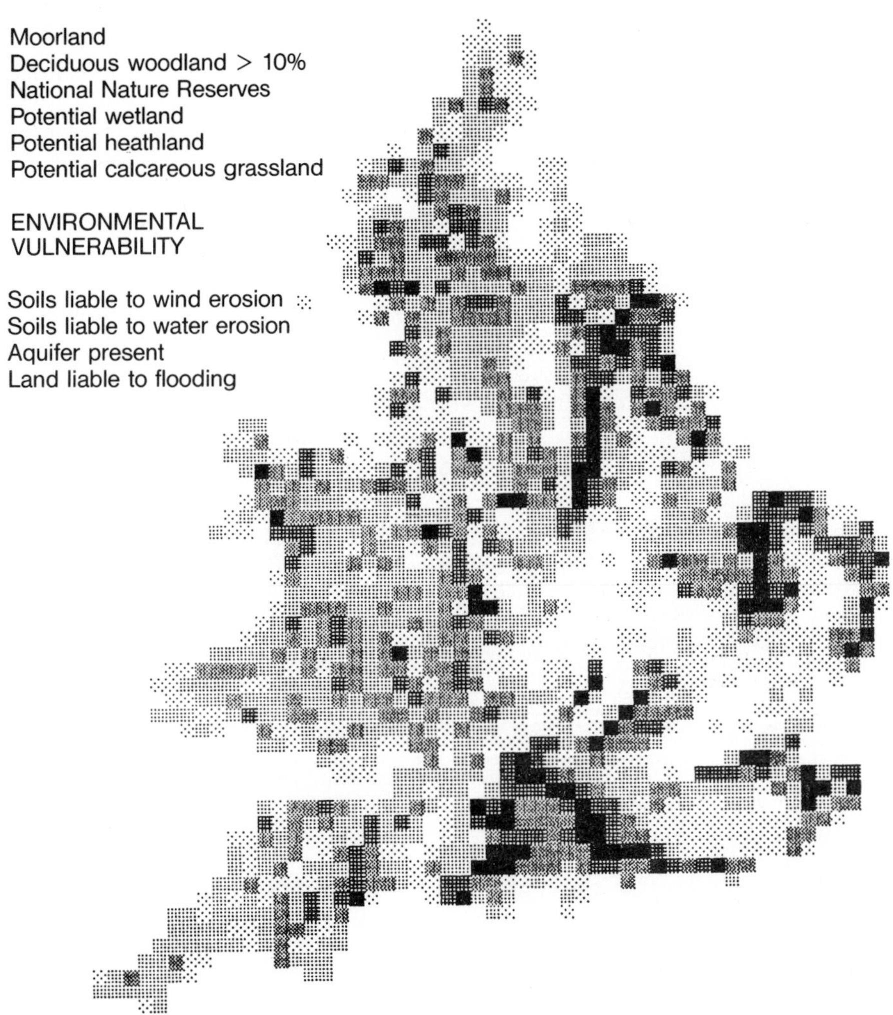

Figure 8.1 Evaluation scheme for environmental factors favouring selection as a 'set-aside target area'.

Table 8.3 Landscape and habitat characteristics of Environmentally Sensitive Areas in England and Wales

ESA	Area (ha)	Landscape characteristics	Habitats	Threats
Anglesey	18,500	Rocky outcrops Small woodland Varied coastline	Semi-natural lowland habitat and wetlands	Intensification Upgrading of peat to pasture
Breckland[b]	51,200	Heathland and calcareous grassland Valley grasslands	Grazed heath Calcareous grassland Wetlands	Ploughing Tree invasion Draining of wetland
Broads[a]	29,870	Grazing marshes Fen and fen carr Flat valleys and wet marshes	Drained marshland Marsh/dyke Fen/carr woodland	Ploughing Stock reduction Lowered water tables, nitrogen run-off
Cambrian Mountains[a]	72,800	Varied grasslands Oak woods on valley sides	Rough grazing Oak woodland Haymeadows	Burning and moorland afforestation Land improvement
North Peak[b]	32,000	Open moorland Heathland	Peat moorland Grasslands Heather heath	Insufficient moorland burning Overgrazing Grazing of woodlands
Lleyn[b]	40,200	Intricate field system, heather, grass and gorse Valley woodland	Coastal headlands Valley wetlands	Intensification Afforestation Abandonment
Pennine Dales[a]	15,900	Varied grassland Valley bottom meadows Dry stone walls and field barns	Wet pasture Haymeadows Woodlands	Drainage Fertilisers Neglect

Somerset Levels and Moors[a]	26,970	Open landscape divided by rivers, ditches and rhynes	Wetland	Drainage Grassland improvement Ploughing
South Downs (East)[a] South Downs (West)[b]	26,600 22,800	Open downland Chalk grassland Scattered trees and scrub cover	Species-rich chalk turf	Ploughing Underdrainage Spray damage
Suffolk River Valleys[b]	20,890	Valleys with rivers Grasslands, hedges and reedbeds	Reedbeds Wet grasslands Rivers and ditches	Drainage Ploughing Fertilisers
Test Valley[b]	2,300	Rivers, trees, carr Reedbed and pasture	Chalk stream communities Chalk grassland	Ploughing Fertilisers Increased stocking
Welsh Marches[b]	103,800	Moorland Intricate field system	Rough grazing Valley woodlands Hedgerows	Intensification Ploughing Woodland grazing Hedgerow demise
West Penwith[a]	7,200	Rocky coastline Intricate field system, moors	Coastal heathland Inland heathland Scrub and wetland	Reclamation of scrub Wall neglect Lane widening Construction

Notes: [a]designated in March 1986; [b]designated in May 1987.

Source: after Potter (1988)

the East Coast states of the USA or read their enormous ecological literature on old field succession to appreciate the rich ecosystems and beautiful landscapes that can be spontaneously developed on abandoned farmland. Rank grass, scrub and trees may be undesirable at the expense of heath or downland, but not at the expense of arable or improved grassland.

There is now an unprecedented opportunity to do this in Britain and restore some of the habitats and species lost through post-war agricultural intensification. The Countryside Commission has imaginatively proposed the establishment of a major new forest of 40,000 ha in the Midlands and the creation of other large urban fringe woodlands. These proposals have been well received and, with appropriate targetting, could be achieved using the new policy measures. The CAP early retirement proposals, if implemented, would help greatly. Farmers over the age of 55 would be given a pension to abandon their land. It has been estimated in Germany that this measure could take 7.4 per cent of all farmland out of production at less cost than present support (Schopen, 1987).

Conclusions

The post-war planning and management of the British countryside has been largely determined by socio-economic factors. In particular, price incentives have been used to shape the structure of the agricultural industry, with little thought as to what effect this would have on the pattern of land use and resource exploitation. European food surpluses and the consequent major changes in the Common Agricultural Policy which are now being proposed offer the opportunity to redevelop countryside planning and management onto a more resource-orientated foundation. Conservation is thus going to have to change from a predominantly negative, protectionist, planning activity to a much more positive, practical, managerial one: indeed a land-use in its own right. There will be great challenges to restore land to new types of cover. Farmers and foresters are practical men used to clear objectives and sound technical and financial advice as to how to implement them. They are beginning to demand this of conservationists. The great challenge now facing applied ecologists and the conservation movement is to set out what kinds of countryside it wants to see and develop, codify, cost and promote the management techniques needed to achieve them.

References

Bell, M., 1987, *The future use of agricultural land in the United Kingdom*, Agricultural Economics Society Conference, 1987.

Best, R.H., 1981, *Land use and living space*, Methuen, London.

Buckwell, A., 1986, *Controlling Cereal Surpluses by Area Reduction Programmes*, discussion paper on Agricultural Policy, **86/2**, Wye College, University of London.

Burnham, C. P., Green, B. H., Potter, C. A., 1987, *Set-aside as an Environmental and Agricultural Policy Instrument*, Set-aside Working Paper No. 3, Wye College, University of London.

Countryside Commission, 1986, *Monitoring Landscape Change*, Countryside Commission, Cheltenham.

Daniels, J.L., 1988, 'America's Conservation Reserve Program: rural planning or just another subsidy?' *Journal of Rural Studies*, 4, 405–11.

Edwards, A. M., Wibberley, G. P., 1971, *An Agricultural Land-budget for Britain 1965–2000*, Wye College, University of London.

Edwards, Angela, 1986, *An Agricultural Land Budget for the United Kingdom*, Set-aside Working Paper No. 2, Wye College, University of London.

EEC, 1985, *A Future for Community Agriculture*, (Brussels), Com **85**(5).

Ervin, D. E., 1987, 'Cropland diversion in the US: are there lessons for the EEC set-aside discussion?' in D. Baldock and D. Conder (eds.), *Removing Land from Agriculture*, CPRE/IEEP, London, pp. 53–63.

Lawrence Gould Ltd., 1986, *Changes in Land Use in England, Scotland and Wales 1985 to 1990 and 2000*, Nature Conservancy Council.

MAFF, 1987, *An Extensification Scheme*, Ministry of Agriculture, Fisheries and Food, London.

Marsh, J., 1987, *Directions for Change: Land Use in the 1990s*, National Economic Development Office.

Potter, C. A., 1988, 'Environmentally Sensitive Areas in England and Wales: an experiment in countryside management', *Land Use Policy*, **5**, 301–14.

Potter, C. A., Gasson, Ruth, 1987, *Set-aside and Land Diversion: the View from the Farm*, Set-aside Working Paper No. 6, Wye College, University of London.

Rackham, O., 1980, *Ancient Woodland: its History, Vegetation and Uses in England*, Edward Arnold, London.

Rackham, O., 1986, *The History of the Countryside*, Dent, London.

Schopen, W., 1987, 'Removing land from agriculture: the German view', in D. Baldock and D. Conder (eds.), *Removing Land from Agriculture*, CPRE/IEEP, London, pp. 35–39.

Wells, T.C.E., Sheail, J., Ball, D. F., Ward, L. K., 1976, 'Ecological studies on the Porton Ranges: relationships between vegetation soils and land-use history', *Journal of Ecology*, **64**, 589–626.

9

Reconstruction of habitats on farmland

J.M. Way

Introduction

In discussing opportunities for reconstruction of habitats in farmland in the twentieth century and beyond, we are usually concerned with habitats in the man-made landscapes that are familiar today. There are powerful pressures for the preservation of these landscapes and the wildlife habitats that they contain, but patterns of land-use and the economics of the countryside change. We need to discuss not only what we should be doing to reconstruct habitats in the existing older landscapes, but in considering new landscapes, and creation of new habitats, to take advantage of the patterns of land-use of our time and for the future, so far as we can predict them.

Whether we should describe the latter as reconstruction or as creation may seem to be a question of semantics, but it highlights the distinction between repair or reinstatement of existing or recently lost areas of habitat, and the creation of new ones. The extensive creation of new landscapes and new habitats will not be an instant process; nevertheless, whilst existing landscapes and habitats have, until recently, evolved slowly from gradual changes in land-use, in the future both the rate of change and the degree of positive direction seem likely to increase (e.g. Countryside Commission, 1987a).

In reconstructing or creating habitats on farms results can be achieved by management or by new works, or both. For instance it will often be quicker to establish woodland on a new site by planting than to allow one to develop naturally; whilst it may be cheaper and quicker to manage agriculturally improved downland to encourage diversity and herb-richness rather than sow new grassland.

It is axiomatic that any programme of habitat reconstruction must be accompanied by subsequent management, and that the resource costs and long-term nature of this management should be understood by the farmer from the start. In addition, proposals will be more attractive to farmers where management for wildlife conservation, sport, amenity or recreation can be used to generate an income. Management may have to include control of (mainly) vertebrate pests (especially rabbits) for which the proposed new forms of land use may provide attractive habitats. Management has also to be taken into account in the design of schemes, for instance where machinery is likely to have to be used in the maintenance of a pond, or for cutting grass without awkward manoeuvres or damage to trees.

Opportunities for reconstruction of farm habitats

Opportunities for habitat reconstruction on farms arise from the various motivations of the farming community. Actual operations are carried out under the direction of, if not actually by, farmers. The interest of farmers is therefore focal if integration between agriculture and wildlife conservation is to be widely practised.

Factors that influence individual farmers include Government policies both for agriculture and the countryside, public opinion, sporting interests, and the personal knowledge and concern of farmers for the wildlife and landscape of their farms. In addition, conservation works can have a beneficial effect on land values.

Costs, advice and personal factors

Interwoven with these factors are questions of costs. Basically the costs may be wholly met by the farmer, or may be ameliorated by income generated from conservation-related activities, e.g. from sport or sale of fuel wood. Other sources of support are available from national and local government, including grants from the Agricultural Departments, Countryside Commission (CC) via Local Authorities, and the Forestry Commission (FC) for hedges, walls and other features, for tree planting and woodland schemes, and from many Local Authorities for these and other environmental purposes management agreements are available. In statutorily protected areas e.g. National Parks, Sites of Special Scientific Interest (SSSIs), to compensate farmers for loss of income arising from environmental modifications to their normal farming practices. In the Environmentally Sensitive Areas (ESAs) (see Figure 9.1) the concept of management agreements has been considerably extended, and agreements have now been made covering a high proportion of the eligible land in each of the ESAs in England and Wales.

Advisory support for farmers is provided through the Nature Conservancy Council (NCC), CC, FC and particularly through the countrywide network of Agricultural Development and Advisory Service (ADAS) officers of the Ministry of Agriculture, Fisheries and Food (MAFF) in England and Wales. Further advice is available from many Local Authorities, who employ landscape architects and ecologists. In the voluntary sector advice is available from the county Farming and Wildlife Advisory Groups (FWAG's), from such groups as Project Silvanus in England and Coed Cymru in Wales on woodlands, from the Game Conservancy and the British Association for Shooting and Conservation on sport, from the County Trusts for Nature Conservation, from the Royal Society for the Protection of Birds (RSPB) and others.

Consequently there are many ways in which farmers can receive help and financial support for conservation activities on their farms. However, rarely are the costs met outside SSSIs, other protected areas and ESAs, so that the reconstruction of habitats on farmland in the general countryside is always going to be paid for to a greater or lesser extent by farmers.

There is a great deal of goodwill. The Country Landowners Association (CLA) and the National Farmers' Union (NFU) have given support to conservation policies (NFU/CLA, 1977). Leaders in the farming community set up the County

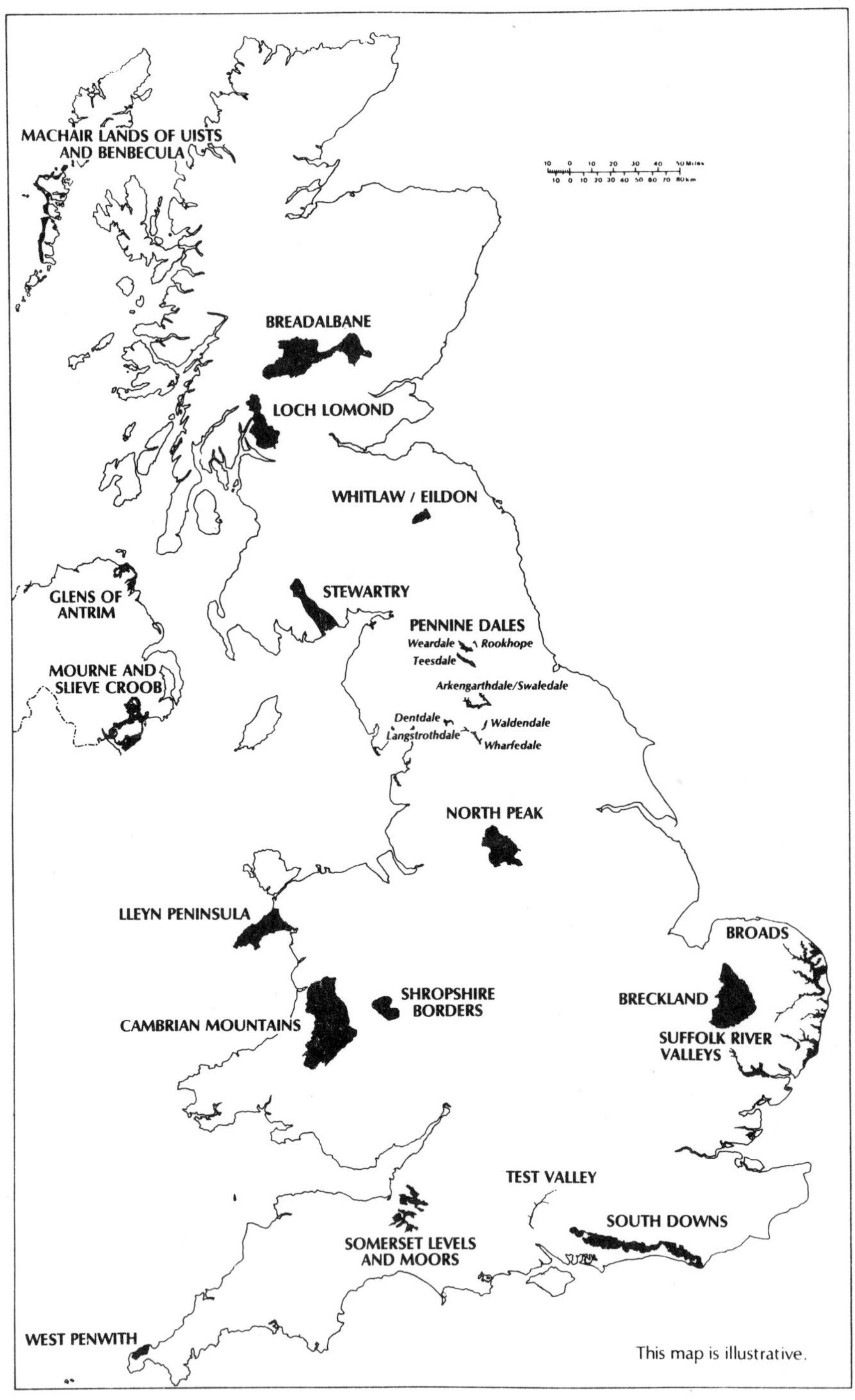

Figure 9.1 Environmentally sensitive areas in the United Kingdom. (Reproduced with permission. Crown copyright 1987 and MAFF Cartographic Unit.)

FWAG's of which many individual farmers are now members (as they are of other countryside organisations). The interest of farmers is demonstrated in requests for advice. It has not been possible to extract meaningful figures for advice provided by ADAS, because much conservation advice has been given at the same time as advice on other aspects of farm management. However, for FWAG's in 1986/87, about 30 Farm Conservation Advisers in England and Wales made 3,174 farm visits (including follow-ups) (Farming and Wildlife Trust, 1987). There was no shortage of requests for advice, and in those counties with a backlog of visits, there was a problem in dealing with the work. Further resources would probably be needed for the national advisory organisations to service a greater number of requests for on-farm conservation advice than is being met at the moment.

Cost and possible (financial) benefits influence farmers on the types of habitats that they select. Advisory requests to ADAS have been principally for woodlands, often in connection with a Forestry Commission or Local Authority grant, and next for ponds. Hedges, shelterbelts, amenity plantings and field corner planting advice have been in demand, but there has been considerably less interest in grasslands, wetlands and watercourses. A similar pattern applied to FWAG's (*op. cit.*) for advice during 1986/87. Interest in the management of arable field margins is likely to increase, both as a result of the work of the Cereals and Gamebirds Research Project (e.g. Rands and Sotherton, 1987, Boatman *et al.*, in this volume), and options under the UK Set-Aside Scheme for the reduction of cereal acreages. Some of this advice has been given in response to specific enquiry, and some as part of conservation appraisals of whole farms.

Changing opportunities in the countryside

Government policies for agriculture and for the countryside are changing. Diversification schemes, farm woodland and afforestation schemes, the designation of Environmentally Sensitive Areas, set-aside and proposed extensification schemes, and the introduction of the Farm and Conservation Grant Scheme, all provide opportunities for reconstruction of habitats on farmland in parallel with other objectives. Many organisations are anticipating these opportunities. There is no lack of ideas and constructive discussions are being held in many quarters.

Farm advice

A farm advisor may be able to give conservation advice *either* on an opportunity (farm-gate) basis, *or* in response to a specific request for project advice on a particular habitat such as a wood or a pond, *or* in the course of drawing up a farm plan.

Opportunity advice

Much very useful day-to-day conservation advice is given by ADAS advisers in the course of advisory farm visits made for other purposes. The discussion of wildlife conservation topics is necessarily limited in scope and in most cases is unlikely to lead to any significant proposals for habitat reconstruction, but on occasion may lead to subsequent requests for more detailed advice. Advisers from other organisations are not under the same statutory obligation to balance their agricultural advice with environmental considerations (Agriculture Act 1986, s.17; HMSO, 1986).

Conservation project advice

Most commonly advisers from any of the wide range of organisations (ADAS, NCC, FC, Local Authorities, National Park Authorities, FWAG's, consultants and voluntary bodies) may receive enquiries from farmers and landowners for advice on a specific project (e.g. woodland or tree planting, or pond construction), and here there is a great deal of scope for proposals for habitat reconstruction. Success with one project may lead to further enquiries from the farmer, and sometimes from his neighbour(s).

Farm plans for wildlife

Farm plans may range from single interest wildlife conservation plans, to various combinations of interests (e.g. wildlife *and* landscape), to plans incorporating all interests. In all of these plans assessments are made of the wildlife interest on the whole farm and opportunities for habitat reconstruction identified. Clearly, every opportunity needs to be taken to promote whole farm plans, which have much more value for the countryside of the future than individual projects on isolated habitats.

Whole farm single interest plans can be drawn up quite rapidly and cheaply. It is more costly if all interests are considered: assessments have to be made of agricultural aspects and the farm business, forestry, wildlife conservation, landscape, sport, recreation and amenity, archaeological and historical features, and their relative values on a field-by-field basis weighed against each other in order to arrive at an optimum solution. The benefits of this holistic approach are considerable in both the short and the longer term, and pressures for production of whole farm plans are likely to grow in connection with changes in agricultural land-use and other environmental policies. In this connection, ADAS has developed a methodology for the production of Integrated Land Management Plans taking all interests into account in arriving at optimum land-use and management solutions. A service for the production of these plans is available to farmers and landowners. General guidelines on the principles and practice of preparing multipurpose management plans, using a systematic approach, have also been published by the Countryside Commission (1986), and by consultants for the Countryside Commission Demonstration Farms, with accounts of the implementation and subsequent monitoring and management of selected habitat proposals (Countryside Commission, 1987b).

Production of farm plans require consideration of both the implementation of proposals and financial aspects. Discussion between the authors of the plan and the farmer on implementation should be agreed and recorded in the first instance and for any subsequent changes, and advice on sources of materials, procedures and other aspects given as appropriate. Similarly, plans must provide estimates of costs of implementation of environmental aspects, including costs of subsequent management, and of sources of income from non-agricultural activities, information on available grants, tax savings and other tangible benefits.

Farm habitats

Although farm habitats are ecologically interdependent, their management, and consequently opportunities for habitat reconstruction are different, and usually independent. The major farm habitats are (1) arable land, including fallow and field boundaries; (2) grasslands and boundaries, including moorland and rough grazing; (3) trees, woodlands and scrub; (4) open water and wetland habitats; (5) artefacts, including buildings and structures, farm roads and tracks, quarries and pits.

Arable land

Arable land is not normally regarded as a natural habitat but nevertheless its wildlife content is considerable but decreasing (Ratcliffe, 1977). The interest is generally ephemeral. However, despite their artificiality, arable habitats are attracting attention at the present time, focused on the increasing rarity of a number of 'weed species', on the open ground bird fauna, and on the results of the Game Conservancy's Cereals and Gamebirds Research Project (e.g. Perring, 1974; Botanical Society of the British Isles, 1987; Rands and Sotherton, 1987).

Both 'weeds' and non-competitive wild plants find their greatest opportunities at the early growth stages of arable crops but most are capable of establishing whenever open ground (ruderal) situations occur. In modern agriculture a first step in restoring wildlife populations depends primarily on a reduction in the intensity of arable land management. That populations of individual species, occurrences of apparently rare species and diversity of species can recover from intensive management (particularly following reductions in the use of pesticides), sometimes in surprisingly short periods of time, has been demonstrated by the Cereals and Gamebirds Research Project (Boatman *et al.*, in this volume).

It seems likely that significant areas arable land taken out of production under the set-aside scheme will be managed under the fallow options. Where traditional rotational fallows are practised involving green cropping or occasional cultivations, not only may some of the rarer weeds become re-established, but copious food sources for insects, birds and other animals will become available with the wide range of often abundantly flowering and seeding annual plants that appear.

Opportunities for habitat reconstruction also occur at the margins and boundaries of arable fields, e.g. Way and Greig-Smith (1987). At its fullest expression a lowland field margin may incorporate a woody hedge, a herbaceous bank, a ditch, a track, a grassed or sterile strip and an unsprayed crop margin or headland. Each of these reinforces the other, so that a hedge with a ditch is more valuable for birds (Arnold, 1983; Lack, in prep) than either alone. However, whilst research is in progress, and more is needed, the principal effort at the present time needs to be in getting farmers to accept existing ecologically sound field boundary practices aimed at the establishment and management of perennial vegetation.

Grasslands

There has been a considerable loss of grassland habitats on farmland since the Second World War (NCC, 1984; Fuller, 1987; Wells and Sheail, 1988), but opportunities for their reconstruction are variable. Many farmers do not appreciate the environmental conservation interest of 'natural' grassland in the same way as for woods, hedges or ponds, nor see any economic advantage in them. Sometimes an adviser has to advise the farmer that existing grassland *per se* is of more conservation value than the creation of some other form of wildlife habitat.

As with other habitats, grasslands can be promoted by a range of techniques, from changes in management regimes to sowing with seed mixtures containing wild flowers appropriate to the soil conditions (Wells *et al.*, 1981, and in this volume). The most extensive opportunities for grassland habitats in productive areas seem to be with management agreements of various kinds, associated with schemes such as the Broads Grazing Marsh Scheme (Countryside Commission/MAFF, 1988), and in ESAs. In less agriculturally productive areas, the danger is that lack of management will not encourage the diversity of species that is usually regarded as the prime ecological objective.

Grassland habitats of ecological value are usually associated with long-standing agricultural practices (NCC, 1977), so that we need to look at these historical practices. The principal elements are time, inputs and stock management.

The element of time is well established, although sometimes it is over-estimated with the assumption for instance that downland grassland or herb-rich hay meadows once destroyed cannot be replaced. This should be qualified. There are instances of quite exceptionally diverse plant communities on railway land, much of which is not more than 150 to 200 years old, whilst Wells *et al.* (1976) recorded similar periods of time for sheep walk at Porton Down for the most diverse communities, and shorter periods for others. On the Somerset Moors elements of herb-rich hay meadow communities have rapidly recovered following only two or three years of relaxation of intensity of agricultural use, including raising of water tables. Heather can re-establish in a matter of years following reduction of grazing pressures on moorland and re-establishment of positive management practices, including controlled burning. In all cases the presence of propagules is critical, but so is the continuity of management over time that allows species to aggregate and stabilise into communities. We need to know more about the element of time in grassland.

It is tempting to identify usage of inorganic fertilisers as the most important single cause of loss in recent years of the ecological interest of agricultural grasslands. Whilst this grossly over-simplifies the problem, reductions in fertiliser usage may well be the key to the diversification of many grassland habitats. For instance other factors damaging ecological interest including drainage, silage making, herbicide usage, and intensive management of stock depend on the additional use of fertilisers to optimise benefits.

There is a very considerable body of knowledge about the agricultural management of grasslands both in the uplands and the lowlands, but less about management for ecological objectives. In particular, in the drawing-up of conservation management agreements there has been a lack of scientific information to back up proposals for prescriptions on levels, and timing, of applications of both organic and inorganic fertilisers, of cutting dates for hay and of stocking rates under a variety of different circumstances. At the same time the pool of knowledge and experience available from the generation of farmers whose farming methods sustained valued grassland habitats is diminishing. Consequently, both ecological research and the pooling of experience on the agricultural management of grassland (including rough grazings, moorland and heaths) for nature conservation objectives, are urgent for the provision of advice on grassland habitat reconstruction on farms.

Specific grassland habitats of concern include:

1. All grasslands on base-rich strata.
2. Semi-natural, often neutral, wet grasslands, where wet meadows/pastures provide habitats for wading birds and plants, including many areas where the ecological interest (e.g. the Broads Grazing Marsh) is dependent on the continuance of low-intensity grazing regimes by store cattle. In some instances, where production falls below economic levels, there may be opportunities to diversify for nature conservation purposes (perhaps under management agreements) by seasonally flooding the land, or by raising water-levels in drainage systems.
3. Belts of rough grassland along hedges, ditches and margins of arable fields, and of areas of rough grassland in uncropped field areas, on banks and similar places.
4. Reversion from arable, particularly in traditional sheep-rearing areas, aiming for low-intensity systems and zero fertiliser inputs compatible with the development of diverse/herb-rich swards.
5. Heaths and moorlands, and regeneration of heather.

Woodlands, shelter-belts, trees and shrubs

Reconstruction of farm woodlands may be achieved by reintroduction of management regimes (e.g. coppicing) that have lapsed, by new forms of management in an existing wood or by the planting of a wood on a new site. New plantings will be especially valuable where they link up existing isolated woods and copses, and at moorland margins to diversify habitats.

Literature and advice is available from the FC, MAFF (e.g. MAFF/FC, 1986), NCC, CC, Local Authorities, FWAG's, the RSPB (Smart and Andrews, 1985) and many other organisations including the BTCV's (1980) *Practical Handbook for Woodlands*. However, farmers are rarely trained as woodland managers and are more likely to call in contractors than to undertake woodland operations themselves. Alternatively they may join a co-operative or management agency of one form or another, such as Project Silvanus, presently operating in Devon and Cornwall and providing a comprehensive service from the preparation of management plans to marketing of the crop, including woodland habitat conservation.

Many woodland species and plant communities are very slow to colonise (see Buckley and Knight, in this volume). There seems therefore to be no chance of reconstructing ancient (Rackham, 1980; Peterken, 1981) or even old woodland habitats *de novo*, and limited opportunities for doing so by managing existing woodlands. Thus reconstructed farm wood habitats will not generally serve the highest wildlife conservation objectives. However, some species, notably of birds and mammals, may appear or reappear surprisingly quickly once a favourable woodland structure has been restored. In addition, newly planted woods pass through a period of some years during which they are becoming established and over this time present a succession of structures and habitats of changing wildlife value.

Scrub arises more often by default than intent and is rarely managed for its own attributes. Nevertheless, it is an important wildlife habitat, although this may depend upon what other habitats it displaces. Where it is invading valued habitats such as chalk or limestone grasslands, it is only tolerable up to a certain level of cover, and its subsequent removal may be very disturbing to the turf. The most effective way of controlling scrub in these situations is by some programme of management, minimally by annual cutting or appropriate grazing regimes. Where scrub is not competing with other habitats it can be managed passively by leaving it alone, or positively, in anticipation of its succession to woodland, by cutting of rides and creation of glades. Although some land falling out of agricultural production will provide opportunities for development of scrub habitats, there are unlikely to be any incentives for its deliberate establishment. Nevertheless, it will be worthwhile in farm plans to identify areas that might be allowed to go to scrub and to plan for it.

Open water and wetland habitats

One of the problems with wetland habitats, as with grasslands, is to prevent new conservation works causing damage to areas of existing ecological importance.

On farmland, farmers are restricted in their management of streams (including rivers) which may be under the control of Water Authorities, of Internal Drainage Boards, or of other statutory bodies.

Greater opportunities for habitat reconstruction occur for off-stream lakes and ponds, and farm reservoirs (even though statutory permissions may be required), for which there is considerable interest at the present time. Open water areas on farms are important for conservation, for landscape and for sport (mainly duck shooting and fishing), and still have a use for fire-fighting, stock watering, and irrigation. Considerable expense can be involved in pond construction or reconstruction but sometimes it is possible to employ machinery that is available on the farm for some other purpose (e.g. CC, 1987b), leading to reduced costs, and such opportunities are worth encouraging. MAFF Farm and Conservation grant is available, and grants may also be obtained from CC, NCC, National Park Authorities or Water Authorities for pond construction or reconstruction, or work on adjoining land.

Ditches have the same linear habitat characteristics as hedges, and are contemporaneous with many hedges which were planted or arose naturally on the spoil thrown up from ditch digging. In some low-lying areas such as the East Anglian Fens, the Somerset Moors, Romney Marsh, the Yorkshire Ings and the Lancashire Mosses, drainage ditches serve in the absence of hedges as field boundaries (wet hedges), and are usually permanently water filled, and of year-round aquatic interest in contrast to field ditches in other parts of the country which commonly dry out. Whilst there do not seem to be great opportunities for new ditch construction, there are advisory opportunities for the sympathetic management of existing ditches to encourage their wildlife interest.

Opportunities for habitat reconstruction in wet grasslands have been discussed above, and must depend on the willingness of farmers to permit water tables to rise or to allow periodic flooding to occur. Where farmers already have wet grassland there is probably little existing use, and consequently little scope for reduction of fertilisers, lime or pesticides. The nature conservation interest usually depends on the continuance of summer grazing and traditional haymaking practices.

To the extent that lowland swamps, marshes and fen mires, and lowland and upland bogs, are subject to agricultural management, there may or may not be opportunities for habitat reconstruction. The most obvious ecological effects arise from the agricultural lowering of water tables of surrounding land, and from pollution and nutrient enrichment. When opportunities to control these occur there will be a considerable contribution to the continuing survival of these habitats.

From an advisory point of view there are publications on aquatic habitats produced by MAFF (1986 a,b,c), BTCV (1976), NCC (1982), NCC (Newbold *et al*, 1983), RSPB (Lewis and Williams, 1984), and CC (1980, 1982, 1987b and others). While a great deal of advice has been given on the construction and wildlife value of farm ponds and reservoirs, and many have been made or 'reconstructed', it would be useful to have more reports on the work, and of monitoring of the results to record successes and failures. The same is true for other habitats.

Buildings, farm roads, quarries and other artefacts

Farm buildings provide habitats for a wide range of plants and animals, some of which are welcome and some are not. The characteristics that separate farm buildings from other buildings in the countryside lie in the openness of many structures (allowing birds and other animals access), and in their environs. Where a building is open there are opportunities to encourage nesting of birds and especially barn owls (*Tyto alba*), but farmers will not want to attract species that foul or damage the contents of the building. In the Pennine Dales ESA farmers are required to maintain and repair field barns as part of their agreements. If this saves these structures from disappearing then there will be wildlife as well as landscape benefits; otherwise a decrepit barn will often provide an equally good habitat for many species.

Opportunities for habitat conservation along farm roads and tracks is limited by the fact that they generally receive very little management. However, where there are banks and other areas of grassland, wildlife value can be enhanced by annual grass cutting. Grassy banks have special value for partridge nesting and this may provide many farmers with an additional interest in maintaining them as herbaceous vegetation.

Some farms have old railway lines crossing them that have been acquired by the farmer, and these, together with quarries and other similar artefacts, present opportunities for programmes of habitat conservation and development ranging from minimum disturbance to different forms of active management. Problems arise with rubbish dumping and this, together with uncontrolled development of scrub, often lead to unsightly degeneration, which needs to be controlled.

Discussion

In giving advice about habitat reconstruction on farms, advisers need to be thinking more widely. It is important to see individual habitats as part of the whole farm, and the whole farm in the context of the surrounding countryside and landscape into which it fits. This demands an ecosystem approach. Although individual habitats are identified and managed separately, their ecological interdependence has to be recognised, and although we compartmentalise them for convenience this should not constrain us from considering them together. Such an approach for birds has been discussed by O'Connor and Shrubb (1986), who describe how structural changes in habitats on farmland Common Bird Census plots have interacted to affect species composition and populations on farms.

When one comes to the level of advising on 'whole farms', and extends this discipline to surrounding areas, new opportunities for habitat reconstruction arise when it becomes possible to integrate management of a habitat on one farm with that of its neighbour. Thus, with some habitats such as woods, there may be part ownership in the hands of several owners, and watercourses generally cross farm boundaries, whilst other habitats on neighbouring farms, although discontinuous, nevertheless contribute to and reinforce each other. As more and more farmers become interested in conservation management so the opportunities for cross-boundary advice will increase, and there will be possibilities for proposals for one farm to be integrated with those for its neighbours.

We are now seeing extensive changes in agricultural policies in England and Wales and these will be reflected in changes in land-use and management. For instance, in the ESAs there is essentially a reversion to low input farming, whilst in the set-aside scheme arable land will be taken out of agricultural production altogether. When land is taken out of production it may be diversified into other income-generating uses. But, in the wider countryside there is a limit to the area of land that can be diversified into non-agricultural products (with the exception of woodland) and of the markets to sustain them. Consequently, there will be significant areas of farmland either being turned over to woodland, to fallow, or in failing to find an alternative use, they will be at risk from not being managed at all. It is this latter category of land where there will be major opportunities for habitat reconstruction, generally by management rather than by new works.

One of the management options may be quite deliberately to do nothing, but this should be a positive decision in the knowledge of the likely developments and how they will integrate with other habitats. This option of allowing areas to revert to wilderness is an important one for ecological and amenity objectives, and should not be discarded. However, for the greatest area landowners and farmers will want to manage their land. The challenge for land management advisers in the fields of nature conservation, landscape, recreation, amenity and sport, and for socio-economists, will be to identify objectives and make practical proposals, acceptable to Government landowners and to farmers, for national and local financial support for the management of areas that fall out of agricultural production. In turn, this will provide opportunities for the sort of integrated approach to the reconstruction of habitats over broad areas of countryside, cutting across farm and estate boundaries, suggested above.

Acknowledgements

I should like to acknowledge the encouragement given to me by MAFF in the preparation of this paper, and for the help and comments of my colleagues. However, it must be emphasised that the opinions are my own, and do not represent the views of the Ministry. I am also grateful for the help of colleagues in other organisations noted in the text for information and comment.

References

Arnold, G.W., 1983, 'The influence of ditch and hedgerow structure, length of hedgerows, and area of woodland and garden on bird numbers on farmland', *Journal of Applied Ecology*, **20**, 731–750.

Botanical Society of the British Isles, 1987, BSBI Arable Weed Survey, *BSBI News* No. 45, p. 6.

British Trust for Conservation Volunteers, 1976, *Waterways and Wetlands*, British Trust for Conservation Volunteers, Reading.

British Trust for Conservation Volunteers, 1980, *Woodlands*, British Trust for Conservation Volunteers, Wallingford.

Countryside Commission, 1980, *Farm Ponds*, Countryside Conservation Handbook, Leaftlet No. 5. Countryside Commission, Cheltenham.

Countryside Commission, 1982, *Conserving Farm Ditches and Watercourses*, Countryside Conservation Handbook, Leaflet No. 12, Countryside Commission, Cheltenham.

Countryside Commission, 1986, *Management Plans*, CCP 206, Countryside Commission, Cheltenham.

Countryside Commission, 1987a, *Shaping a New Countryside*, CCP 243, Countryside Commission, Cheltenham.

Countryside Commission, 1987b, *Conservation Monitoring and Management*, CCP 231, Countryside Commission, Cheltenham.

Countryside Commission/MAFF, 1988, *Broads Grazing Marshes Conservation Scheme 1985–1988*, CCD 20, Countryside Commission, Cheltenham.

Farming and Wildlife Trust, 1987, *Summer Bulletin 1987*, Farming and Wildlife Trust, Sandy.

Fuller, R.M., 1987, 'The changing extent and conservation interest of lowland grasslands in England and Wales: a review of grassland surveys 1930–84', *Biological Conservation* **40**, 281–300.

HMSO, 1986, *Agriculture Act 1986*, Her Majesty's Stationery Office, London.

Lewis, G., Williams, G., 1984, *Rivers and Wildlife Handbook*, Royal Society for the Protection of Birds, Sandy.

Ministry of Agriculture, Fisheries and Food, 1986a, *Farm Ponds: Design and Construction*, ADAS leaflet 3026, MAFF, Alnwick.

Ministry of Agriculture, Fisheries and Food, 1986b, *Farm Ponds: Management and Maintenance*, ADAS leaflet 3025, MAFF, Alnwick.

Ministry of Agriculture, Fisheries and Food, 1986c, *Field Drainage and Conservation*, ADAS booklet 2522, MAFF, Alnwick.

Ministry of Agriculture, Fisheries and Food/Forestry Commission, 1986, *Practical Work in Farm Woods*, Leaflets P 3017–24, MAFF, Alnwick.

National Farmers' Union/Country Landowners Association, 1977, *Caring for the Countryside*, NFU, London; CLA, London.

Nature Conservancy Council, 1977, *Nature Conservation and Agriculture*, NCC, London.

Nature Conservancy Council, 1982, *The Conservation of Farm Ponds and Ditches*, NCC, Shrewsbury.

Nature Conservancy Council, 1984, *Nature Conservation in Great Britain*, NCC, Shrewsbury.

Newbold, C, Purseglove, J, Holmes, N., 1983, *Nature Conservation and River Engineering*, NCC, Shrewsbury.

O'Connor, R.J., Shrubb, M., 1986, *Farming and Birds*, Cambridge University Press, Cambridge.

Perring, F.H., 1974, 'Changes in our native vascular plant flora', in D.L. Hawksworth (ed.), *The Changing Flora and Fauna of Britain*, Academic Press, London, pp. 7–25.

Peterken, G.F., 1981, *Woodland Conservation and Management*, Chapman & Hall, London.

Rackham, O., 1980, *Ancient Woodland*, Edward Arnold, London.

Rands, M.R.W., Sotherton, N.W., 1987, 'The management of field margins for the conservation of gamebirds', in J.M. Way and P.W. Greig-Smith (eds.), *Field Margins*, British Crop Protection Council Monograph No. 35. BCPC, Thornton Heath, pp. 95–104.

Ratcliffe, D.A., 1977, *A Nature Conservation Review*, Vol. I, Cambridge University Press, Cambridge, p. 332.

Smart, N., Andrews, J., 1985, *Birds and Broadleaves Handbook*, Royal Society for the Protection of Birds, Sandy.,

Way, J.M., Greig-Smith, P.W., 1987 (eds)., *Field Margins*, British Crop Protection Council Monograph No. 35, BCPC, Thornton Heath.

Wells, T., Bell, S., Frost, A., 1982, *Creating Attractive Grasslands Using Native Plant Species*, NCC, Shrewsbury.

Wells, T.C.E., Sheail, J., 1988, 'The effects of agricultural change on the wildlife interest of lowland grasslands', in J.R. Park (ed.), *Environmental Management in Agriculture*, Belhaven Press, London, pp. 180–20.

Wells, T.C.E., Sheail, J., Ball, D.F., Ward, L.K., 1976, 'Ecological studies on the Porton Ranges: relationships between vegetation, soils and land-use history', *Journal of Ecology* 64, 589–626.

10

Common-sense approaches to the construction of species-rich vegetation in urban areas

Jane Smart

Introduction

Over the last ten years it has become widely accepted that a considerable diversity of wildlife habitats occurs in towns. Bunny Teagle's exhaustive survey of the West Midlands (Teagle, 1978), and the London Wildlife Habitat Survey (commissioned by the Greater London Council for planning purposes and carried out by the London Wildlife Trust in 1984), produced comprehensive information for the first time. These surveys and other work (Goode, 1986; Smyth, 1987) have revealed a wealth of urban wildlife, many of which are semi-natural habitats caught within the urban environment: verges and embankments of roads and railways, canals and canalised rivers, sewage farms and the freshwater and canal-feeder reservoirs; and areas of vacant and derelict land awaiting the next cycle of development (Cole, 1983).

An urban nature conservation 'movement' has developed to press for the conservation and appropriate management of many of these urban sites, part and parcel of which has been a surge of interest in 'habitat creation'. Through research into reclamation techniques (Bradshaw and Chadwick, 1980) and into the creation of species-rich swards, we are now developing the technology to create 'naturalistic' replicas of many semi-natural habitats. *Creating Attractive Grasslands Using Native Plant Species* (Wells, Bell and Frost, 1981), based on over eight years of research at the Institute of Terrestrial Ecology's Monks Wood research station, was a milestone in the habitat creation literature. Produced as a response to the tremendous loss of the old herb-rich grasslands stemming from the intensification of agriculture in the last forty years, the book describes habitat management techniques based on detailed research on the ecology of the species and communities concerned. This work is referred to in many other guides and handbooks which describe practical techniques for the development of species-rich grasslands (e.g. Baines and Smart, 1984; Emery, 1986). The common objective is to provide floristically-diverse swards which will both enhance the aesthetic quality of the environment and increase the diversity of wildlife, in areas such as roadside verges, land for amenity purposes in towns, country parks, nature reserves and gardens.

Hand in hand with the development of habitat creation techniques has been the response of landscape designers to issues concerning the conservation of the natural world. The landscape profession has perceived that the poor quality of living environments of most modern cities can be redressed by adopting an 'ecological approach' to landscape design (Tregay, 1982; Goode and Smart, 1986). Techniques developed in Holland over the last 50 years have provided a stimulus for many landscape designers to construct, using basic ecological principles, replicas of traditionally managed habitats such as woods and haymeadows in towns.

But some habitats are easier to construct than others. Wetlands, from small artificial ponds to large lakes developed from restored gravel workings, are readily planted with native species and have proved extremely attractive to wildlife (Harrison, 1974) as have woodlands established as nature-like plantations (Tregay, 1986). Grasslands, however, have proved extremely difficult to develop. Despite their popularity, and the boom in wildlife gardening (Baines, 1985), together with the mass production of native wild flower seed mixtures, species-rich swards constructed 'from scratch' are a rare sight. This is surprising because there is certainly plenty of room in towns and cities. Formal parkland held by local authorities has long been seen by nature conservationists as a prime candidate for the development of species-rich communities.

Whilst there is no doubt that, given sufficient horticultural attention, attractive diverse swards can be developed (the show gardens of the seed manufacturers are prime examples), it is not generally appreciated just how difficult it is to develop a meadow on nutrient-rich soils (but see Hodgson, in this volume), for example in formal parks. Experience suggests that the main problems are unsuitable choice of site, inadequate site preparation and inappropriate management. Thus, in attempting to generate species-rich grassland communities 'from scratch' in urban areas, it seems that we spend much time and effort fighting against nature instead of working with it (McHarg, 1969; Bradshaw, 1986). If habitat generation is based on natural systems and uses natural processes to achieve desired ends we are more likely to produce resilient and self-sustaining solutions (Bradshaw, in this volume).

This article is concerned with more 'natural' approaches to the generation of species-rich vegetation in cites, with particular reference to London. The habitats which have developed so successfully on industrial 'wasteland' and vacant urban land illustrate what is possible. They are a good guide since most of the species involved are already adapted to the artificial conditions of towns and cities. These are the species which do best and most of them are very attractive (Goode, 1987). Cities are often the home of rare and localised native herbaceous plants as well as a wide variety of common species and a range of exotic garden escapes.

Methods which encourage the natural colonisation of sites include a spectrum of techniques from *laissez-faire*, manipulation of development ('creation through management'), to restoration, rehabilitation of habitats and positive construction. Clues to common sense techniques for the generation of communities are gleaned here from four sites in London on which artificial habitats have developed; one by design and three by default.

Natural approaches to the generation of species-rich vegetation

Natural colonisation: wastelands

Many sites in cities, now highly valued for nature conservation, have in fact developed naturally, by default rather than by design. These 'wastelands', clothed in unsown, unmanaged vegetation are becoming regarded as the true urban countryside (Greater London Council, 1986). Their ecological characteristics, regional variation and cultural value have been described by Gilbert (1981, 1983 and 1984) who has named them 'urban commons'. They often have a natural history that is both rich and unique to a particular area: a blend of native plants with naturalised exotics. Since wasteland is not traditionally recognised as valuable wildlife habitat its qualities have hitherto been underrated.

There is a wide variety of wasteland in London: abandoned wharves, disused railway sidings, abandoned gardens and allotments. These sites are usually temporary, and at any moment may be developed into a new car park or office block. They tend to be young in comparison with many other habitats and almost always result from disturbance and feature bare ground. When the site is a result of dereliction or demolition the substrate is artificial: brick, glass, tarmac and rubble. Semi-natural substrates such as gravel, clay and topsoil are more often produced by excavation, construction work and tipping (Greater London Council, 1986).

The primary colonising species of wasteland are determined by the nature of the substrate, the composition of the buried seed bank and local seed sources. Plant succession takes the vegetation of these areas through a series of phases culminating in woodland. Annuals and short-lived perennials move in initially; some years later tall herbs tend to take over. The less hospitable the substrate, the slower this process is. The proportion of grasses gradually increases so that the community typically becomes grassland with pockets of tall herbs. The process is quicker in abandoned gardens and allotments where there is a soil covering, and the seedlings of trees and shrubs are often present from the beginning. Whatever the substrate, successions tend to converge towards secondary woodland, usually passing through a phase of scrub development.

The acceptability of wastelands

To many inner-city dwellers, wastelands are the nearest thing to conventional countryside, reflecting seasonal change in a way rarely experienced in towns. They cater for a wide range of 'unofficial' and impromptu activities which are not allowed anywhere else: children's adventure play, bonfires and the grazing of horses and goats. Although wastelands are very popular with naturalists and children, large areas of raw derelict land are still viewed officially as unsightly and depressing. The usual local authority reaction, when faced with an 'overgrown' urban site to 'landscape' is to clear it, smother it with topsoil and reseed it.

With a small management input these popular areas can soon be made 'respectable'. When the vegetation dies down in the autumn rubbish tends to become visible. Clearly this needs to be removed, and the road frontages kept tidy. Some means of preventing fly tipping is necessary, such as a ditch or a line of bollards. Footpaths following desire lines should be retained – a covering of bark chippings or even hoggin will increase their acceptability. Turning such areas into nature parks and promoting active educational use also results in social respectability. For small sites this approach is now established in certain western European cities such as Berlin.

Even with the tidying up described, however, wasteland vegetation can look drab for over half the year, being dominated by dead vegetation. This is unacceptable to many citizens who have grown up with city gardens full of organised flowers and evergreens, and who associate 'weeds' on derelict sites with bomb damage of the last war. These problems, however, can be eased by careful manipulation of the vegetation.

Manipulation of existing swards

The acceptability, and species-richness, of existing swards can be increased by simple management techniques whether the sward is recently developed, such as on derelict land, or developed from original meadow pasture. At Warrington New Town, diverse swards have been developed by management of existing vegetation (Tregay and Gustavsson, 1983). One site, located near housing, was rather featureless apart from its cover of existing grassland and scrub. To enhance the look of this area, the sward was cut using a hand-held rotary cutter and the debris raked off, while young shrubs such as willow and gorse were retained. Grass walkways based on trodden paths were mown through the area, the design evolving gradually as a response to vegetation succession and to the way the site was used by people.

Many urban grasslands are already quite rich in species but, being kept closely shaved, we tend not to be able to see this. This applies particularly to grasslands in cemeteries, hospital grounds and other areas where some species remain from the 'original' semi-natural grassland. As Baines and Smart (1984) point out, a simple change in management can result in a quick transformation to a colourful wild flower meadow. A community which is rich to start with can be enhanced with a gap of around four to five weeks in the mowing regime, some time in early summer. This allows the flowers present to bloom and set seed. By removing the clippings, the nutrient status of the soil is reduced which tends to favour wild flowers rather than coarse grasses. Although extremely simple, this kind of management regime is rarely utilised. In places where regularly mown grassland contains a variety of species, this is the most effective first step in encouraging species-richness.

Assisting natural colonisation

Where species-richness is the primary objective and the time-scale is of no consequence, it is of course possible to allow diverse swards to develop through natural colonisation. For example, several areas for natural colonisation were incorporated into the landscaping at Warrington (Tregay and Gustavsson, 1983), starting with areas of undisturbed subsoil, and controlling colonisation and development by a flexible and responsive management regime.

More often, time is limited and trampling pressure dictates that some 'instant' green cover is necessary. In this case a well-tried, sensible approach is to create an open sward into which herbs can establish by natural colonisation (Corder and Brooker 1981). For this, fescue-based grass mixes sown at low rates work best. On one such meadow in Warrington a *Festuca-Agrostis* sward on a sandy subsoil was naturally colonised by 65 other species after only three years (Moffatt, 1986).

Such achievements depend on the precise timing and height of all cutting operations and the presence of a local seed source (Scott *et al.*, 1986). For the first two years, regular mowing is essential to eliminate competitive species which tend to exclude the more attractive meadow flowers. After this establishment phase, mowing is reduced to allow natural diversification to occur. A cut in late August or early September prevents dead grass litter building up and maintains the open nature of the sward (Tregay and Gustavsson, 1983).

Examples of species-rich vegetation by design and by default

The flora of a created nature park is compared here with that of three abandoned railway sidings in London. The areas of railway land have all been the subject of planning inquiries to decide their future development and may not end up as natural parks. Indeed, abandoned railway land in London is becoming scarce as it undergoes redevelopment, and a study is currently taking place to reassess the quality of that which is left. They are described here to illustrate the nature of the vegetation which has developed on them.

William Curtis Ecological Park

The story of the development of the William Curtis Ecological Park, which existed near Tower Bridge, London, from 1978 to 1985, is well known (Cole, 1983, 1986). This pioneering venture involved the creation of a range of habitats on a small (*c.* 1ha) derelict plot in the London Borough of Southwark (see Figure 10.1). Voluntary labour was used to plant thickets of native trees and shrubs as well as to create a pond, meadow areas, and a sand dune. Areas of rubble were also incorporated. Soil was obtained by opening the site as an inner city subsoil dump; 350 lorry loads of subsoil were brought in over a period of three weeks.

William Curtis Park was remarkably successful, not only in catering for local school children who would otherwise have had little or no contact with nature, but also as one of the first innovative habitat creation ventures in an urban area. The project provided very useful data on habitat creation and management as the Ecological Parks Trust, who developed and managed the site, carried out detailed monitoring throughout the seven-year period that the park was open, to assess the development of the created semi-natural communities (Cole, 1986).

Shakespeare Road Sidings

Shakespeare Road Sidings, just half a mile from Brixton town centre, are immediately adjacent to the London to Dover railway line. These abandoned sidings (*c.*

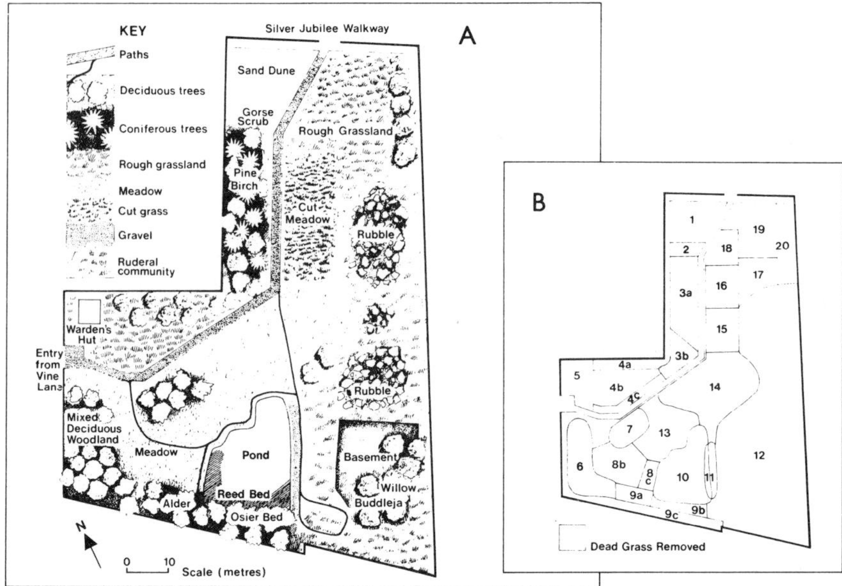

Figure 10.1 Plan showing habitat areas established at William Curtis Park in 1982 (A) and position of original compartments in 1978 (B).

Note: Key to original compartments is as follows: 1: sand dune; 2: gorse scrub; 3a: birch/pine woodland; 3b: birch/oak woodland; 4a: London pioneer species; 4b: species-poor deciduous woodland; 4c: hedgerow; 5: hut compartment; 6: species-rich deciduous woodland; 7: copse; 8a: perennial wayside weeds; 8b: wet peat meadow; 8c: tall sledge community; 9a: alder carr; 9b: willow bed; 9c: pre-existing stand of sycamore, sallow etc.; 10: pond with marginal vegetation; 11: gravel pond margin; 12: area of natural succession on a variety of substrates; 13: wet clay meadow; 14a: copse; 14b: sown grassland; 15: short pasture grasslands; 16: dry meadow; 17: cornfield; 18: chalk grassland; 19: self-sown meadow and scrub; 20: hedge. (After Ecological Parks Trust, 1982).

2ha) fell into disuse around 1960; from then the site remained vacant and by 1986 had developed a cover of birch woodland and grassland. Following a decision by Lambeth Council to refuse planning permission for housing the vegetation on the sidings was bulldozed, with some burning. The sidings are a man-made embankment about three metres high constructed over London clay. The fill consists of chalk, limestone and granite hardcore covered by a variable surface layer of ash, brick, concrete, tarmac, metal and glass.

Bricklayers Arms

Bricklayers Arms (*c.* 12.5ha), alongside the Old Kent Road in Bermondsey, was a busy goods yard and coal depot from the late 1800s until its closure in 1962. Most of the tracks were left for many years and only removed in the 1980s. Over this period plants had grown up between the sleepers and on the embankments. Until its recent development, a medley of grassland, scrubland and wasteland survived amongst the clinker and ballast of the old railway tracks.

Feltham Marshalling Yards

This disused railway siding complex (*c.* 30ha) extends for over a mile beside the railway running along the southern edge of Hounslow Heath and not far from Heathrow Airport. The yard closed in 1968, since when it has been colonised by a great variety of plants. Birch and willow thickets are interspersed with open patches full of flowers.

Plant community development

A great many plant species have been recorded from all four sites. At William Curtis Park, the best studied of the areas, 348 vascular plant species were recorded by 1982. Less detailed studies of Feltham Marshalling Yards have revealed over 200 species. Table 10.1 shows that the flora of the sites includes many attractive herbaceous species, several of which are rare in the Greater London area. Corn chamomile (*Anthemis arvensis*), for instance, which was found at Shakespeare Road Sidings and Feltham Marshalling Yards, is recorded in only 0.5 per cent of the 4 km^2 grid (tetrads) covering Greater London (Burton, 1983). Blue fleabane (*Erigeron acer*) found at Bricklayers Arms, has been recorded from only 6.2 per cent of London's tetrads and kidney vetch (*Anthyllis vulneraria*), recorded from William Curtis park and recorded in only 2.7 per cent of tetrads, is also scarce.

Interestingly, most of the species at William Curtis Park were not introduced. Some colonised from wind-borne seed and some appear to have been introduced with the imported subsoil. On the railway sidings, many of the plants may have been introduced by the trains themselves. Seed of species characteristic of chalk grassland such as musk thistle (*Carduus nutans*) may have been transported to Shakespeare Road via trains coming through chalk areas of the South Downs. Feltham Marshalling Yards are further from the centre of London than the other three sites and perhaps benefited more from local sources of seed; all three sites would have had much greater local sources 100 years ago.

On Shakespeare Road Sidings the soil pH ranges from 4.6 to 7.6, no doubt reflecting the diverse range of materials making up the sidings and indicating a variable soil chemistry. This has allowed many species of plants to coexist: eyebright (*Euphrasia officinalis* s.1.) and bee orchid (*Ophrys apifera*) reflect the chalky soil as do blue fleabane (*Erigeron acer*) and pale toadflax (*Linaria repens*) on other sites (see Table 10.1). Sheeps sorrel (*Rumex acetosella* agg.) indicates poorer acid conditions. Heathland and sand dunes are also represented on the sites by, amongst others, tormentil (*Potentilla erecta*) and hares-foot (*Trifolium arvense*) respectively.

Common to all the sites is the abundance of legume species. As many as 27 species were recorded from William Curtis Park, of which only three were purposely introduced, and the same number was found on Feltham Marshalling Yards. Species include grass vetchling (*Lathyrus nissolia*) from William Curtis Park, birdsfoot (*Ornithopus perpusillus*) from Feltham Marshalling Yards and hares-foot (*Trifolium arvense*) from both Shakespeare Road and Feltham Marshalling Yards. Birdsfoot trefoil (*Lotus corniculatus*), introduced at William Curtis Park, was also found at the other three sites.

Table 10.1 Occurrence of specified herbaceous plants on some sites in London

Latin name	English name	WC	SR	BA	FM	% occurrence in London
Achillea millefolium	Yarrow	·	·	·	·	100.00
Ajuga reptans	Bugle	·			·	23.75
Anthemis arvensis	Corn chamomile		·		·	0.50
Anthyllis vulneraria	Kidney-vetch	·				2.75
Ballota nigra	Black horehound	·	·	·	·	98.50
Cardamine hirsuta	Hairy bitter-cress	·	·		·	37.50
Cardaria draba	Hoary pepperwort	·	·	·	·	73.75
Carduus nutans	Musk thistle		·		·	3.00
Centaurea nigra	Hardheads	·*			·	89.50
Centaurium erythraea	Centuary	·			·	11.25
Chenopodium album	Fat hen	·	·	·	·	99.00
Cirsium vulgare	Spear thistle	·	·	·	·	99.75
Clinopodium vulgare	Wild basil				·	9.25
Cymbalaria muralis	Ivy-leaved toadflax	·*				56.00
Daucus carota	Wild carrot	·	·		·	41.50
Digitalis purpurea	Foxglove	·*				44.25
Dipsacus fullonum	Teasel	·*			·	51.50
Dipsacus pilosus	Small teasel			·		0.50
Echium vulgare	Viper's bugloss				·	2.50
Epilobium adenocaulon	American willowherb	·	·			80.75
Erigeron acer	Blue fleabane			·		6.25
Erodium cicutarium	Common storksbill	·	·	·	·	8.75
Euphorbia peplus	Petty spurge	·	·	·	·	97.00
Euphrasia officinalis s.1.	Eyebright		·		·	5.00
Galium verum	Lady's bedstraw				·	44.00
Geranium dissectum	Cut-leaved cranesbill	·			·	67.25
Geranium molle	Dove's-foot cranesbill		·			70.00
Geranium robertianum	Herb robert	·	·	·	·	52.50
Geum urbanum	Wood avens	·*			·	57.75
Hypericum hirsutum	Hairy St John's wort				·	12.25
Hypericum perforatum	Common St John's wort		·	·	·	52.50
Hypochaeris radicata	Cat's ear	·	·	·	·	98.00
Knautia arvensis	Field scabious				·	15.25
Lathryus pratensis	Meadow vetchling	·			·	72.75
Lathyrus nissolia	Grass vetchling	·				9.00
Leontodon autumnalis	Autumnal hawkbit	·	·	·	·	90.25
Lepidium heterophyllum	Smiths pepperwort				·	1.00
Linaria repens	Pale toadflax			·	·	2.25
Linaria vulgaris	Yellow toadflax	·	·	·	·	75.75
Lotus corniculatus	Birdsfoot-trefoil (common)	·*	·	·	·	85.00
Malva sylvestris	Common mallow	·	·	·	·	96.75
Medicago lupulina	Black medick	·	·	·	·	96.25
Medicago sativa	Lucerne	·		·	·	9.25
Melilotus alba	White melilot	·	·		·	33.00
Melilotus officinalis	Melilot	·	·		·	62.25
Mercurialis annua	Annual mercury	·	·	·	·	74.00
Odontites verna	Red bartsia	·				14.00

Latin name	English name	WC	SR	BA	FM	% occurrence in London
Onobrychis viciifolia	Sainfoin	·*				2.25
Ophrys apifera	Bee orchid		·			4.50
Ornithopus perpusillus	Birdsfoot				·	7.00
Pentaglottis sempervirens	Alkanet	·	·	·	·	22.00
Picris hieracioides	Hawkweed ox-tongue	·	·	·		20.00
Polemonium caeruleus	Jacob's ladder	·*				0.50
Potentilla erecta	Common tormentil				·	23.50
Prunella vulgaris	Self-heal	·*				85.00
Pulicaria dysenterica	Fleabane				·	19.25
Ranunculus repens	Creeping buttercup	·	·	·	·	99.50
Reseda lutea	Wild mignonette		·	·	·	34.50
Rumex acetosa	Sorrel		·		·	88.50
Rumex acetosella agg.	Sheep's sorrel	·	·	·	·	66.00
Sanguisorba minor	Salad burnet				·	8.25
Senecio squalidus	Oxford ragwort	·	·	·	·	99.00
Silene alba	White campion	·	·		·	76.00
Silene dioica	Red campion	·*	·	·	·	59.50
Silene nutans	Nottingham catchfly				·	0.00
Silene vulgaris	Bladder campion	·		·	·	45.00
Sisymbrium officinale	Hedge mustard	·	·	·	·	98.75
Solanum dulcamara	Bittersweet	·	·	·	·	98.75
Solanum nigrum	Black nightshade	·	·	·		87.50
Sonchus asper	Spiny milk-thistle	·	·			93.75
Tanacetum parthenium	Feverfew	·	·			73.00
Trifolium arvense	Hare's-foot		·		·	8.50
Trifolium campestre	Hop trefoil		·		·	36.00
Trifolium dubium	Lesser yellow trefoil	·	·		·	82.50
Veronica chamaedrys	Germander speedwell	·			·	63.75
Veronica hederifolia	Ivy-leaved speedwell	·	·	·	·	79.25
Vicia cracca	Tufted vetch		·		·	65.75
Vicia hirsuta	Hairy tare	·	·		·	21.50
Vicia sativa	Common vetch	·	·		·	79.25
Vicia sepium	Bush vetch		·			30.25

Notes: 1. The occurrence of some attractive herbaceous plants on some sites in London: one created park (William Curtis Ecological Park (WC); Ecological Parks Trust 1982) and three abandoned railway sidings: Shakespeare Road Sidings (SR), Bricklayers Arms (BA) and Feltham Marshalling Yards (FM). Species introduced into William Curtis Ecological Park are marked by an asterisk.

2. The species lists are selective and are in no way intended to portray the full botanical diversity of these areas, nor the abundance or time of occurrence of the species on the sites. Because of the unequal searching effort on the four sites, the lists should not be used to contrast their species-richness.

3. The percentage occurrence in London was calculated from tetrads (map squares of side two kilometres, area four square kilometres) on maps in 'Flora of the London area', R.M. Burton (1983), London Natural History Society. These give a broad indication of the distribution of species in the Greater London area. O indicates that the species was previously unrecorded.

After an initial increase, species characteristic of grassland communities at William Curtis Park (some of which were introduced in a seed mix and some of which colonised naturally) tended to decline, following the development of a dense sward of *Festuca rubra* (included in a sown mixture of bents and fescues, and itself encouraged by the nitrogen-fixing legumes). In fact in 1982 the Ecological Parks Trust reported that the increasing homogenity of the vegetation necessitated a change in management. The 20 original habitat compartments had not retained their identities, but had gradually merged to form a few large areas which themselves showed signs of merging (Ecological Parks Trust, 1982).

Extensive amounts of dead grass were removed from 'grassland' compartments in early 1982 in an attempt to encourage herb species. By 1983 it became clear that the park was supporting a much taller vegetation than formerly, and scything and removal of the previous season's dead growth was again carried out to counteract this (Ecological Parks Trust, 1983). At Shakespeare Road Sidings removal of the existing vegetation was by bulldozing. In the first summer since bulldozing over 150 vascular plant species were recorded, ironically a greater number than previously, probably resulting from the germination of buried seed.

Each of the four sites also appears to have its own characteristic flora. For example, Bricklayers Arms is the only site with small teasel (*Dipsacus pilosus*), Shakespeare Road is the only site with bee orchid (*Ophrys apifera*) and Feltham Marshalling Yards the only place where salad burnet (*Sanguisorba minor*) and field scabious (*Knautia arvensis*) were recorded. In the same way, red bartsia (*Odontites verna*) was found only at William Curtis Park where it was not introduced. Clearly any of these species might at some time appear at one of the other sites, or could even be introduced. Nevertheless, the fact that each site has its own particular complement of species certainly adds to their interest and popular appeal.

Lessons for a sensible approach to the development of species-rich vegetation

At William Curtis Ecological Park a diverse grassland-like sward was generated by a combination of manipulation of the existing landform, soil and water, with planting, seeding and management. At Shakespeare Road Sidings, Bricklayers Arms and Feltham Marshalling Yards diverse swards developed through natural colonisation of the existing substrates. Some accidental 'management' occurred at Shakespeare Road Sidings when the vegetation was bulldozed. Several key 'lessons' emerge from these examples.

The substrate

The nature of the substrate can be assumed to be a prime factor accounting for the variety of vegetation at these sites. Firstly, the variability of materials utilised in the construction allowed a high number of species to survive. By importing a wide range of substrates such as chalk, limestone, sand and clay, species characteristic of several different semi-natural habitats will be encouraged to grow through natural colonisation or introduction.

Secondly, substrates often created stressful conditions: subsoil and other hardcore materials are low in available plant nutrients and are subject to moisture stress because of their freely draining nature. It is surprising that this relationship between inhospitable habitats and species-richness is not utilised more in the design of new habitats (Grime, 1986), as the required materials are cheap and easily obtainable.

The conservation value of railway land has been highlighted by Sargent (1984). At the sidings sites described here botanical composition will reflect the composition and construction of the cess and ballast. The cess is usually defined as the freely draining area of cindery material over which ballast (the track bed) and rails are laid. The cinder is usually exposed between tracks and in station and shunting yards (Sargent, 1984). These conditions and others which relate to the location and past use of the railway sites would be hard to re-create in their entirety. Nevertheless, the materials from which they were constructed could be used almost anywhere to encourage the development of species-rich vegetation as the history of William Curtis Park demonstrates (see Table 10.1).

Management

As with existing swards, management is crucial to the maintenance of newly developed vegetation. It is of course required to prevent the development of early-successional herbaceous vegetation into scrub and thence to woodland. Many of the plants found on the sites described here are ruderals. These have rapid growth rates, abbreviated life-spans, and prolific reproduction, all of which allow the intervals between disturbances to be effectively exploited (Grime, 1986). To maintain this kind of vegetation repeated disturbances must be created through management. Alternatively, vegetation dominated by ruderal species can be allowed to develop into a short rich turf of mainly grassland species. Thereafter, cutting or mowing regimes with flowering gaps appropriate to the species on site (followed by removal of clippings) will maintain and enhance the diversity of the vegetation.

Hardcore substrates have the additional advantage that their very nature arrests succession; maintenance is made relatively straightforward. They also have the merit of making good footpath material and are resilient to trampling pressure. Whatever the substrate, the long-term survival of any seral vegetation type will be dependent on management. Ideally this should be flexible and exploit the dynamic qualities of vegetation processes, using them to creative ends. As such management should be an ongoing part of the development of a site.

Time

The management problems that were seen at William Curtis late in its seven-year life suggest that the ultimate character of that site would be somewhat different from the railway sites. The latter were all much older, having been abandoned for at least two decades, but the pace of change on them has been very slow and management needs minimal. There is a considerable value in this stability, in both aesthetic and scientific terms, which is at best anticipated, but not yet realised, in places like William Curtis Park. Thus when presented with a choice of retaining a site of some age and existing value, or accepting a newly created alternative, other things being equal, one would always prefer the existing site (see Newbold, in this volume).

Encouraging the development of species-rich vegetation

As long as there are no safety problems (e.g. contaminated land) there is a lot to be said for manipulating the 'urban commons' which already exist. Such areas speak for themselves in the sense that it is clear that the plants growing in them can withstand the environmental conditions imposed by a harsh city environment. These places tend to be popular and have their own local or cultural significance (Gilbert, 1983). But to enhance their appearance *some* 'tidying up' of these areas is often desirable. This can range from neatening the edges and erecting a sign, to full-scale restoration which might involve the removal of all tall vegetation at the end of the season, followed by continuing appropriate management.

For the creation of new swards, natural colonisation can often do the job, given sufficient time. Usually, however, it is important to attain an instant show of colour in a new nature park. This is particularly so on small-scale sites where the emphasis needs to be on obviously attractive and interesting habitats. As Lyndis Cole (1986) points out, a small piece of Wimbledon Common transported to a small inner city gap would look remarkably uninteresting. Inoculating the site with species that gave an instant, albeit temporary, colourful and species-rich meadow community was an essential part of convincing the sceptics and increasing the faith of all those involved with the William Curtis project. In addition, of course, a great deal was learnt. The imported substrate was not only a medium for planting; it was also a means of bringing more species and, because of its particular composition, a means of encouraging other colonisers to become established.

A meadow of cornfield annuals can easily be developed on chalk, limestone and other hardcore substrates; these will flower in the first season and look extremely attractive. Once a site is established (both ecologically and socially) it will probably not matter a great deal if the ruderal 'weeds' invade and some of the wild flowers from the seed packet are ousted. At this point all concerned will probably have realised that some of the 'weeds' that come in are really quite attractive and interesting. Another way of giving a 'new' site a start in life is to develop an open sward of bents and fescues into which species can colonise naturally. The success of this gradual diversification depends on an appropriate management regime and a local seed source, but it is often not viewed positively as a means of creating species-rich swards.

All the sites described here are popular with people. This has been demonstrated by public support at the various planning inquiries held to consider their future. This support has shown a clear case for both the retention of the sidings sites and the construction of new areas of species-rich vegetation in established parks and public open space. Their construction may involve a combination of several of the techniques reviewed here. Their longevity will depend on the awareness and use of ecological principles and the application of well-known (but often ignored) management techniques. If successful, we can move towards more widespread culture of species-rich swards whilst making better use of popular 'wasteland' sites.

References

Baines, J.C., 1985, *How to Make a Wildlife Garden*, Elm Tree Books, London.

Baines, C., Smart, J., 1984, *A Guide to Habitat Creation*, GLC Ecology Handbook No. 2, Greater London Council.

Bradshaw, A.D., 1986, 'Ecological principles in landscape', in A.D. Bradshaw, D.A. Goode and E. Thorp (eds.), *Ecology and Design in Landscape*, 24th Symposium of the British Ecological Society, Blackwell Scientific Publications, Oxford, pp. 15–36.

Bradshaw, A.D., Chadwick, M.J., 1980, *The Restoration of Land*, Blackwell Scientific Publications, Oxford.

Burton, R.M., 1983, *Flora of the London Area*, London Natural History Society, London.

Cole, L., 1983, 'Urban nature conservation', in A. Warren and F.B. Goldsmith (eds), *Conservation in Perspective*, Wiley, London, pp. 267–86.

Cole, L., 1986, 'Urban opportunities for a more natural approach', in A.D. Bradshaw, D.A. Goode and E. Thorp (eds.), *Ecology and Design in Landscape*, 24th Symposium of the British Ecological Society, Blackwell Scientific Publications, Oxford, pp. 417–32.

Corder, M., Brooker, R., 1981, *Natural Economy. An Ecological Approach to Planting and Management Techniques in Urban Areas*, Kirklees Metropolitan Council.

Ecological Parks Trust, 1982, *Second Report*, Ecological Parks Trust, London.

Ecological Parks Trust, 1983, *Third Report*, Ecological Parks Trust, London.

Emery, M., 1986, *Promoting Nature in Cities and Towns: a Practical Guide*, Croom Helm, London.

Gilbert, O.L., 1981, 'Plant communities in an urban environment', *Landscape Research,* **2** (3), 5–7.

Gilbert, O.L., 1983, 'The wildlife of Britain's wasteland', *New Scientist,* **97**, 824–9.

Gilbert, O.L., 1984, 'The urban common, New Directions 7', *Landscape Design,* **6**, 35–6.

Goode, D.A., 1986, *Wild in London*, Michael Joseph, London.

Goode, D.A., 1987, 'Nature in the city', *Urban Design Quarterly*, **24**, 12–14.

Goode, D.A., Smart, P.J., 1986, 'Designing for wildlife', in A.D. Bradshaw, D.A. Goode and E. Thorp (eds.), *Ecology and Design in Landscape*, 24th Symposium of the British Ecological Society, Blackwell Scientific Publications, Oxford, pp. 219–35.

Greater London Council, 1986, *A Nature Conservation Strategy for London: Woodland, Wasteland, the Tidal Thames and two London Boroughs*, Ecology Handbook No. 4, GLC, London.

Grime, J.P., 1986, 'Manipulation of plant species and communities', in A.D. Bradshaw, D.A. Goode and E. Thorp (eds.), *Ecology and Design in Landscape*, 24th Symposium of the British Ecological Society, Blackwell Scientific Publications, Oxford, pp. 175–94.

Harrison, J., 1974, *The Sevenoaks Gravel Pit Reserve*, WAGBI, Chester.

McHarg, I.L., 1969, *Design with Nature*, Natural History Press, New York.

Moffat, D., 1986, 'The natural approach to open space', in R. Brooker and M. Corder (eds.), *Environmental Economy*, Spon Ltd., London.

Sargent, C., 1984, *Britain's Railway Vegetation*, Institute of Terrestial Ecology, Natural Environment Research Council, Cambridge.

Scott, D., Greenwood, R.D., Moffat, J.D., Tregay, R.J., 1986, 'Warrington, New Town: an ecological approach to landscape design and management', in A.D. Bradshaw, D.A. Goode and E. Thorpe (eds.), *Ecology and Design in Landscape*, 24th Symposium of the British Ecological Society, Blackwell Scientific Publications, Oxford, pp. 143–60.

Smyth, B., 1987, *City Wildspace*, Hilary Shipman, London.

Teagle, W.C., 1978, *The Endless Village*, Nature Conservancy Council, West Midlands Region.

Tregay, R., 1982, 'Nature and an ecological approach to landscape design', in A.R. Ruff and R. Tregay (eds.) *An Ecological Approach to Urban Landscape Design*, University of Manchester.

Tregay, R., 1986, 'Nature-like plantations', A.D. Bradshaw, D.A. Goode and E. Thorpe (eds.) *Ecology and Design in Landscape*, 24th Symposium of the British Ecological Society, Blackwell Scientific Publications, Oxford, pp. 275–84.

Tregay, R., Gustavsson, R., 1983, *Oakwood's New Landscape: Designing for Nature in the Residential Environment*, Sviriges Lantbruksuniversitet and Warrington and Runcorn Development Corporation, Warrington.

Wells, T.C.E., Bell, S., Frost, A., 1981, *Creating Attractive Grasslands Using Native Species*, Nature Conservancy Council, London.

11

Opportunities for habitat enhancement in commercial forestry practice

M.A. Anderson

Introduction

Blanket planting with conifers has been a 60-year explosion. In the North-Temperate zone millions of hectares of ligniculture (Everard, 1974) and commercial plantations now clothe land hitherto or primevally covered with deciduous trees. Broadleaved communities have dwindled, been isolated or lost.

Public distaste has followed. This has been well marked in Britain, which had the extreme condition of a few exotic species massively planted on land that had already lost most of its tree cover (Nature Conservancy Council, 1986), and in Scandinavia, where mixed forest on wet ground is being replaced intensively with conifers (Gamlin, 1988). National policies and planning are now reflecting this concern. Encouragement is being given to reinstating broadleaf habitats and communities. Britain, in particular, gives substantial grants for new broadleaved planting (Forestry Commission, 1988).

The precise objectives of habitat reconstruction and enhancement depend a great deal on the communities remaining in the conifer plantations. These aims are shown in Table 11.1. Where there is little present in the way of native communities they must be encouraged to return, perhaps sometimes even translocated. This aim is termed **Import**. Where there are substantial, though perhaps small, communities, they must be sustained, and this is termed **Maintenance**. Finally, established and re-established populations must be arranged and encouraged so as to be able to invade a great deal of the adjacent forested land when light or soil conditions permit. This last aim is termed **Export**. All three objectives embrace the notion that component species should eventually maintain themselves. To start with, this may be difficult in plantation forests; it will require their managers to cycle or arrest successions artificially. Plant and animal communities are dynamic.

Commercial foresters might be expected to seek the best return for the least expenditure (but for a contrasting view see Dannatt, 1987). In this paper I summarise some topical ideas about ready-made opportunities for reinstating native deciduous communities with minor deflections from production silviculture. It is not a comprehensive review. I do not examine all the possible opportunities. But opportunities can arise because of new legislation, because of changes in conventional practice or because they were there all along, and I consider some of each.

The paper divides opportunities into those at the scale of the forest's mosaic or

Table 11.1 Aims of habitat reconstruction in British conifer plantations

	Name	Description
1.	IMPORT	Encourage return of previous native species or communities
2.	MAINTENANCE	Stabilise and sustain remaining native species or communities
3.	EXPORT	Establish colonising fronts of native species

patchwork, those inside individual compartments of trees and those that occur along edges. It is largely botanical, animals being expected to follow their food. British conditions are used, rightly or wrongly, as a model for others.

Opportunities arising in the forest's patchwork

Restructuring

Conifer plantations are usually divided into management units or 'compartments'. Often each is several hectares in area. On uniform ground they are also usually rather evenly sized. Forest managemént practices, such as thinning and fertilising, are generally applied uniformly to a whole compartment.

Figure 11.1A shows three such compartments in Kielder Forest, England. This is an upland area with much uniformly planted Sitka spruce (*Picea sitchensis*) Bong. (Carr.). The compartments were all planted at the same time, about 50 years ago, and are now due for commercial felling. The figure shows underlying features, such as a stream, previous paths and roads, which possess little in the way of plant communities because the trees are 20m tall, only 5m apart and shade most of the open ground.

At clear-felling an opportunity might conceivably arise to vary the grain and variety of the patchwork, and to release areas of native communities that have survived. Until recently, however, clear-felling in these areas has been done by 'rolling up the carpet'. Harvesting starts on a line and moves steadily across the scenery, leaving little woody growth behind. Adjacent compartments are usually harvested in sequence. Replanting is done in rather the same way, although there is likely to be more variety in species because foresters will tend to put more productive crops on ground that supported better growth in the first rotation. And winds may have blown gaps in the forest so that there are patches of different age. But there is still a risk that the next rotation will be almost as uniform as the first, and that native species will become shaded out once again.

Figure 11.1B shows an ecologist's perception of the same scene. The open rivers, wetter ground and edges all assume greater importance. They are able to support a range of vegetation types as well as being likely places in which to find relict communities.

The aim of Maintenance should direct managers towards establishing and encour-

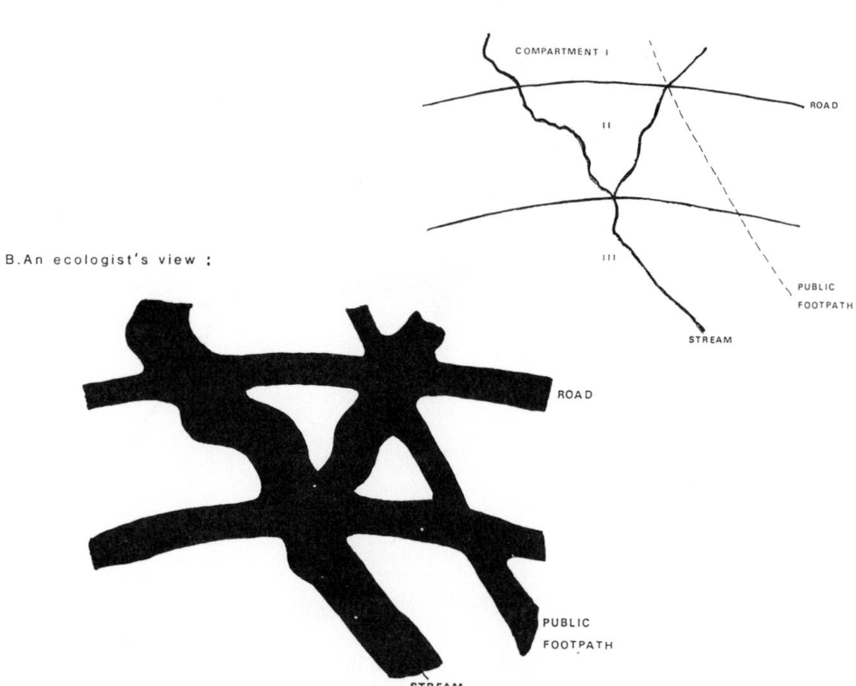

Figure 11.1 Patchwork and edges in a single-species tree plantation, Sitka spruce in Kielder Forest, England.

aging light and broadleaved growth along these features. Import might also be satisfied, as conditions here will suit many invading species, always assuming that they are there to invade. Export, invasion potential, could be satisfied by forming smaller compartments with permanent rims of native communities. Variety of tree species amongst compartments would give a greater range of species a chance of a foothold, even if it is temporary. Variety in growth stage between compartments would reduce invasion distances.

Are there opportunities for this to occur? Hibberd (1985) describes the principle of forest 'restructuring'. In this the broad-scale practice of 'rolling up the carpet' is replaced by one with a fine patchwork that includes permanent broadleaved margins for purposes of silvicultural and landscape improvement. Figure 11.2 shows this applied to the same scene as in Figure 11.1. The permanent broadleaved rims, some perhaps planted with the minimum five per cent of broadleaved trees now officially requested for all British conifer planting, make the plantation edges windfirm. Snaky belts of broadleaves forming a net can approximate to many suggested improvements in the landscape. Variety in age or crop species also gives an element of landscape diversity.

Integrated with this there are opportunities arising from changes in other unrelated practices. Streamsides, for example, will no longer be planted with conifers because of their suggested deleterious action, through acidity, on fish. And management of deer often requires that a number be shot; this is often done by attracting

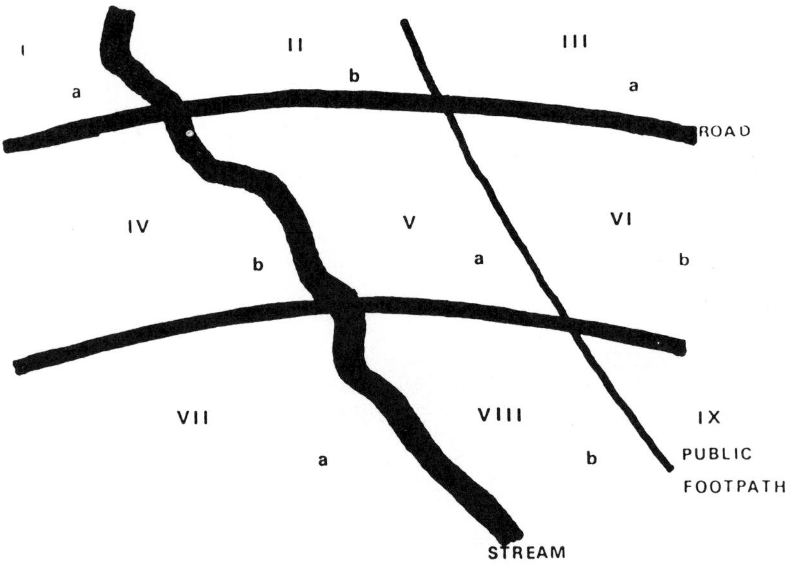

Figure 11.2 Forest 'Restructuring', Kielder Forest, England. Roman numerals indicate the new compartments formed for the second and all subsequent rotations. The broadleaf margins shown in heavy outline will be permanent also.

them to specially made 'glades', where they can be seen and dealt with (Ratcliffe, 1985). A combination of glades and unplanted streamsides may form ecologically continuous 'string-of-pearls' patchworks with a range of light conditions and possible habitats.

'Restructuring', or elements of it, is likely to be used in much of upland Britain. It appears to provide a number of ready-made opportunities for habitats to re-form, perhaps with a little help in the form of special translocations.

Will it have the desired effect? Some idea can be gained from studies in areas where this sort of patchwork has been set up for some time, although perhaps originally arranged for quite different reasons. One example is in Radnor Forest, Wales (see Figure 11.3). Here narrow strips of native broadleaves were allowed to spring up from previous woodland between the planted crops of spruce. Some strips were alongside the rides and roads. Others ran through the tree crops, separating the compartments. During botanical assessments in the 1970s these became known as 'strip banks', describing their supposed role as stores of native plants. The connected network of broadleaf rims is well arranged for colonisation of the compartments after felling.

Figure 11.4 shows the degree to which the aims of Import, Maintenance and Export were satisfied in the Radnor study. The results are from 14m × 14m sample areas in belt transects across 14 banks. The data are means. Twelve plant species, on average, remained in the strip banks five years after the conifer crop was clear-felled. This averaged 86 per cent of the species present previously, and indicates a

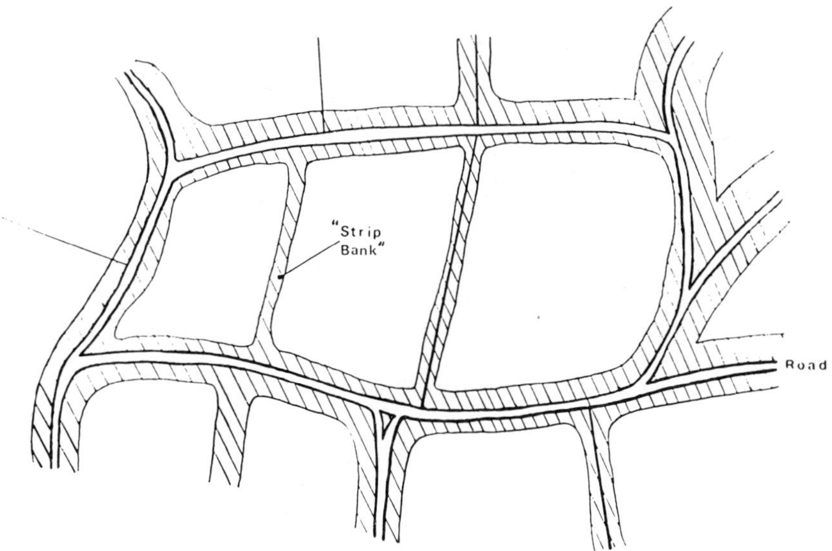

Figure 11.3 A conifer plantation with a net of broadleaved trees and scrub, Sitka spruce in Radnor Forest, Wales. 'Strip banks' are the ribbons in the net, shown hatched, which run across the compartments.

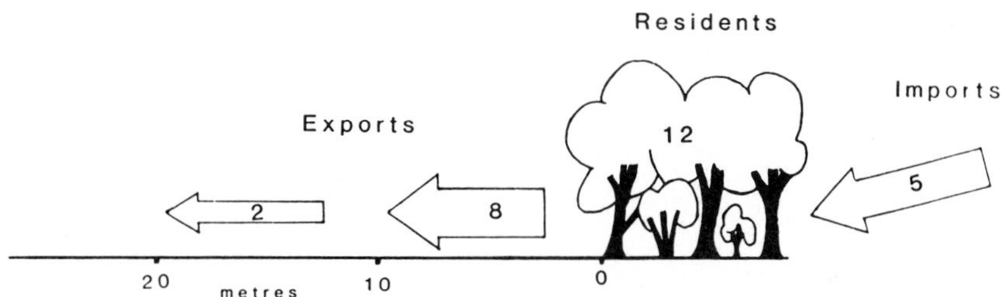

Figure 11.4 Flux and reservoir of plant species at a permanent division, formed of broadleaved trees, between compartments of Norway spruce, Radnor Forest, Wales. Fluxes are numbers of plant species five years after clear-felling. Residents are previous inhabitants still present after five years.

mean loss of around two species, of which unfortunately one was a slow-colonising plant of old woodland. Five new species appeared in the strip banks, though all were wind-dispersed or widespread opportunists, such as fireweed, *Chamaenerian angustifolium*. On average eight species appeared in the cleared areas, having previously only been in the strip banks. Only about two of these reached as far as 20 metres into the open ground and most were wind-dispersed.

Although the figures are disappointingly small for a previously wooded area, they largely indicate changes in the desired direction. Perhaps it is most depressing that the movements were mostly of widespread wind-dispersed colonisers while immobile woodland ones were lost. This prevalence of anemophilous species is in line with many other British results (Hill, 1979). But many of the apparent movements may be deceptive: Brown and Oosterhuis (1981) have shown that plant species may move in and out of soil seed banks with alternating light and shade.

Blocks of different tree species

The five per cent or more of broadleaves may be planted in blocks, large or small. Larger ones in a mosaic perhaps provide big enough areas for deep woodland species to have a chance of maintaining themselves somewhere (Peterken, 1981). But the broadleaved compartments will probably be on a rotation too short (<100 years) to allow a really good opportunity for extreme conditions to develop.

The extra broadleaves are very likely to be used in small blocks, strips and patches; these will provide fine habitat opportunities if they can be laid out with some contiguity. The contiguity should help colonisation and the long edge formed by small patches or strips will provide many sites for marginal habitats.

Uneven-aged patchwork

The ravages of wind, fire, planting failures and landscape fellings knock holes in the fabric of commercial forests. But it is only on waterlogged soils and the most exposed hills that these amount to more than a few per cent of the area. It takes several rotations for a highly uneven-aged mosaic to develop. Nonetheless, a long edge will eventually develop, separating the patches of contrasted stature. This and the presence of closely adjacent growth stages will provide enhanced habitats for some raptors (Petty, 1987).

It is a fairly minor deflection from current practice in commercial forestry to vary the date of clear-felling in adjacent compartments by just a few years. A difference of two to five years will not make a vast difference to long-term revenue. But it could be laid out so as to give a very long edge and very great local contrasts in height. Again, though, it will take at least two or three rotations before many of the local contrasts are of more than a few decades.

Opportunities arising inside forest compartments

Thinning

Thinning lets light into dark plantations. It can also temporarily decrease the thickness of litter carpets (Anderson, 1987). Both of these effects will be beneficial to many plant communities. Commercial plantations are usually thinned at planned intervals. The exceptions are on large areas of wet, shallow soils in windy places. These remain dark throughout the rotation. Fast-growing crops on the best sites may be thinned at intervals as short as five years, starting at about the twentieth. The pulses of light can last half the interval between thinnings. Unfortunately these crops will have closed canopy at least 10 years before the first thinning. Many plants will have been shaded out altogether in this period.

Foresters have some latitude in the way that the thinning is done. Frequency, proportion of trees harvested and pattern can be played off against each other on many sites. Heavy and frequent thinning provides opportunities for many plant species to spring up from dormant seed or to invade, establish and grow. In particular, thinnings more than about 10 per cent greater in intensity than is commercially conventional can promote the formation of a shrub layer.

In general, though, heavy thinnings in conifer plantations only increase the amount of growth and the influx of wind-dispersed species, or quantities of speedy colonisers, and plants with seeds that can remain viable for long periods in the soil (Hill, 1983). For other plants the damage will have been done during the period of gloom after canopy closure. Generally speaking, the immobile species characteristic of old broadleaved woods and forests do not have seeds with great longevity and are lost in the decade of dark. Some might survive as rhizomes, bulbs or similar bud banks, particularly if they can function 'saprophytically', like some orchids (Summerhayes, 1964).

Spruce crops are probably the worst offenders. They close canopy early and do so quickly after all but very heavy thinnings. An increase of 10 per cent or so in thinning intensity or frequency may satisfy Import, Maintenance and Export aims, but mostly for common rapid colonisers or for those woodland plants capable of lying dormant for decades in the soil. Habitats for most broadleaved woodland plants will not be reconstructed under spruce.

Pine crops are quite different. Some fast growing pine plantations close canopy quickly like spruces. But most do not. On fairly poor soils the dark periods are quite short and a 10 per cent increase in frequency or intensity here allows sunlight to fall fairly continuously on the forest floor (Anderson, 1979). In consequence, heavily-thinned pine crops often have continuous shrub layers and habitats for many woodland plants and butterflies. But there is still the problem of the initial dark period before first thinning. If this can be got down below ten years there is an increasing chance of habitats being provided for the spring plants of broadleaved woodland.

Some observations in line-thinned crops in southern England support this. Planted on limestone soils pines soon suffer from canopies thinned by chlorosis. This shortens the dark period to a few years. Primroses (*Primula vulgaris*) and wood spurge (*Euphorbia amygdaloides*) have been found on the thinned 'racks' where lines of trees have been harvested. They could not be found in the shady parts between. Both species are relatively characteristic of long-established broadleaved woodland. They may have survived here because the canopy would not have been dark for too long when thinning was done. It would be a minor deflection of conventional practice to start thinning pine crops a few years earlier so as to create these conditions more widely.

It is the pattern, or type, of thinning that provides some of the most interesting opportunities. Figure 11.5 shows the pattern of first thinning used in upland spruces in Britain. Typically 25–40 per cent of the crop is harvested. In the basic pattern rows of trees are removed, often about one in three. Chevron and herringbone alternatives differ from this in having fewer rows ('spines') removed and in having short side rows ('ribs') attached to each spine. Chevron patterns have opposite ribs and herringbone ones have staggered ribs.

Where the ribs join the spines a glade is formed. This becomes especially large if the timber stems are dragged out with only average care and attention. Some corner trees will be lost and the glades will be made much bigger, perhaps six to eight

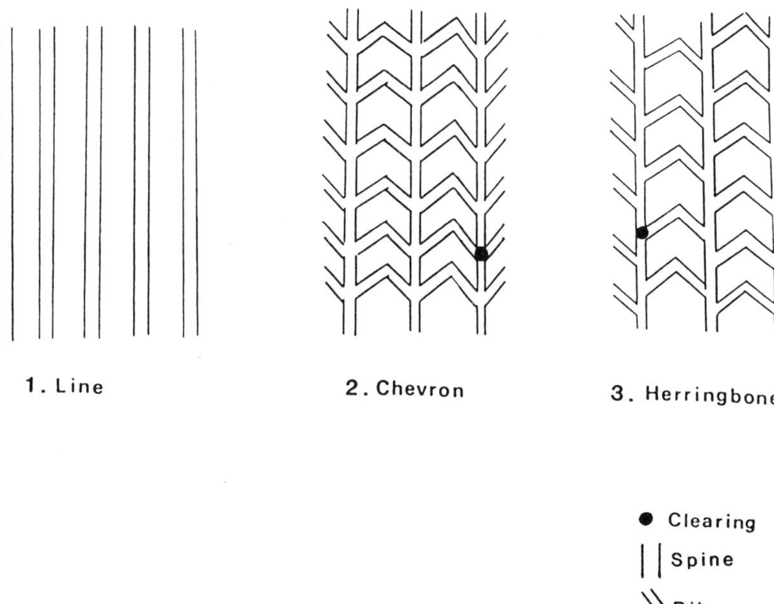

Figure 11.5 Patterns of light penetration after thinning in conifer plantations. These patterns form from three different types of first thinning, carried out at 20–25 years. Open clearings or glades are formed where the ribs of chevron and herringbone patterns join the spines or 'racks'.

metres across. A gap this size may not close up before the next thinning is done. And the rutting caused by large logs turning where the ribs join the spines may expose mineral soil for colonisation.

Figure 11.6 shows some effect of chevron thinning on plant communities. The data are from a designed experiment with thinning patterns in Sitka spruce grown in Rheidol Forest, Wales. The figure shows a stand ordination using the first two axes of a DECORANA analysis (Hill, 1979, method as in Page, 1981). The data were cover estimates done in 4m × 4m quadrats five years after thinning.

The plant communities in the unthinned part of the plantation are clustered together in the ordination and hence are rather similar in several aspects. They mostly consist of a few bryophytes, notably *Isopterygium elegans*, and the odd fern. The stands on ribs are distinct from these in having a flowering-plant flora that includes some opportunist plants notorious for their seeds' longevity in the soil, such as foxglove (*Digitalis purpurea*). Most of the rest are anemophilous, fireweed being well represented once again. But a few woodland plants are present, such as bramble (*Rubus fruticosus* agg.) although their seed might have been brought in by birds.

The stands in the glades are variable in composition, overlapping that of stands on the ribs. This variability probably reflects that in glade size, which depends on

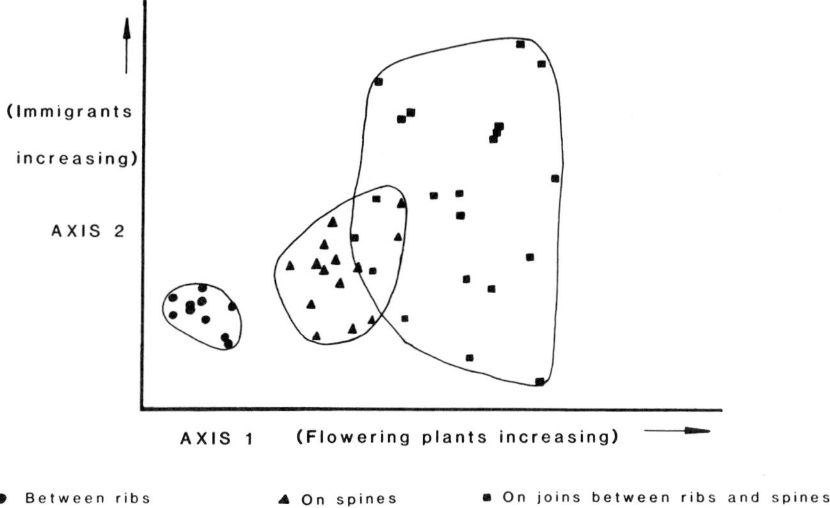

Figure 11.6 Stand ordination of vegetation in chevron thinning, Sitka spruce, Rheidol Forest, Wales, five years after first thinning. The ordination, of percentage cover, shows the dissimilarity in the composition of vegetation from unthinned plantation, closed circles; ribs, closed triangles; and spines, closed squares. The ordination was done using DECORANA, Hill (1979).

whether corner trees get knocked down, and on how close the ribs are to the end of their spines. Ground near the access end of the spines is likely to be more disturbed than that further back. It receives more passes from extraction tractors and so has more logs dragged across it. It is consequently more rutted and has more bare soil. Plants with wind-dispersed seed tend to be more prevalent here. Eight plant species were found, in such places, that did not occur on ribs or in the unthinned crop. One of them, wood sorrel (*Oxalis acetosella*), may have persisted from previous woodland, which was oak coppice heavily mown by sheep. Some glades had almost no plants in them at all. These were in places where rutting and puddling had caused temporary ponds.

Use of chevron thinning, in preference to line thinning, clearly has some potential to reconstruct habitats for a few broadleaved woodland plants, though it is not clear how many will be suited or how permanently. And there is still the bottle-neck before first thinning to contend with in spruce. Hill (1983) has suggested that we may have to accept that little is to be gained by special thinning in upland spruce crops. Chevron and herringbone patterns cannot be used in windy areas or on waterlogged soil anyway because they worsen the risk of windblow. But they might be valuable in combination with early thinning in existing pine crops on previous broadleaved woodland sites, or for reconstructing habitats for marginals and opportunists, in wind-firm spruce crops.

Mixed-species crops

Commercial foresters rarely convert single-species plantation stands to intimate mixtures before clear-felling. This reduces immediate opportunities for habitat reconstruction through introducing broadleaves or other species. Only old crops can seriously be considered. The exceptions to this rule, of course, are crops being underplanted in advance of felling, group-felling treatments, prematurely windblown patches and land that is being planted for the first time. Here, much might be achieved for Maintenance and Import of native woodland habitats by planting a proportion of the trees of the right species. These will usually be deciduous and native. On some sites these may, however, be slow to establish themselves, particularly in the uplands where they are often not well adapted to the exposure. On occasion, therefore, conditions for native woodland plants might be more easily established in the first place by planting quick-establishing exotic broadleaves. In Britain, the South American beeches of the genus *Nothofagus* have been shown to support the floras rather similar to some found under native oaks (Wigston, 1983; Anderson, 1979).

Opportunities for establishing more mixed-species compartments have arisen in Britain with the introduction of recent policies, in state-run woods, and with grant schemes requesting five per cent of broadleaves. Their use and effectiveness will depend much on the terrain and its previous woodland history. The proportion of broadleaves in mixture with an upland conifer crop could put it at risk from the wind. This is because broadleaved trees will generally not establish or grow as fast as the conifer and will therefore leave a rough canopy for the wind to tug at. This will not happen in lowland areas at less risk from the wind. It is here that more intimate mixtures of conifer and broadleaved trees may help to Maintain and Import habitats by reducing the gloom and the litter mattress of pure conifer stands.

The planting pattern of the added tree species is critical. Dispersed small patches of broadleaved trees satisfy small birds (Bibby, 1987). It is not clear whether they satisfy small plants. Very small patches of a few trees will not maintain permanent high light levels on the forest floor. The conifers alongside will shade them out. The patches probably need to be at least 15 metres across and a few of these will quickly use up more than five per cent of the area. A few large, widely-spaced patches in a compartment may satisfy the aim of Maintenance but not Export. Small edge:area ratios and wide spacing reduce the chances of colonisation being able to take place very far after the conifer is felled.

Clumps, small or large, within compartments may be looked upon unfavourably by the commercial forester. They could hinder thinning and extraction of conifers. Foresters will be more inclined to use them on the scale of the forest patchwork than inside compartments.

Extended rotations

Late stages of conifer crops can be well lit and the forest floor may have ameliorated (Anderson, 1979). For this to happen heavy thinning is needed. But many crops are not thinned at all. And fast growing stands of trees close over quickly even after late thinnings. Hill (1983) pointed this out as an unfortunate paradox; he suggested that dense upland crops are best managed on short rotations so that the period of viability of seeds in the soil is not at risk of being exceeded. He also suggested that short rotations in the lowlands may be beneficial to plant habitat formation by simulating coppice cycles.

Nevertheless there will always be some crops that have been heavily thinned on wind-firm sites, perhaps for landscaping. And others with established floras might be allowed to grow on for a few years without a great loss of revenue. They will rarely support habitats for many plants of our native woodlands, though, since most of these will have been shaded out before the first thinning.

Opportunities arising along the forest's edges

Cutting corners

Plantation forests contain roads, rides, paths and outer edges. All of these are better lit than the interior. They usually support some plant growth (Anderson and Carter, 1987). Sometimes, however, these features are completely obscured as the conifer crop closes overhead.

At thinning there are opportunities to reconstruct and sustain habitats for plants and butterflies in particular. The edge of the forest is often cut back, particularly at corners, as temporary timber landings and loading places for lorries. Even if they are not cut back for these purposes it is a minor addition to forestry practice to do so at the same time as thinning the crop.

Cutting-back, particularly at corners (intersections of roads and rides) can satisfy all three aims of Import, Maintenance and Export. This can provide room and light for belts of woodland plants, scrub and grassland to occur in varied proximity. Relics of previous woodlands can survive here and reinvade the area behind when the crop is finally harvested. Habitats for incoming plants and butterflies can be established along an edge of graded stature simulating several stages in one succession. This is provided by tall shrubs standing behind shorter ones with grass sward in front (Carter and Anderson, 1987).

Repeated mowing and coppicing is needed to maintain this arrested succession or to cycle it. This is expensive. It may not have to be done for any other forestry purpose and so only the initial cutting-back is really a ready-made opportunity.

Spaced edge glades

Loading places, stacking areas and the mouths of thinning racks are all forms of local ride-widening. They form bays or glades that let in light. They provide space for broadleaved woodland, scrub and grassland habitats to be maintained or formed at intervals along the front of the crop. If spaced closely enough these features form a continuous invasion 'bridgehead' from which the crop behind can receive Exports.

There is more shade in a rectangular edge-glade than in a cut-off corner of the same size. Ready-made glades must be surprisingly large to provide enough light for sward as well as shrubs. Figure 11.7 shows the basis of one widely used criterion designed to set the minimum depth of widened edge or edge-glades. This criterion is the ride-width needed to let light reach the edge during the growing season; this is approximated by the width of ride that just lets sunlight fall on the edge at noon on the equinoxes. The figure shows this for a south-facing edge. In Britain the length of shadow at this instant is about the same as the height of the tree casting it. As a minimum, therefore, glades need to be deep enough for the distance to the tree opposite to equal its height, making roughly a 1:1 ratio.

Using this 1:1 ratio one row of trees needs to be removed in a 10-metre tall crop, assuming a standard British forest road of 10-metre width. This is shown in Figure 11.7A. Two rows need to be removed when the crop is 20 metres tall as in Figure 11.7B. This is about the average height of many late-stage conifer crops in Britain.

The same rule applied along the ride dictates that the glade should be at least 20 metres long. Two rows × 20 metres is about the size of a small loading bay and so

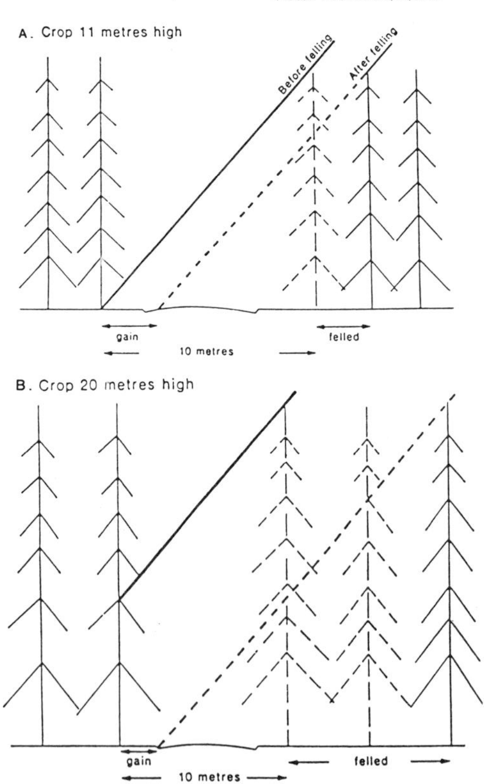

Figure 11.7 Felling needed to form widened rides or ride-side glades as edge habitats for plants and butterflies. The diagram shows the shadow limits of tree crops at the start of the growing season (equinoctial noon). Dashed trees are the ones that will have to be removed if the criterion of light just touching the lit edge at this instant is to be satisfied. The ride shown is aligned east-west.

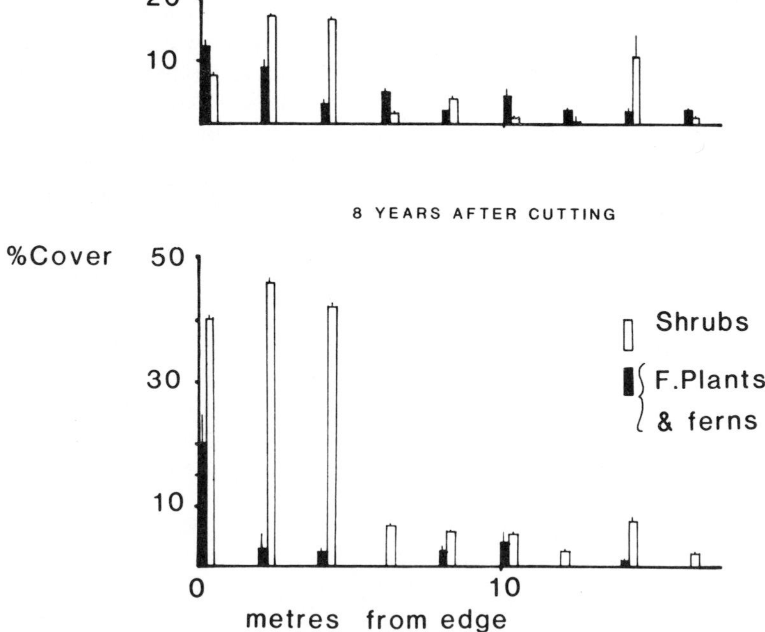

Figure 11.8 Plant regrowth after ride-widening in Corsican Pine, Bedgebury Forest, England. The histogram bars are percentage cover estimated in metre-square transects in June. Standard errors are shown. Distances are reckoned from the original plantation edge before it was cut back three metres. This cutting satisfied the criterion illustrated in Figure 4.11.

some opportunities of the right size already exist. But bell-mouths at the end of thinning racks are likely to be less than half this size and hence too small. Bigger glades are needed also on north-facing edges or on north–south rides. If two-shrub depths of edge belt are needed then the glade must be wider still. And so it may be necessary to form the glade specially in these circumstances. Doing so by cutting out trees between racks or loading places at the same time as doing the thinning is a minor modification to a practice that has to go on anyway, and so thinning time provides a fine reconstruction opportunity. Leaving some of the broadleaved growth to become trees benefits birds and a few buttterflies, and contributes to a wind-firm edge, as in 'restructuring'.

Do edge-glades actually develop the required vegetation? Figure 11.8 shows results from a transect survey in metre-square transects across an edge in a plantation of Corsican pine (*Pinus nigra* var. *maritima*) in Bedgebury Forest, England. The site had previously held an old woodland flora under coppice of European Sweet chestnut (*Castanea sativa*). When the crop was 18m tall its edge was cut back three metres, which was slightly less than the minimum dictated by the 1:1 rule. The figure shows percentage cover for plants along 16 transects extending 20m into the stand of trees from its previous edge. Before cutting, a sparse edge cover (<15 per cent) of shrubs overhung a two to ten per cent cover of old woodland and marginal plants. The shrub cover was not appreciably greater than at some points in the plantation, although the cover of plants was smaller and at least twice as high on the outer edge as inside. This may have been because of high light intensities on the plantation edge in spring, before the shrub canopy was fully expanded.

Eight years after cutting, the edge belt of shrubs covered the ground three times more fully at the edge than it did before and the outermost belts of marginal plants had doubled in percentage cover (Anderson and Carter, 1987). But the cover of woodland plants was more than halved for the next six metres in, where shrub cover exceeded 40 per cent at ground level. This suggests suppression of the smaller woodland plants. But the design of this sort of survey prohibits the definite conclusion that shrubs did any suppressing. And these plants may be present dormant as seeds and buds underground anyway. Nonetheless, one of the plant species affected was the violet (*Viola riviniana*) whose leaves are the food of the larvae of Fritillary butterflies. At this site violets were present in the expanded vegetation at the outer edge as well as inside and so the caterpillars would not run short of something to eat. Elsewhere, though, they might be lost by the compaction and scraping of the soil caused by lorry and log movements.

The 1:1 criterion is crude, and refinements are at hand. Sunlight hours and energy input may fit species' requirements more closely, and autecological studies of exact light requirements may be needed. Dent (1986, reported in Spellerberg, 1987) has largely achieved this for some British reptiles. Recent developments of fast algorithms to forecast light and heat in glades of all shapes and sizes (for example, Yallop *et al.*, 1988) should make available the other side of the equation.

Even using the sample 1:1 criterion some of the subtleties of selection of aspects

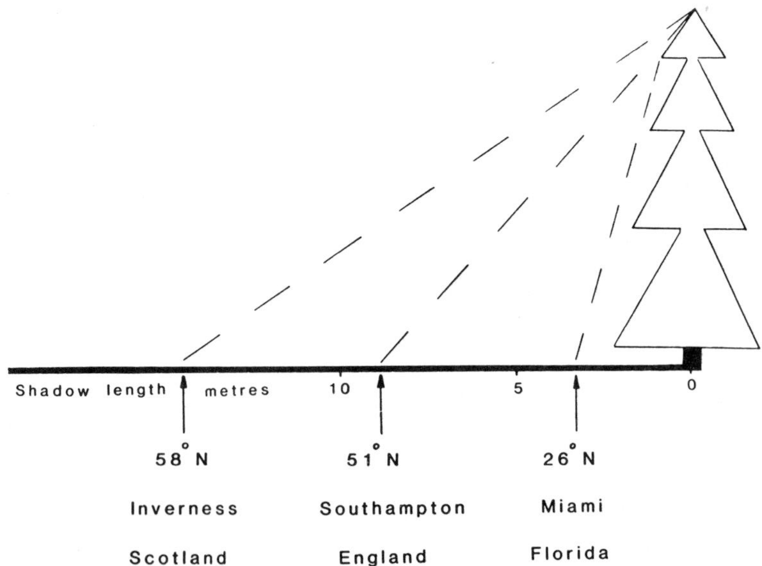

Figure 11.9 Lengths of shadow cast by a 10-metre tall tree at noon, equinox for three latitudes. The shadow at the lowest latitude is influenced by the breadth of the tree as well as its height.

and crops suitable for glade management can be glimpsed. Figure 11.9 shows shadows cast by a 10-metre tree at three different latitudes. In north Scotland a much deeper glade is needed (14m) than the 8-metre one that will suffice in southern England. In north Scotland glades may need to be cut so deep that no realistic opportunities exist on a flat ground and it may be necessary to concentrate on north–south rides where at least there is always a spark of light around noon, and on sloping ground where the sun can peep over the opposite crop and shine in. Further south, as shown by the shadow cast in Florida, the breadth of the tree becomes a major factor as well as the distance between trees. But here, of course, there will hardly ever be a need for glades.

Successionally-cut glades

Where edge-glades are close together it is possible to cut their bushes and swards in sequence so that many stages in the succession are close together. This gives a chance for Export of many different species along a concentrated front into the area behind after it is felled. This arrangement is probably the best combination of ecotectural design (Anderson and Carter, 1987) to satisfy all three aims of Import, Maintenance and Export.

But it is also likely to be almost unmanageable. It requires a fixed timetable of coppicing and mowing. It is only likely to find an opportunity where a forester does repeated thinnings, extracting the timber at points staggered successively along a ride. In practice forestry is not this neat.

Uniformly widened rides

One hundred to two hundred metre lengths of a ride are occasionally widened uniformly as stacking grounds for very large amounts of timber. More informally they may also be treated like this for landscape improvement. These long belts might be expected to satisfy all three aims of habitat reconstruction if wide enough.

In practice, however, there are drawbacks. A great deal of coppicing and mowing is needed, although it may be more efficiently done in a straight line than in indentations. Alignment with prevailing winds can cause butterflies to be blown away. And foresters might be reluctant to accept the hindrance of not being able to drag timber freely out from the crop behind if all the edge is to remain undisturbed by heavy vehicles or logs. Spaced glades are less hindrance because timber can be extracted around and between them if necessary.

Dannatt (1987) has argued that spaced glades may provide the forester with opportunities. They could provide him with ready-made access points for thinning. This only applies in those circumstances where the edge habitats are of a type that will benefit, or at least not suffer, from traffic. Heathland communities, complete with reptiles may be an example of this. They need periodic opening-up and ground-scraping for regeneration. In this sense glades may be a forester's boon, not his bane.

Discussion and conclusions

Many of the opportunities described in this paper have one awkward factor in common: they need continued management. Glades and scrubby edges, however they have been formed, generally need subsequent and repeated coppicing or mowing (Carter and Anderson, 1987). This is likely to be expensive and difficult.

But if ready markets for coppice shoots and rough grass are at hand then the prospects are much brighter. In some areas of Britain and Europe the sale of coppice poles continues, albeit on a fairly small scale. And there are small markets for the wild flower seed contained in mowings from forest rides. But without expansion in these markets there is a risk that some of the present enthusiasm for edge habitat management will wane.

There is a major distinction between opportunities in lowland and in hilly areas. In Britain particularly, many upland conifer forests are on sites that had little or no broadleaved woodland cover before planting took place. Many lowland plantings have, in contrast, been on land that had on it ancient woodland, that is woodland with more than about 400 years' continuous cover of trees. The latter, but not the former, started off with the full panoply of sensitive and sluggish forest greenery. These areas started with a lot but may lose it, while those in the uplands started off with much less, will not easily acquire it and will then lose it readily again. The possibilities for encouraging new communities in upland plantations are fewer because wind-blow often reduces the latitude that a forester has with his management. And the difficulty of getting broadleaves apart from birches to start growing and establish in cold, wet mountain country is daunting.

Thoughts such as these have led to arguments that there is little point in trying to reconstruct broadleaved woodland in many of these difficult areas. Rather we are urged to concentrate on safeguarding and developing the remaining examples of old semi-natural habitats in long-established woodland. But formation of streamside glens and deer glades will go on anyway and will certainly provide opportunities for limited habitat reconstruction if well planned with ecologists' assistance.

From present data and projections I conclude that:

1. Restructuring conifer forests as patchworks with broadleaf fringes is a new but existing opportunity for survival and spread of some native organisms.
2. Government support for increased broadleaf planting will help habitat reconstruction if the trees are laid in contiguous patchworks or similar ecologically-designed arrangements.
3. Use of patchy thinning practices may enhance woodland habitats in some pine crops.
4. Opportunities to form and maintain edge habitats exist widely, but need fine tuning to get the light regimes right.
5. Markets for small coppice wood and cut rough grass may need to expand if edge habitats are to be maintained for long in many areas.

References

Anderson, M.A., 1979, 'The development of plant habitats under exotic forest crops', in S.E. Wright and G.P. Buckley (eds), *Ecology and Design in Amenity Land Management*, Wye College, University of London, pp. 87–109.

Anderson, M.A. 1987, 'Conserving soil fertility: applications of some recent findings', in R.J. Davies (ed.), *Proc. ICF Conference: Forestry's Social and Environmental Benefits*, Institute of Chartered Foresters, Edinburgh.

Anderson, M.A., Carter, C., 1987, 'Shaping ride-sides to benefit wild plants and butterflies', in D.C. Jardine (ed.), *Proc. ICF Conference: Wildlife Management in Forests*, Institute of Chartered Foresters, Edinburgh, pp. 66–80.

Bibby, C.J., 1987, 'Management in commercial conifers for birds', in D.C. Jardine (ed.), *Proc. ICF Conf: Wildlife Management in Forests*, Institute of Chartered Foresters, Edinburgh, pp. 60–5.

Brown, A.H.F., Oosterhuis, L., 1981, 'The role of buried seed in coppice woods', *Biological Conservation,* **21**, 19–38.

Carter, C., Anderson, M.A. 1987, 'Enhancement of forest ridesides and roadsides to benefit wild plants and butterflies', *Forestry Research Information Note*, 126, HMSO, London.

Dannatt, N., 1987, 'Wildlife management in forests – a summary', in D.C. Jardine (ed.), *Proc. ICF Conf: Wildlife Management in Forests*, Institute of Chartered Foresters, Edinburgh, pp. 118–19.

Dent, S., 1986, 'The ecology of the sand lizard (*Lacerta agilis* L.) in Forestry plantations and comparisons with the common lizard (*Lacerta vivipara* Jacquin)', Ph.D. thesis, Southampton University, England.

Everard, J., 1974, 'Ligniculture – modern silviculture in France', *Forestry and Home-grown timber*, Feb/March 1974, 24–6.

Forestry Commission, 1988, *Woodland Grant Scheme*, Forestry Commission, Edinburgh.

Gamlin, L., 1988, 'Sweden's factory forests', *New Scientist*, 28 Jan, 41–7.

Hibberd, B., 1985, 'Restructuring of plantations in Kielder Forest District, *Forestry* **58**, 119.

Hill, M.O., 1979, 'The development of a flora in even-aged plantations', in E.D. Ford, D.C. Malcolm and J. Atterton (eds.), *The ecology of even-aged forest plantations*, Institute of Terrestrial Ecology, Cambridge, pp. 175–92.

Hill, M.O., 1979, 'DECORANA – FORTRAN program for detrended correspondence analysis and reciprocal averaging', *Ecology and Systematics*, Cornell University, New York.

Hill, M.O., 1983, 'Plants in Woodlands', in E.H.M. Harris (ed.), *Centenary Conference on Forestry and Conservation*, Royal Forestry Society, Tring, pp. 56–68.

Nature Conservancy Council, 1986, *Nature Conservation and Afforestation in Britain*, Nature Conservancy Council, Peterborough, England.

Page, H., 1981, 'Tree species influence in plantation ecology', M.Sc. thesis, University College, London.

Peterken, G.F., 1981, *Woodland Conservation and Management*, Chapman & Hall, London.

Petty, S.J., 1987, 'The management of raptors in upland forests', in D.C. Jardine (ed.), *Proc. ICF Conf: Wildlife Management in Forests*, Institute of Chartered Foresters, Edinburgh, pp. 7–23.

Ratcliffe, P., 1985, 'Glades for deer control in upland forests', Forestry Commission Leaflet 86, HMSO, London.

Spellerberg, I., 1987, 'Management of forest habitats for reptiles', in D.C. Jardine (ed.), *Proc.*

ICF Conf: Wildlife Management in Forests, Institute of Chartered Foresters, Edinburgh, pp. 83–91.

Summerhayes, V., 1964, *Wild Orchids of Britain*, New Naturalist Series, Collins, London.

Wigston, D.L., 1983, 'The ecological, landscape and amenity implications of the introduction of exotic trees into British forestry. II. The example of *Nothofagus* Blume', *Arboricultural Journal,* 7, 3–13.

Yallop, B., Carter, C., Anderson, M.A., 1988, 'Duration of daylight and levels of energy input to sunlit ridesides', *Forestry Commission Annual Research Report, 1987*, HMSO, London.

12

Habitats for education: developing environmental resources on school grounds

K. Funnell

Origins

Our programme of environmental initiatives within school grounds began almost by accident as a by-product of technical experiment. During the period 1979–81 the Landscape Branch of Kent County Council (KCC) was engaged in direct tree-seeding trials, chiefly on highway verges. One of the biggest problems proved to be the acquisition of local tree and shrub seed at reasonable cost. A solution was to ask local schools to collect seed for the trials, envisaging that they would retain sufficient supplies to grow their own trees and shrubs from seed and supply the surplus for use in direct seeding.

Growing trees from seed in school was nothing new, but this initiative aroused considerable interest, particularly at the primary level. It was supported by a teachers handbook *Trees from Seed* launched for National Tree Week 1982, the theme being 'For Every Child A Tree'. Two interrelated schemes were developed: The Kent Tree Seed Collection Scheme and The Kent New Woodlands Project. The aim was quite simply for children to collect and germinate seed, to grow on seedlings for two or three years and then to plant them out either on their own school grounds or on other suitable sites in the locality. Gradually these schemes developed a wider remit until eventually the registered schools, of which there were by now over 90, were invited to become involved in the development of other resources and habitats within the school grounds.

From this small beginning, (two schools participated in 1984–5), demand has mushroomed to the point where over 120 schools are now at various stages in developing diverse habitats within their grounds. Projects are based on partnership, both between landscape staff and the Environmental Education Inspectorate, and between these agencies and the schools. Approaches from schools arrive via local landscape staff or the Inspectorate, and at all stages of sophistication from the speculative enquiry to the submission of detailed, integrated schemes. They may be for a specific habitat type, woodland, meadow or most especially a pond, or they may encompass plans for the whole site.

Philosophy

A tentative division can be made within the conservation movement between the 'determinists' and the 'possibilists'. The former display a strictly puritanical line, believing habitat creation and change should be limited to climax vegetation types determined by soils, microclimate and so forth, however theoretical this might be. The latter recognise that every site presents a range of possibilities that may even be enhanced by landscape engineering works such as soil importation. Both have drawbacks in relation to school grounds. Determinism would severely restrict the range of habitat types, while possibilism might lead to confusion about the reason and logic behind habitat types and vegetation patterns.

In practice our approach has generally been flexible and broadly possiblist, while avoiding excesses such as laying chalk downland on acid heathland by means of soil importation, if for no other reason than inordinate expense. More diverse school grounds save expensive and time-consuming excursions, and provide appropriate resources or opportunities within reach during lesson time. The artificial nature of some created habitats is made clear if it is not immediately evident in, for example, the nature of pond construction.

Although on a broader scale habitats may be created for their own sake, habitat creation within school grounds is designed specifically for *use*. As such it is important to be fully aware of user demands related to nature conservation, and to the site as a whole.

User surveys

Two user surveys have been undertaken by the Landscape Branch of KCC, first on primary schools in 1985 and subsequently on secondary schools in 1987. The former highlighted a dramatic upturn in the demands on the school grounds for environmental education. There was evident frustration with the all too familiar 'green desert' of gang-mown grass contained by chainlink fences whose sole function was for informal play and organised games. Use of the outdoor classroom as a resource for the curriculum was by no means limited to natural history but included maths, science, art, language, craft, design and technology.

Contrast with secondary schools could not have been more dramatic. Yet the advent of the new and more practical GCSE exams, with the need to carry out individual pieces of investigative practical work, has focused more interest in the grounds, especially in relation to sciences, geography, art and home economics. Although there was a general lack of awareness in both primary and secondary schools of the opportunities for habitat creation, our surveys showed that there was a demand for a range of habitat types – woodland, scrub, hedgerow, pond/marsh, meadow and specific features/artefacts e.g. butterfly garden, log pile, dry stone wall and ditch. Constraints of space and other demands on the site meant that they had to be small – 'pocket' habitats – and robust enough to sustain use and abuse. It was also important not to confine them to one area, in order to encourage the concept of the universality of conservation, to maximise use and experience of the site and to develop a wider sense of responsibility and ownership. As a result of this feedback, and the developing partnership with inspectors and teachers, our brief has now broadened to encompass the provision of resources for outdoor learning. It extends across the curriculum, throughout the year and everywhere within the site boundaries, and accepts that play is an essential part of learning and development.

Involvement of the school

Without clear definition of user demands in the past, apart that is from the requirement for playing fields, grounds maintenance staff have tended to maintain the grounds to their own satisfaction and in their own style, essentially cosmetic, tidy and formal. Whilst machinery, herbicides and related safety regulations militate against direct involvement by teachers and pupils, a lack of awareness and indeed of confidence has acted as a break on local initiative. Consequently, the early schemes were largely designer-led, with a plan of proposed habitat types being prepared after an initial site meeting.

The approach has gradually evolved to the point where an ideas plan is usually prepared for the teachers and pupils to develop and amend in relation to their own aspirations and needs. Teaching staff are now more fully involved, although their reluctance to include the children as part of the consultation process is sometimes surprising. Only rarely is the school the prime mover in the development of ideas and plans, but this is the ultimate goal and trends in this direction must be positively encouraged.

Initiatives and techniques

Most frequently the initial approach is narrow, limited to one type of habitat, especially a pond, or one particular area of the site. Habitat creation may not even be the school's prime motive, but merely the argument for quite another objective, such as the removal of a redundant swimming pool or a new hedge to prevent random access. The latter highlights the multi-purpose nature of many initiatives.

An analysis of the types of habitat asked for and created within school grounds shows the popularity of ponds, although there appears to be increasing awareness of the delights and opportunities of wild flower meadows (see Figure 12.1). Declining school rolls in Kent mean only two new schools are built each year, so the opportunities for soil management, earth modelling and consequent habitat creation are limited. Nevertheless, specific techniques have been developed and much achieved on these new sites. Enrichment and diversification of existing sites is much more difficult due to the predominant, aggressive and often nutrient rich close-mown sward. The techniques employed to create various habitat types are described and appraised below:

Woody vegetation

Most tree and shrub planting is programmed for National Tree Week in late November/early December. Besides direct benefits to schools in being part of this

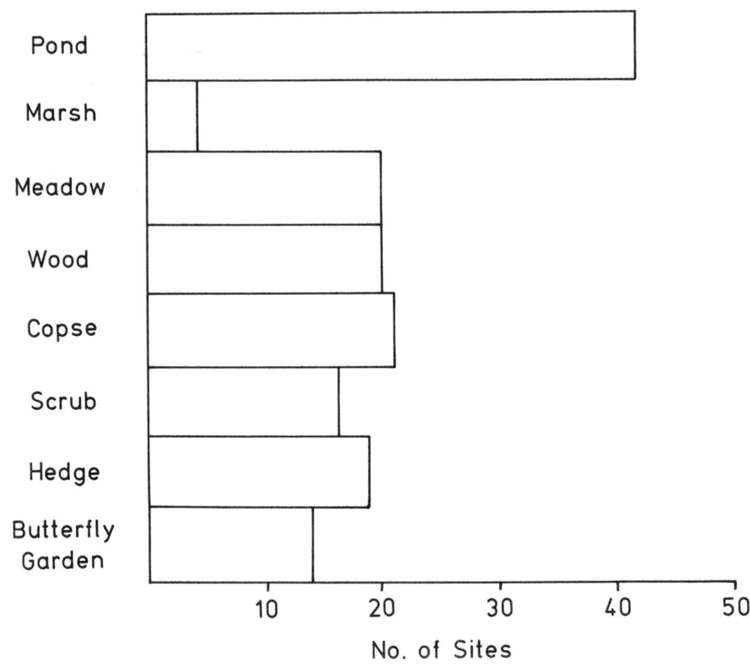

Figure 12.1 Numbers of new habitats created on school sites in Kent from 1983 to 1987.

event, all experience to date points to greater success in establishment with schemes completed before the end of the calendar year. Nursery stock has either been grown by the school or is supplied as forestry transplants in size ranges 300–450mm or 450mm–600mm. Species choice is difficult since teachers may be looking for diversity, even unique characteristics in form, height, bark, leaves, berries/nuts or flowers. Yet every effort is made to create some ecological cohesion, and mixtures and proportions are based on recognised types (Peterken, 1981). Proportions of trees and shrubs vary from 70 per cent to 30 per cent, dependent on the site and the objective; centres are normally two metres, avoiding regular spatial pattern, and species are set in random groups of 1 to 10.

Habitat design is too frequently contrived, regular and excessively logical. The concept of pioneer species (see Buckley and Knight, in this volume) has not been pursued since these are invariably non-native, and long-term management cannot guarantee that they will be removed in sufficient time to prevent their dominance of the woodland. Not merely is care taken to avoid throwing together a cocktail of species likely to be successful on a given site, but occasional species of one to five per cent of total mixture are included. Too often woodland mixtures consist of four to six common species, and less frequent but often appropriate species such as lime (*Tilia cordata*), hornbeam, (*Carpinus betulus*), wild cherry (*Prunus avium*) or chequer tree (*Sorbus tominalis*) are frequently forgotten. Even with hedgerows a basic thorn mixture of 70–90 per cent is enriched by two to four occasional species dependent on the site and the use.

Meadows

This is more difficult to achieve and sustain on existing school sites. Topsoil stripping and removal, or conversely importation of subsoil, is neither practical nor financially viable on most sites. Killing of the existing sward, shallow cultivation of the soil and sowing of a proprietary wild flower mixture has met with some success. However, the raising and planting of suitable pot-grown wild flowers into swards has been more widely adopted (see Wells *et al.*, in this volume).

My colleague, Dr J. Frankton, developed an initial list of 13 species that met the main criteria – varied flower/leaf form, colour and seasonal display during the school terms – and could survive and hopefully spread within such a sward. The approach was consequently as much populist as ecological. However, this encouraged the planting of some species outside their normal habitats and a further refinement was necessary to produce four broad wild flower categories that could be found or created on most school sites (see Table 12.1).

In this latest development only certain of the species have been supplied to schools, who are encouraged to collect or buy seed, germinate seedlings, grow on and plant out pot-grown wild flowers. Collection of many species is frustrated by the summer holidays, but it is hoped to develop a 'Wild Flowers From Seed' scheme to complement that of 'Trees from Seed'. On new school sites or where development such as a building extension, playing field construction or even a pond allows for soil manipulation to create an impoverished growing medium, other techniques have been tried. These include natural regeneration, seeding with a proprietary wild flower mixture and late-cut hay as a seed source, the latter being most successful.. The technique utilises hay bales from a local nature reserve, and initial results suggest greater success with autumn, rather than spring sowing. After light hand raking of the subsoil, bales are opened, spread and moved about with hand tools to encourage seed to work into the ground and stalks to act as a covering mulch. It is currently the preferred method on bare earth sites on the grounds of both cost and success.

Experimentation continues. Plans are in hand to assess the spreading of late-cut hay on rotovated close-mown sward; the seed is incorporated by means of a pedestrian contravator. The latter was designed for remedial work to cricket squares by rotovating narrow slits in the sward and sowing seed in one operation. It is used in two or more directions. The technique, akin to slot-seeding (see Wells *et al.*, in this volume) is nothing new, but the attraction of this pedestrian machine is that it can be used cost-effectively on small areas where mini-meadows are required. Comparison can then be made between this technique and killing plus rotovation of the entire sward.

Wetlands

Ponds are universally popular, to the point where there appears to be no direct correlation between the quality of the resource provision and its use within the cur-

Table 12.1 Wild flower mixtures developed for four school habitat types

Habitat type	Wild flower mixture
Well-drained open meadow (cut mid-July onward)	*Achillea millefolium* (yarrow) *Leontodon hispidus* (rough hawkbit) *Leucanthemum vulgare* (ox-eye daisy) *Lotus corniculatus* (birdsfoot trefoil) *Potentilla reptans* (creeping cinquefoil) *Prunella vulgaris* (self-heal) *Trifolium pratense* (red clover) *Veronica chamaedrys* (germander speedwell)
Persistently damp open meadow (cut August onward)	*Ajuga reptans* (bugle) *Cardamine pratensis* (milkmaid) *Centaurea nigra* (knapweed) *Filipendula ulmaria* (meadow sweet) *Lotus corniculatus* (birdsfoot trefoil) *Ranunculus acris* (field buttercup) *Rumex acetosa* (sorrel) *Tragopogon pratensis* (goat's beard)
Hedgerow base (some open ground maintained by summer shade)	*Alliaria petiolata* (Jack-by-the-hedge) *Fragaria vesca* (strawberry) *Galium cruciata* (crosswort bedstraw) *Glechoma hederacea* (ground ivy) *Lathyrus pratensis* (yellow vetchling) *Primula vulgaris* (primrose) *Stellaria holostea* (greater stitchwort) *Vicia sepium* (bush vetch)
Young plantations (some open ground maintained by summer shade)	*Anemone nemorosa* (wood anemone) *Geranium roberianum* (herb robert) *Geum urbanum* (wood avens) *Hyacinthoides non-scriptus* (bluebell) *Lamiastrum galeobdolon* (yellow archangel) *Primula vulgaris* (primrose) *Silene dioica* (red campion) *Stellaria holostea* (greater stitchwort)

riculum. Nevertheless they are easy and quick to construct, and offer immediate benefits to teacher and pupil. Almost all recent construction has utilised butyl liners on a sand base. Two sizes of liner have been used, 8m × 5m and 6m × 4m.

Early designs incorporated a marsh area within one corner of the pond, but its small size, the precision required in achieving accurate levels, and the fluctuation in water level all led to the construction of separate marsh areas abutting the pond. This is done using a cheap 4 × 4m polythene pond liner on sand, covered with 150mm of a soil/peat mix. Ideally one or two pools of water up to 300mm deep are created towards the centre of the marsh. The marsh takes any overflow from the pond, and where possible a surface water source of supply, such as rainwater from adjacent downpipes, is fed into the pond to reduce the need for topping up.

Concrete paving slabs 600 × 600mm are laid along one pond margin to facilitate access for study purposes; while regrettably utilitarian, they are ideal for use with a liner. Thus direct access is precluded around at least half the perimeter. The 'wild' area containing the pond is usually demarcated by an agricultural fence backed by new hedging for the purposes of supervision and consequent safety. There has proved to be little difference in cost between excavation and preparation by mini-digger and excavation by hand. Planting ledges were originally 225mm wide in specification, but were often too narrow for a planting basket when constructed. The ledge has now been specified at 450mm to achieve a realistic flat area of 300mm. Occasionally ponds have been raised up on one (sloping site) or all sides to make access for children easier. The consequent retaining walls make this type of construction exceedingly expensive and unattractive for amphibians. Where vandalism is likely, rigid liners have been installed, made of durable plastic rather than glass-fibre, with reinforced edges and floor. These are of comparable cost to a 8m × 5m butyl liner, but are only 3.5m × 2m × .45m deep. Sometimes the problem can be overcome more simply by locating the pond within a courtyard area. In any event, location close to the building complex is preferred for ease of access, supervision and maintenance.

Management

On a large scale, management of habitats, such as meadow, can be achieved by utilising or amending agricultural techniques. Within the school grounds new 'pocket' habitats can only realistically be managed using the horticultural machinery available to the grounds staff or contractor. Removal of cuttings after mowing is fraught with particular problems, and without a mini-forage harvester costs are prohibitive. Ground staff should receive specific training in the objectives of such management and the techniques involved, since long grass and weeds, rotting logs, soil impoverishment and poor drainage are contrary to all past training and experience. Tidy-mindedness is a formidable obstacle, and a frequent source of frustration through lack of communication between user and manager.

After-care of newly created habitats is too often insufficient, neglected or totally forgotten. In the past it has on occasion been left to the school, particularly where habitats were in a defined area, and this has sometimes quickly become overgrown and inaccessible. Major maintenance can and should be undertaken by ground staff or contractor and presents no problem in principle for either specification or implementation. However, it fails to command the attention of either landscape manager or ground staff, since emphasis is usually on building surrounds and playing fields. It is assumed that when the season is at its peak the 'wild areas' will get by, on the basis that no maintenance is good maintenance, and no lasting harm will be done.

Teachers and pupils can undertake limited maintenance tasks, and derive educational benefits from so doing. A partnership where the ground staff or contractor take on major works and the school selects support or follow-up tasks, including regular inspections, offers far greater scope. One example is where the ground staff cut a mini-meadow and the school rake up and collect hay and consequent seed.

Costs

Small-scale experimentation in habitat creation rarely pays great attention to financial implications. In a totally managed landscape such as school grounds, the costs of implementation, establishment and maintenance of new habitats must be borne by existing and frequently hard-pressed budgets. Furthermore, comparison must usually be made with costs of 'conventional' maintenance to horticultural standards. Distinction between capital and revenue budgets for new works and day-to-day running costs respectively mean it is difficult to find expenditure for major initiatives such as planting of woodland. Even in this context it is important to be aware of the eventual 'pay-back' period.

Whilst involvement of children, parents and volunteers and the acquisition of grant aid may ease the financial burden and be important for support from the local authority, it generally makes but a small contribution to the cost of new works and none whatsoever to the cost of ongoing maintenance. The latter must be substantially undertaken by ground staff or landscape contractor, and funding of this work must consequently bear comparison with conventional techniques.

Costs for both new works and maintenance depend on the specification and the degree and quality of supervision to ensure this is achieved. Size of nursery stock (see Table 12.2) spacing, type of guard and planting technique all have a direct bearing on the cost of a new woodland. The type of mowing machine used depends very much on the size, shape, slope and accessibility of the grass area. Too frequently it may also depend on the whim of the grounds maintenance officer, yet there are dramatic variations in the cost of grass cutting using different types of machinery.

Accepting these variations it is possible to make some comparison between different new works and maintenance regimes with both an amenity and a conservation bias. At the same time it is essential to be aware of further comparison with the cheapest forms of hard landscape where maintenance implications are minimal.

Table 12.2 Planting numbers and coverage obtainable for different nursery stock with a £100 budget

	£ Unit	Number	Cover m²	Spacing
Seeds	.0025	40,000	400	100/m²
Bulbs	0.14	750	37.5	20/m²
Seedlings (Notch)				
Transplants	1.50	67	290	2m
Transplants (Pit)	2.00	50	225	2m
Whips				
Small Feathered	4.40	23	225	3m
Standards	15.00	7	44	5m
Ex. Heavy St.	37.00	3	32	8m
Groundcover	1.80	56	7	8/m²
Shrubs (Orn)	3.50	28.5	9.5	3/m²

Thus to establish an ornamental shrub border is usually comparable in expense to *in situ* concrete or tarmac. A new woodland is nearly half as expensive, while grass is cheaper still, seeding being substantially less than turf. Where wild flower seeding is used on a new site, the very high cost of seed supply, up to eight times that of amenity grass for a given area, is more than offset by a reduction in certain cultivations and fertiliser. Where it obviates the need to return topsoil to the area far greater savings can be made. Conversely enrichment of an existing sward represents additional cost that may need to be recouped by savings in subsequent maintenance.

The school user surveys highlighted headteachers' desire for hedges as boundary treatment, as much for their functional and amenity value as for the additional wildlife habitat created. Accepting that on most sites a new hedge demands a low cost protective fence during establishment, it is still cheaper to establish than the all too familiar chainlink. Yet as soon as it requires cutting after, say five years, it quickly becomes more expensive than chainlink. Thus hedge planting may need to rely on savings from more cost-effective techniques on other new habitats within the grounds.

These may well accrue, but not necessarily from the savings in a meadow cutting regime that might immediately be assumed (see Figure 12.2). Given a large compact area, gang-mowing is incredibly cheap. With very little formal bedding or herbaceous borders, savings can be made by reducing the area of rose and/or shrub borders in favour of grass or ground cover, provided this is desirable and acceptable. However, there is a more complex relationship between 'meadow' and standard amenity grass maintenance. As a rule of thumb it has been accepted in the past that

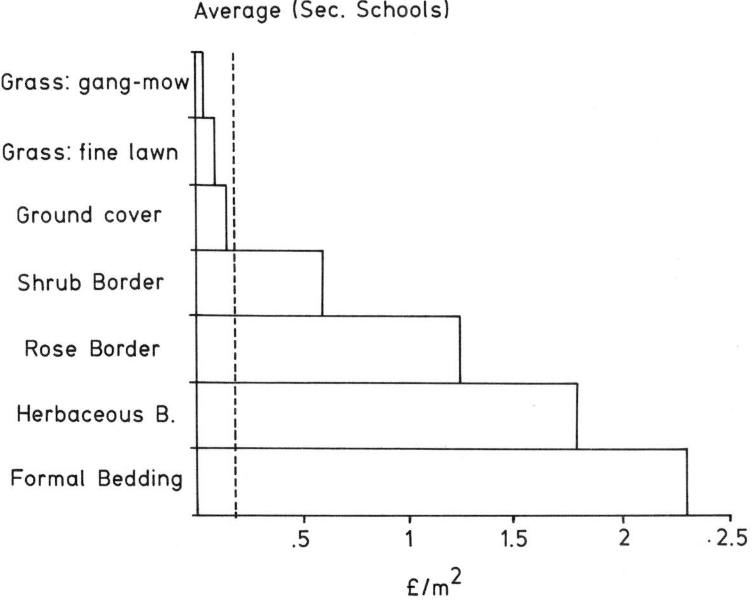

Figure 12.2 Comparative landscape maintenance costs for established soft landscape areas per 1,000m^2 per annum.

an autumn meadow maintenance cut in late August, and removal of cuttings, is broadly equivalent to the cost of a season's low cost gang-mowing.

But there are different types of wild flower meadow maintenance e.g. 'spring' meadows (late-season cutting), 'summer' meadows (early-season cutting), autumn cutting, etc. (Emery, 1986), with consequent variation in machinery and cutting frequency. Bearing in mind that choice of machinery is crucial to cost, it is possible to highlight these differences and compare them with conventional techniques (see Figure 12.3). Size of area is central to this equation, as indeed is the removal of cuttings. It is too simplistic merely to compare these rates with the cheapest gang-mower rate (for areas over 6,000m^2). The inclusion of more expensive rotary and cylinder mowing can make some types of meadow maintenance more attractive financially. There is most argument for a meadow replacing small difficult areas mown by pedestrian machine, rather than extensive gang-mown playing fields. This fits in very well with the realities of the situation and demands on the site.

The desire to create new woodlands, copses and tree groups on school sites is growing. This is an area where a more wildlife-conscious design and management of school grounds may have longer-term cost advantages as well as undoubted amenity, resource and conservation benefits. The pay-back period, when the total of annual mowing costs exceeds the expenditure necessary to establish a woodland area, has

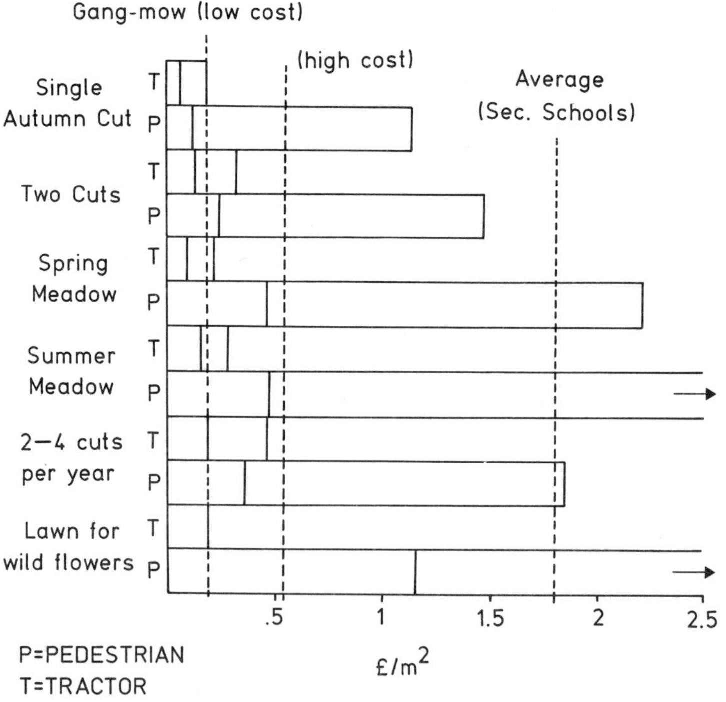

Figure 12.3 Comparative maintenance costs for different meadow regimes (see Emery, 1986) using pedestrian or tractor machinery. High and low-cost options are dictated by small and large areas, respectively.

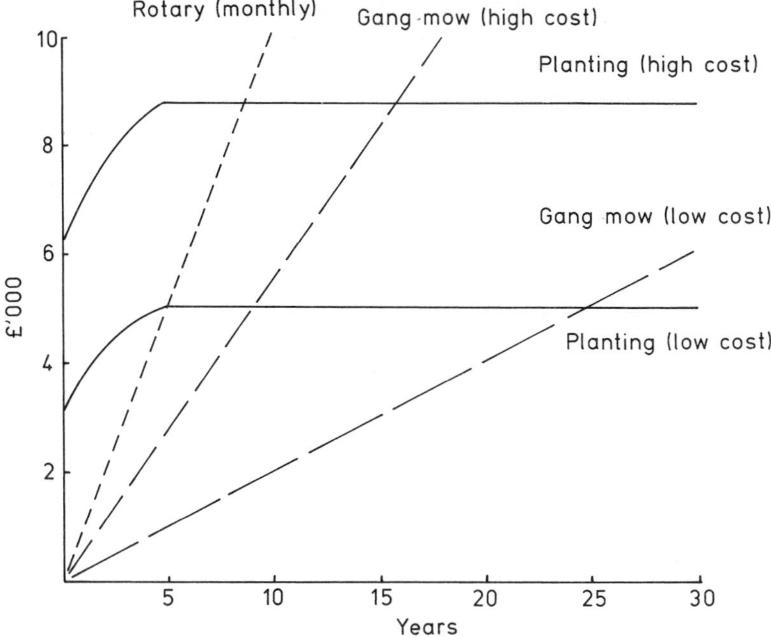

Figure 12.4 Pay-back periods for conversion of conventional, close-mown grass to woodland per 1,000m². High and low-cost options are dictated by size of area.

been held to be as little as six years. Yet this figure is entirely dependent on the cost-effectiveness of the mowing regime and the specification of techniques for planting and after-care. In reality there can be no single 'cross-over' point, but bands contained by low and high mowing and planting respectively (see Figure 12.4). With low-cost planting and high-cost mowing, pay-back periods can be as short as five years, or with the converse equation, as long as 43 years. A more probable time-span is in the region of 10 to 20 years.

Conclusion

Habitat creation on school grounds must take place within the unique context of local site conditions and invariably be set in a multi-use situation. Woodlands, meadows and ponds are being developed for study, appreciation and enjoyment. It is hoped they will lead to an understanding of, and support for, conservation issues and objectives by future generations (Baines, in this volume). They should take pressure off the more fragile local, regional and national nature reserves. In the same way that many old mineral workings figure in the latter categories, there is no reason why such man-made habitats should not achieve longer-term conservation value in their own right, by design rather than by accident. While generally not cheaper than conventional horticultural techniques, it is likely that more emphasis on habitat creation would not involve additional cost, but could be contained within existing budgets. The contribution towards the learning resources of the school is inestimable.

References

Emery, M., 1986, *Promoting Nature in Cities and Towns: a Practical Guide*, Croom Helm, London, p. 264.

Peterken, G.F., 1981, *Woodland Conservation and Management*, Chapman & Hall, London.

SECTION 5
Creating new habitats

After appraising the broad opportunities for habitat reconstruction, the detailed task of designing, planning and specifying works at an individual site level begins. Traditionally in landscaping schemes, habitat 'creation' relies mainly on herbaceous seeding with separate tree and shrub planting. But in the habitat construction there are many other operations to consider, perhaps including transplanting and diversifying existing habitats as part of the overall scheme – as many of the following contributions testify. Transplantation and diversification techniques are, however, dealt with in more detail in Sections 6 and 7 respectively. The main purpose here is to explore *de novo* habitat reconstruction.

At the start, decisions must be taken about the best way in which to develop the site. Penny Anderson suggests several criteria for planning the initial habitat design. First, it must be determined whether the new habitat is appropriate to its surroundings, and whether it bears any resemblance to the original semi-natural communities of the area, past or present. If existing semi-natural habitats are present, these can be used as the blueprint and built on: but if they are to be destroyed by development, it might still be possible to 'rescue' them as part of the scheme, by methods such as transplantation. Perhaps most difficult of all is deciding how to *integrate* a number of different habitats within a single scheme, and how to determine the optimum proportions of each. This is a question which takes us back to ecological theories such as the relationship of viable population sizes to the dynamics of immigration and extinction, where the guidelines are still surprisingly few.

When eventually the habitat design is complete, the next challenge is to translate this on to the ground. For large schemes, there is no escaping proper job specifications, bills of quantity, legally binding contracts procedures and, perhaps most important of all, close site supervision during the operation itself. Seemingly straightforward tasks, such as planting *Typha* or *Phragmites* at the edge of a new lake, require unambiguous yet detailed instructions to be relayed to landscape contractors, as Bickmore and Larard show, if habitat reconstruction is to be successful. This is particularly so as many habitat 'creation' procedures are far removed from conventional landscaping or agricultural practice.

Habitat 'creation' is clearly more difficult for some habitats than for others. Sowing a seed mixture containing the 'correct' plants is on the face of it an easy way to 'create' attractive communities. Brown reviews the upsurge in demand for such wild flower mixtures (in which the customer is still guided by cost rather than ecological appropriateness) and the provision made for this within the horticultural industry. Suppliers naturally aim to avoid disappointment by prescribing mixtures which are even simpler, more attractive and increasingly reliable. The alternative, especially as the wild flower market is oversubscribed, is to carefully select and collect local seeds for the particular situation, as Kaule and Krebs do for arable habitats such as field edges and headlands.

The success of meadows, the most popular of the plant communities regularly created by *a priori* sowing, is by no means guaranteed. Much depends on the particular site conditions, the occurrence of other species already present as vegetation or in the soil seed bank, and the management regime adopted. All these factors are taken into account by Davis and Coppeard, who describe the development of sown meadow communities on an infertile landfill site over several years. Under the right conditions, and with appropriate measures to control invasive species, it is clear that seeding can, in the short term at least, determine and maintain the desired species composition against the forces of natural colonisation, while creating genuinely new opportunities in terms of other wildlife such as butterflies and invertebrates.

Creating ponds and lakes is almost as popular a pastime as sowing meadows. Water areas require expensive engineering to make the most of the habitat, but they are soon productive in terms of the wildlife which they quickly attract and in terms of their high educational and aesthetic value. In contrast, scrub and woodland may be less attractive a form of habitat 'creation' because the results are inevitably longer-term. Until recently, landscapers have thought it sufficient to plant native trees and shrubs, conveniently forgetting the understorey plants and other components of the community. But there is no reason why these should not be included as part of the process of habitat 'creation', although maintaining the correct species composition requires extremely subtle canopy management.

The same principles are true of other engineered habitats which are not discussed here. Like the examples given, their successful implementation will depend on careful matching of the new habitat to the particular features of the site, the correct choice of the community stereotypes, and a long-term commitment to management and maintenance.

13

Creating new habitats in intensively used farmland

G. Kaule and *S. Krebs*

Species diversity in the farm landscape

In landscapes dominated by farming, species diversity is greatest at habitat interfaces: forest edges, hedges, ditches and herbaceous strips between the fields. Investigations in different parts of Southern Germany have shown that almost 45 per cent of the flora exists facultatively or exclusively in edge-habitats which cover not more than 8–10 per cent of the landscape (Kaule 1986). Such habitats are thus of great interest to nature conservation (Schoenichen 1954).

But we need to ask whether, given their size, and the pressures of the environment surrounding them, edge-habitats can be viable and independent as island ecosystems. Reductions in species diversity will be bound to occur at the edge as the field ecosystem itself is increasingly simplified, due to the use of fertilisers and pesticides. Nevertheless, weed populations may survive for many years in edge habitats when they have disappeared from the arable land in between. Eventually, however, they may become extinct by the forces of ecological succession (see Bradshaw, this volume).

Species conservation and edge habitats

Traditionally, edge biotopes were managed by mowing, coppicing, or even by using them as depositories for stones from field cultivations. This resulted in a high diversity of micro-habitats, corresponding to high species diversity. Although the habitats changed with time, the various successional stages were always present in the overall landscape pattern.

Today, neither the immigration of species from the surroundings to the edge, nor the habitat diversity of the edge itself, occurs in the same way in the modern landscape. Improving the chances for species survival on farmland edges is therefore an important part of conservation strategy. In the network of semi-natural habitats, edges are refuges (at least for a transitional period) and are stepping stones for the colonisation of new habitats, as well as being important habitats in themselves. In general, there are four main conservation strategies which can be applied to farmland, in the following order of priority:

1. Protective conservation of existing habitats.
   The highest priority should be given to the conservation of existing habitats where all other strategies fail to prevent species loss. (See Newbold, this volume).

2. Habitat creation through natural succession.
   Natural succession of a habitat may be a successful means of habitat creation if there is sufficient immigration from the surroundings. As Bruns (1987) shows, the diversity of a new habitat such as a clay or gravel pit largely depends on the diversity of those habitats directly adjacent to it (see also Hodgson, this volume).

3. Transplantation of ecosystems.
   If habitat destruction cannot be avoided and space for new habitats is available, complete transplantation may be a successful method.

4. Planting and sowing local genetic varieties of plants.
   If the species diversity is already heavily reduced, the best way of enriching the habitat is to directly introduce species.

It is obvious that strategies (1) and (2) should be the normal procedure, but in many cases species loss is so advanced that others must be applied. But for many ecosystems, strategies (3) and (4) are ineffective or even unsuccessful. Some habitats are immensely complex and have no chance of being satisfactorily restored or reconstructed (Kaule (1986): these include forests with long-developed soils, oligotrophic peat bogs with thick peat layers and old meadow ecosystems. For some species of these ecosystems new habitats may be designed, but this is species conservation rather than habitat reconstruction.

Transplanting habitats

Hedges

Land consolidation during the last century caused much arable land to be cleared. Hedges, herbaceous strips and trees disappeared, while new hedges were planted, destroying the original populations of plants and animals. To prevent this process occurring today, methods for transplanting hedges have recently been developed (Unger, 1981; Kaule, 1986). Before transplantation, the hedge must be coppiced. The second step is to dig a ditch at the new site. The hedge is then transplanted carefully using a specially equipped caterpillar shovel, which moves it directly to the new site without intermediate steps such as loading it on to a lorry (cf. Park, this volume). Because transplantation tends to cause nutrient mobilisation, nutrient-poor conditions are better guaranteed if the transplanted hedge is placed on the mound rather than in the ditch, at the receptor site.

Recently we began a research programme to observe the species dynamics of carefully mapped 10m segments of hedge prior to transplantation. Large individuals were marked, and will be monitored in the same segment at the new site following transplantation.

Freshwater habitats and fenland

Besides hedges, we have successfully transplanted littoral habitats and fenland. Rhizomes and invertebrates are able to survive in the sludge if the transplantation is done in the winter. Material collected was put in the periodically flooded zone of artificial oxbow lakes, where we documented the development of the resulting communities (Bruns 1988; Kaule 1986), and observed that they achieved their typical patterns within two years.

Habitat reconstruction by sowing plants of local provenance

Seed mixtures which are normally available from horticultural sources are inadequate for habitat reconstruction. Such material is usually uniform, and does not include local genetic strains. The percentage of grasses and legumes in these seed mixtures is also generally too high. To avoid these problems, in our research programme we used local plant material collected within a radius of 10km and growing in environments similar to the areas chosen for habitat reconstruction. Experiments were devised in order to create conditions in which species-rich vegetation would result. The three basic treatments were:

a) bare rock or parent material (complete soil removal)
b) reduced soil layers (removal of topsoil only)
c) the existing arable soil

Treatments a) and b) tend to reduce plant competition, but also have important effects on water relations. Prior to sowing, the germination of seed collected was determined by tetrazolium testing (Bulat, 1961). Germination percentage varied appreciably with species and collection source – for example the rates were between 30 per cent (e.g. *Achillea millefolium, Falcaria vulgaris*) and more than 90 per cent (e.g. *Anthriscus sylvestris, Silene vulgaris*) (Table 13.1).

Field experiments were carried out on arable soils from each of the following geological regions: Jurassic limestone, glacial-fluvial gravel, loess and Keuper marls. The communities to be re-created from the seed mixtures were as follows:

A Control: existing seed bank
B Semi-dry grassland
C Nutrient-rich grassland
D Mesotrophic edge communities
E Pioneer vegetation
F Tall herbaceous vegetation

The species composition of these mixtures is given in Appendix 13.1.

Table 13.1. Percentage germination viability of species collected from edge habitats of different geological regions, as determined by tetrazolium testing. The collection areas were: Kraichgau (K), Hohenlohe (H), Swabian Alb (A), Danube near Hengen (D), Upper Swabian (O).

		Geological Unit				
Species	seeds/g	K	H	A	D	O
Achillea millefolium	6600	28			32	
Anthriscus sylvestris	110	96	96	90	92	91
Agrimonia eupatoria	50	58	63	77	65	
Campanula patula	12000			67		
Centaurea jacea	500	72	73	66	61	70
Centaurea scabiosa	160	75	40		80	
Chaerophyllum aureum	110		82	87	88	
Cichorium intybus	800	83	74		70	
Clinopodium vulgare	5000	90	93	88		
Crepis biennis	1100	62	67	80	52	
Daucus carota	770	95	90	81	87	84
Dianthus carthusianorum	900			97	95	
Falcaria vulgaris	330	25				
Galium mollugo	2000	75	78	82	80	
Galium verum	1700	65	70	69		
Geranium pratense	160	74			84	
Heracleum sphondylium	80	78	80	82		79
Knautia arvensis	210	62	53	57	60	72
Lathyrus pratensis	20	85	80		82	
Leontodon hispidus	850		45		46	
Linaria vulgaris	4000	94				86
Lychnis flos-cuculi	3000					95
Malva moschata	400	84		77	80	
Medicago lupulina	500	21	80	82	89	82
Melilotus officinalis	500	86	82			86
Oenothera biennis	1500	67				
Onobrychis viciifolia	45		64	71	72	
Origanum vulgare	8000	86	86	83		
Papaver rhoeas	9000	60				
Pastinaca sativa	250	87	90	71	71	86
Pimpinella major	1000	69	73			68
Plantago lanceolata	630	85	89	70		
Prunella grandiflora	1200			85	50	75
Rhinanthus alectorolophus	240		64	78	96	
Salvia pratensis	800	40	40	79	56	
Sanguisorba minor	120	90	95	97	91	
Saponaria officinalis	660	86	87			
Scabiosa columbaria	250		66	54		
Silene alba	1300	94		89		
Silene vulgaris	1250	92	96	94	95	97
Torilis japonica	650	77				
Tragopogon pratensis	125			91		

Table 13.1 continued

Species	seeds/g	K	H	A	D	O
Trifolium pratense	300		25			32
Verbascum thapsus	10000	78			82	
Verbena officinalis	2500	73				
Vicia sepium	25	98	98		97	

Preliminary Results

The research programme allows for the continuous monitoring of 10 × 5m plots over 5 years, with an additional recording after 10 years. Results in the first growing season are preliminary, but show that competitive, fast-growing weeds on fertile arable soils exclude the successful establishment of species offering little competition. The same effect is produced by a high percentage of legumes (e.g. *Melilotus*) in the seed mixture. The seed mixture must be selected carefully and success depends on careful preparation of the seedbed.

Removing the soil to expose parent material excludes fast growing weeds, but because these habitats suffer from drought, only early seral species tend to survive. The best results have so far been obtained in existing arable soils with sowings of tall herbaceous vegetation and communities of nutrient-rich meadows. Where the topsoil was removed, sowings of semi-dry grassland species and herbaceous edge communities have responded well, whereas complete soil removal favoured the pioneer, semi-dry grassland and mesotrophic edge communities.

Some sown plots were adjacent to, and in contact with, existing edge habitats, and could be compared with plots isolated within fields. The results for the Keuper Marl area are shown in Fig. 13.1. In this case the adjacent habitat is semi-dry grassland associated with vineyards. The preliminary results suggest that the unsown treatments have the greatest species diversity, but that this is affected both by the existing seedbank and by proximity to neighbouring habitats. On the other hand sowing tends to reduce the species of pioneer communities, as well as the perennial weeds of adjacent habitats.

In summary, it appears that on intensively used arable land the sowing of species from edge communities is likely to be the most successful solution. On very fertile topsoil, species of eutrophic meadows and tall herbaceous communities are preferable (see Hodgson, this volume). The seed mixture should be oriented at the local flora, as natural immigration is heavily reduced by strongly competitive ruderal species.

Topsoil removal treatments in isolated situations tend to be colonised only by species with copious seed production and good dispersal. Partial topsoil removal (of 10–20cm) is a prerequisite for the development of mesotrophic communities. On slopes this is easy to manage, but on receiving sites water runoff and eutrophication will reduce the effect. To avoid this, depressions should be filled with nutrient-poor soil, but in practice it is difficult to manage this on a large scale unless opportunities arise through civil engineering operations in the area such as road construction or mineral extraction.

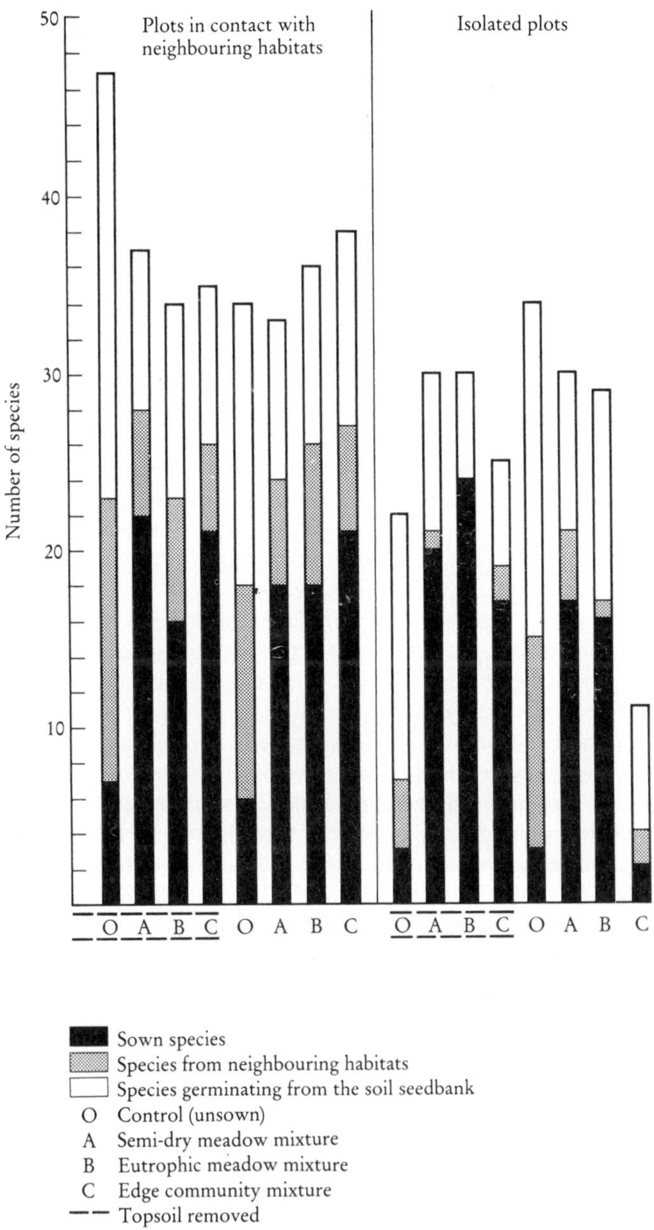

Fig. 13.1 Number of species in plots sown with different seed mixtures, showing the contributions made by sown species, species from the neighbouring habitats, and species from the soil seedbank, including the effect of contact with, or isolation from, adjacent habitats. Results from the loess soils.

References

Bruns, D. (1988): Restoration and Management of Ecosystems for Nature Conservation in West Germany in Cairns, J. (ed): *Rehabilitating Damaged Ecosystems*. CRC Press, Boca Raton, FL, (in press).

Bruns, D. (1987): *Beitrag zur Planung von Ersatzbiotopen gemäß § 8 Bundesnaturschutzgesetz am Beispiel von Sukzessionsflächen auf Lehm*. Dissertation, University of Stuttgart.

Bulat, H. (1961): Reduktionsvorgänge in lebendem Gewebe, Formazone, Tetrazoliumsalze und ihre Bedeutung als Redoxindikatoren in ruhenden Samen. *Prod. Int. Seed Test. Ass.* **26**, (4), p. 685–696.

Kaule, G. (1986): *Arten- und Biotopschutz. UTB große Reihe* (E. Ulmer), Stuttgart.

Lindenbein, W. und Bulat, H. (1960): Grundsätzliches zum Tetrazoliumtest. *Proc. Int. Seed Test. Ass.* **25** (1), p. 449–451.

Schoenichen, W. (1954): *Naturschutz, Heimatschutz. Große Naturforscher Bad.* 16, wiss. Verlagsgesellschaft Stuttgart.

Unger, H.-J. (1981): Verpflanzungen von Hecken und Feldrainen im Rahmen der Flurbereinigung. *Natur und Landschaft* **56** (9), S. 295–300.

Appendix 13.1 Seed mixtures used in experiments on loess soils
a) *Eutrophic meadow* (sowing rate 8.2 g/m^2)

Species	% by weight	viable seeds/m^2	% viable seeds	% cover in first year	
				existing soil	topsoil removed
**group 1					
Achillea millefolium	4.2	647	7.1	37.6	13.7
Cerastium fontanum	0.2	156	1.7	0.5	1.8
Crepis biennis	1.2	81	0.9	0.6	0.3
Galium mollugo	10.3	1360	15.0	2.2	0.8
Geranium pratense	4.2	34	0.4	0.9	0.7
Knautia arvensis	6.1	62	0.7	0.0	0.3
Leontodon hispidus	0.5	17	0.2	0.2	0.3
Leucanthemum ircutanium	2.1	296	3.3	0.7	1.1
Plantago lanceolata	9.1	331	3.6	6.4	1.7
Tragopogon pratense	3.0	28	0.3	1.2	0.5
**group 2					
Daucus carota	6.1	323	3.5	0.0	0.8
Linaria vulgaris	0.6	188	2.1	0.1	0.0
Papaver rhoeas	0.6	255	3.0	4.7	14.7
**group 3					
Lathyrus pratensis	4.2	7	0.1	0.5	0.1
Lathyrus tuberosus	0.6	1	0.0	0.0	0.0
Vicia sepium	1.2	2	0.0	4.5	0.3
**group 4					
grasses	45.5	5259	58.1	14.3	34.0
total	99.7	9047	100.0	74.4	70.9

Appendix 13.1 (contd.)

b) *Semi-dry meadow* (sowing rate 12.6 g/m²)

	% by weight	viable seeds/m²	% viable seeds	% cover in first year	
				existing soil	topsoil removed
**group 1					
Agrimonia eupatoria	19.9	72	0.9	0.0	0.1
Allium oleraceum	11.9	68	0.8	0.0	0.0
Campanula patula	0.1	126	1.5	0.0	0.0
Centaurea jacea	4.0	175	2.1	1.9	0.7
Centaurea scabiosa	1.6	26	0.3	0.4	0.4
Clinopodium vulgare	0.4	225	2.7	0.0	0.1
Galium verum	2.0	298	3.5	0.0	0.0
Hypericum perforatum	0.4	266	3.2	0.0	0.0
Leucanthemum ircutianum	0.4	85	1.0	1.2	0.6
Malva moschata	1.2	50	0.6	0.6	0.1
Origanum vulgare	0.8	688	8.2	0.1	0.1
Picris hieracioides	0.6	42	0.5	0.2	0.2
Plantago media	0.8	196	2.3	0.0	0.1
Salvia pratensis	7.2	288	3.4	0.3	0.1
Sanguisorba minor	2.0	27	0.3	0.2	0.4
Silene vulgaris	4.0	618	7.3	7.7	19.1
**group 2					
Daucus carota	4.0	315	3.7	2.9	2.2
**group 3					
Lotus corniculatus	0.8	43	0.5	1.6	0.9
Medicago lupulina	4.0	52	0.6	52.7	15.7
Trifolium campestre	4.0	90	1.1	0.3	9.0
**group 4					
Grasses	30.0	4656	55.3	14.8	36.1
total	100.1	8406	99.9	84.9	85.9

Appendix 13.1 (contd.)

c) *'Edge' community* (sowing rate 10.2 g/m^2)

	% by weight	viable seeds/m^2	% viable seeds	% cover in first year	
				existing soil	topsoil removed
**group 1					
Agrimonia eupatoria	24.6	72	0.9	0.0	0.2
Allium oleraceum	24.6	114	1.5	0.0	0.0
Clinopodium vulgare	1.0	450	5.8	0.0	0.2
Cichorium intybus	1.2	83	1.1	0.0	0.1
Falcaria vulgaris	9.4	78	1.0	0.0	0.0
Hypericum perforatum	2.0	1066	13.7	0.0	0.1
Linaria vulgaris	4.9	1880	24.1	0.0	0.2
Malva moschata	1.0	34	0.4	0.0	0.3
Origanum vulgaris	1.1	688	8.8	0.0	0.5
Saponaria officinalis	3.4	155	2.0	0.0	0.2
Senecio erucifolius	1.5	234	3.0	0.0	0.7
Silene vulgaris	2.5	309	4.0	0.0	3.6
Tanacetum vulgare	1.0	60	0.8	0.0	1.2
**group 2					
Campanula rapunculoides	0.5	350	4.5	0.0	0.0
Consolida regalis	0.1	4	0.1	0.0	0.0
Daucus carota	9.8	630	8.1	0.0	1.4
Silene pratensis	4.9	564	7.2	1.1	5.4
Oenothera biennis	1.5	151	1.9	0.0	0.0
Papaver rhoeas	0.5	255	3.3	0.0	1.1
Pastinaca sativa	2.5	65	0.8	0.0	0.3
Verbascum thapsus	0.7	492	6.3	0.0	0.0
**group 3					
Melilotus officinalis	1.5	64	0.8	87.8	62.4
total	100.2	7798	32.2	100.1	77.9

Appendix 13.1 (contd.)

d) *'Pioneer community'* (sowing rate 3.0 g/m^2)

	% by weight	viable seeds/m^2	% viable seeds	% cover in first year	
				existing soil	topsoil removed
**group 1					
Cichorium intybus	6.6	132	4.2	0.0	0.0
Daucus carota	16.4	315	10.0	0.0	1.6
Hypericum perforatum	3.3	530	16.8	0.0	0.0
Inula conyza	3.3	182	5.8	0.0	0.0
Linaria vulgaris	1.6	188	6.0	0.1	0.1
Silene pratensis	8.2	280	8.9	0.7	1.9
Oenothera biennis	3.3	100	3.2	0.0	0.0
Papaver rhoeas	3.3	510	16.2	0.0	1.4
Pastinaca sativa	32.8	261	8.3	0.0	0.4
Saponaria officinalis	4.9	84	2.7	0.0	0.0
Senecio erucifolius	3.3	156	4.9	0.0	0.0
Silene vulgaris	6.6	247	7.8	0.0	20.1
Verbena officinalis	1.6	91	2.9	0.0	0.0
**group 2					
Medicago sativa	1.6	17	0.5	0.0	0.0
Meliotus officinalis	3.3	63	2.0	93.5	73.5
total	100.1	3156	100.0	94.3	99.2

Appendix 13.1 (contd.)

e) *'Tall herbaceous vegetation'* (sowing rate 15.2 g/m^2) – existing soil only

	% by weight	% viable seeds	% cover in first year
**group 1			
Anthriscus sylvestris	18.4	5.9	0.0
Daucus carota	6.6	13.9	5.3
Heracleum sphondylium	39.4	8.3	0.0
Linaria vulgaris	2.0	24.9	0.4
Silene pratensis	3.3	12.4	13.3
Pastinaca sativa	19.7	14.4	10.6
Pimpinella major	2.3	7.6	0.3
Torilis japonica	6.6	11.1	1.7
**group 2			
Lathyrus tuberosus	1.3	1.0	0.0
Melilotus alba	0.3	0.5	0.9
total	99.9	100.0	32.6

14

The feasibility of woodland reconstruction

G.P. Buckley and D.G. Knight

The woodland stereotype

Each person's view of the woodland scene is different. For many, the sight of carpets of bluebells and anemones under a leafing canopy in spring is an uplifting experience. Others prefer variety to striking monocultures, and search for local rarities and the plants associated with old, continuously wooded sites. The less romantic find woods impenetrable, gloomy and often wet places, associating them with brambles, stinging nettles and biting insects.

These stereotyped perceptions are hardly representative of the full range of woodland canopy types and ground floras. For a more objective model, the selection of a ground flora can be made on an empirical basis, using any one of a number of vegetation classifications dealing with woodland floras. For example, the recent National Vegetation Classification describes about twenty main semi-natural woodland stand-types in Britain for which detailed species lists are available (Rodwell, 1986).

Suppose we wish to choose a ground flora for a woodland reconstruction on purely aesthetic grounds. A sensible way to go about this might be to short-list attractive, cosmopolitan and shade-tolerant species, following the guidelines for habitat creation recommended by Wells *et al.* (1981). But, on closer inspection, the regular members of real woodland communities are not always what we might wish. According to the National Vegetation Classification, it can be shown that many of the common and widespread species present in semi-natural woodlands are actually far from ideal (see Table 14.1). Although the commonest group of species (Group 1) contain three classic woodland plants, wood sorrel (*Oxalis acetosella*), bluebell (*Hyacinthoides non-scripta*) and dog's mercury (*Mercurialis perennis*), it is hard to imagine anyone taking pains to establish invasive species such as bramble (*Rubus fruticosus* agg.), ivy (*Hedera helix*), bracken (*Pteridium aquilinum*) or wavy hair-grass (*Deschampsia flexuosa*).

Following Peterken (1981), Table 14.1 also attempts a loose classification of the species of all five groups into the categories: (1) poorly colonising plants associated with long continuity of woodland cover; (2) fast-colonising species of recent woodlands; and (3) shade-tolerant species widespread in other habitats besides woodlands, including many ruderals. Although the third category predominates in nearly

Table 14.1 Percentage frequency of selected ground flora species from sample stands*, representing common semi-natural woodland vegetation in Britain

	Frequency	Species	Actual percentage
Group 1	21–41	[2]*Rubus fruticosus*	41
		[3]*Pteridium aquilinum*	33
		[3]*Deschampsia flexuosa*	30
		[1]*Oxalis acetosella*	28
		[1]*Hyacinthoides non-scripta*	26
		[1]*Mercurialis perennis*	26
		[2]*Lonicera periclymenum*	26
		[2]*Hedera helix*	22
Group 2	11–20	[3]*Holcus mollis*	20
		[1]*Viola riviniana*	20
		[2]*Dryopteris dilitata*	20
		[2]*Dryopteris filix-mas*	19
		[2]*Poa trivialis*	17
		[3]*Galium saxatile*	17
		[2]*Urtica dioica*	16
		[2]*Circaea lutetiana*	16
		[3]*Agrostis capillaris*	15
		[3]*Anthoxanthum odoratum*	14
		[3]*Calluna vulgaris*	14
		[1]*Primula vulgaris*	13
		[3]*Deschampsia cespitosa*	12
		[2]*Brachypodium sylvaticum*	12
		[2]*Geranium robertianum*	12
		[3]*Galium aparine*	12
		[3]*Filipendula ulmaria*	11
Group 3	6–10	[1]*Anemone nemorosa*	10
		[3]*Holcus lanatus*	10
		[3]*Agrostis canina*	10
		[1]*Luzula pilosa*	10
		[2]*Geum urbanum*	9
		[3]*Potentilla erecta*	9
		[3]*Ranunculus repens*	8
		[3]*Teucrium scorodonia*	8
		[3]*Veronica chamaedrys*	8
		[3]*Dactylis glomerata*	8
		[1]*Conopodium majus*	8
		[3]*Festuca ovina*	8
		[2]*Glechoma hederacea*	8
		[3]*Ajuga reptans*	7
		[3]*Digitalis purpurea*	7
		[1]*Potentilla sterilis*	7
		[2]*Arum maculatum*	7
		[2]*Rubus idaeus*	7
		[1]*Luzula sylvatica*	7

Table 14.1 continued

	Frequency	Species	Actual percentage
		[1]*Lysimachia nemorum*	7
		[3]*Angelica sylvestris*	6
		[2]*Sanicula europaea*	6
		[1]*Melampyrum pratense*	6
		[1]*Melica uniflora*	6
Group 4	3–5	[1]*Lamiastrum galeobdolon*	5
		[3]*Ranunculus ficaria*	5
		[3]*Heracleum sphondylium*	5
		[1]*Carex sylvatica*	5
		[1]*Chrysosplenium oppositifolium*	5
		[1]*Fragaria vesca*	5
		[1]*Stellaria holostea*	5
		[3]*Molinia caerulea*	5
		[3]*Arrhenatherum elatius*	5
		[2]*Poa nemoralis*	5
		[2]*Rumex sanguineus*	5
		[2]*Stachys sylvatica*	5
		[2]*Silene dioica*	5
		[2]*Bromus ramosus*	5
		[1]*Carex remota*	5
		[1]*Milium effusum*	5
		[1]*Galium odoratum*	5
		[3]*Cirsium palustre*	5
		[1]*Valeriana officinalis*	5
		[3]*Eupatorium cannabinum*	4
		[3]*Anthriscus sylvestris*	4
		[3]*Agrostis stolonifera*	4
		[3]*Phragmites australis*	4
		[2]*Epilobium montanum*	4
		[3]*Caltha palustris*	4
		[3]*Chamaenerion angustifolium*	4
		[3]*Solanum dulcamara*	4
		[1]*Adoxa moschatellina*	3
		[1]*Allium ursinum*	3
		[1]*Veronica montana*	3
		[2]*Tamus communis*	3
		[3]*Prunella vulgaris*	3
		[3]*Succisa pratensis*	3
		[3]*Iris pseudacorus*	3
Group 5	<3	[1]*Platanthera chlorantha*	2
		[1]*Campanula latifolia*	2
		[1]*Carex acutiformis*	2
		[1]*Corydalis claviculata*	2
		[1]*Geum rivale*	2
		[1]*Hypericum hirsutum*	2
		[3]*Festuca rubra*	2

Table 14.1 continued

Frequency	Species	Actual percentage
	[3]*Poa pratensis*	2
	[2]*Festuca gigantea*	2
	[2]*Listera ovata*	2
	[2]*Rubus caesius*	2
	[2]*Viola odorata*	2
	[1]*Campanula trachelium*	2
	[1]*Convallaria majalis*	2
	[3]*Galium palustre*	2
	[3]*Aegopodium podagraria*	2
	[3]*Alliaria petiolata*	2
	[3]*Arctium minus*	2
	[3]*Lapsana communis*	2
	[2]*Iris foetidissima*	2
	[3]*Epilobium hirsutum*	1
	[3]*Scrophularia auriculata*	<1
	[1]*Neottia nidus-avis*	<1
	[1]*Hypericum tetrapterum*	<1
	[1]*Carex pendula*	<1
	[1]*Calamagrostis canescens*	<1
	[3]*Rumex obtusifolius*	<1

Notes: *frequency out of the 2340 sample stands selected for the National Vegetation Classification, distributed among 19 common woodland types in Britain (Rodwell, 1986).
[1] species normally associated with ancient woodland sites
[2] fast-colonising species of recent woodlands
[3] shade-tolerant species widespread in other habitats besides woodlands (after Peterken, 1981)

every group, it is interesting that fast-colonising woodland species are relatively more common in Group 2, while category 1 plants form the majority of Group 5.

This poses a problem for the ambitious restorationist who wishes to reconstruct a particular type of ancient, semi-natural woodland. Many of the attractive and interesting ground flora subjects are also the least widespread and are often associated with continuously wooded sites. These same plants are often poor colonists, and are unlikely to invade a newly made wood by themselves: contrary to the normal ethics of habitat creation, which discriminate against the widespread dispersal of less common species, *they would have to be put there*. At the other extreme category 3 species, which are not necessarily confined to the woodland habitat, may turn out to be invasive, perhaps to the detriment of more desirable species. These objections also apply to some category 2 species, notably *Rubus fruticosus* agg. and nettle (*Urtica dioica*).

Ultimately, then, species choice is a subjective judgement, depending on a number of considerations and guesses. For instance, it must be decided in advance how abundant a species must be before it qualifies as a target species; or how invasive and competitive target and non-target species are likely to be on the site in question. Aesthetic choices have to be made, comparing the attractiveness of different species, or selecting a balance between species diversity and uniformity; and last, but not least, the feasibility of propagating the target species must be determined.

Seed mixtures

An obvious way to create a flora on a new site is to broadcast seed. But in the woodland habitat, the ecological odds are weighed impressively against this, for a number of reasons. Species which persist under stable canopies are typically poor seeding subjects that tend to produce relatively few, heavy seeds, which in turn give rise to robust seedlings capable of surviving the gloom of the forest floor. The same plants are often long-lived perennials, investing less in seed production than in slow vegetative spread. In their stable woodland environment, they have no need to form persistent seed banks, and therefore do not require stimuli such as light, fluctuating temperatures or physical disturbance in order to germinate. Furthermore their seeds may possess a chilling requirement, enabling them to germinate rapidly in spring, a favourable time for seedling development in woods.

In contrast, shade-tolerating species of the woodland edge or more open environments often rely on the efficient dispersal of copious quantities of seed. Being adapted to colonising new sites, this group contains more annuals, biennials and short-lived perennials. Their seed either germinates immediately, or remains dormant in the soil until some stimulus, such as woodland clearance, activates the seed bank (e.g. type III or IV seed banks described by Thompson and Grime, 1979). It seems that only these early and mid-successional species are adapted to accumulate persistent reservoirs of buried seeds (Donelan and Thompson, 1980), and this makes them good subjects for sowing. Late-succession woodland plants have much lower viability and often more exacting germination requirements.

In practice, the species of all groups in Table 14.1 display no particular pattern in reproductive strategy. In Group 1, the common woodland species *Oxalis, Hyacynthoides* and *Mercurialis* are all poor subjects for seeding, as in shade they produce small quantities of seed with low germination success, and may benefit from high summer temperatures or over-winter chilling. This is true of several attractive vernal species of broadleaved woodlands, including wood anemone (*Anemone nemorosa*), ransoms (*Allium ursinum*), moschatel (*Adoxa moschatellina*) and celandine (*Ranunculus ficaria*). Moreover, restorationists will have to wait long periods for flowering to occur from the seedling stage – for example 7–10 years in the case of *Anemone* and arum lily (*Arum maculatum*) (Sowter, 1949; Stirreffs, 1985).

Like the other species cited in Group 1, honeysuckle (*Lonicera periclymenum*) has unreliable germination, while *Pteridium* invades mainly vegetatively. However, the two remaining group members, *Hedera* and *Rubus*, both germinate freely. The success of *Rubus* in particular can be explained by its large persistent seed bank in addition to an efficient method of vegetative spread. Although these characteristics apply equally to other invasive species, there are several visually attractive and colourful species with persistent seed banks which have been successfully exploited by the seed companies. Their great advantage is that they will germinate reliably in the disturbed ground conditions prevailing during habitat reconstruction, and do not require special pre-treatments such as chilling. However, many are short-lived components of stable woodland vegetation.

Commercial shade and woodland seed mixtures

In 1987 we examined samples of seven commercial 'woodland' or 'shade' seed mixes produced by the British horticultural industry. Of a total of fifty-three herb species present in these mixes, only one-third were strictly woodland subjects, the remainder being cosmopolitan, shade-tolerant plants of open habitats such as meadows and waste ground (see Table 14.2). From the point of view of the seed company, such species have four advantages:

1. They may be grown on a large scale for inclusion in other habitat mixes, e.g. the use of ragged robin (*Lychnis flos-cuculi*) and meadow-sweet (*Filipendula ulmaria*) in wetland as well as woodland mixtures.
2. Their seed is more readily produced than from more difficult, less fecund woodland species.
3. Many are attractive or conspicuous species, such as the doubtfully native martagon lily (*Lilium martagon*) and the wasteland species agrimony (*Agrimonia eupatoria*) and mallow (*Malva sylvestris*).
4. Prominent species which germinate readily (e.g. teasle, *Dipsacus fullonum*) and mullein, *Verbascum thapsus*) are good insurance against consumer disappointment.

Table 14.2 Species present in seven commercial 'shade' and 'woodland' seed mixes available in 1987

Species	Number of mixes
1. Herbs	
Silene dioica	7
Digitalis purpurea	7
Geum urbanum	6
Allium ursinum	5
Stachys sylvatica	5
Torilis japonica	5
Hypericum hirsutum	5
Hypericum montanum	5
Teucrium scorodonia	5
Hyacinthoides non-scripta	4
Filipendula ulmaria	4
Lychnis flos-cuculi	4
Betonica officinalis	4
Primula vulgaris	3
Scrophularia nodosa	3
Stellaria holostea	3

Table 14.2 continued

Species	Number of mixes
Primula veris	3
Primula vulgaris	3
Prunella vulgaris	3
Dipsacus fullonum	3
Conopodium majus	2
Geranium robertianum	2
Alliaria petiolata	2
Myosotis arvensis	2
Clematis vitalba	2
Verbascum thapsus	2
Aquilegia vulgaris	1
Campanula trachelium	1
Vicia sylvatica	1
Fragaria vesca	1
Taraxacum officinale	1
Succisa pratensis	1
Agrimonia eupatorium	1
Galium mollugo	1
Lilium martagon	1
Malva sylvestris	1
Mecanopsis cambrica	1
Rhinanthus minor	1
Vicia tetrasperma	1
2. Grasses	
Poa nemoralis/pratensis/trivialis	7
Festuca rubra/tenuifolia/ovina	7
Agrostis capillaris/castellana	6
Cynosurus cristatus	2
Anthoxanthum odoratum	2
Deschampsia cespitosa	2
Deschampsia flexuosa	1
Briza media	1

The inclusion of colourful herb species with excellent germination, such as red campion (*Silene dioica*) and foxglove (*Digitalis purpurea*), together with other reliable species (e.g. hairy and mountain St. John's wort (*Hypericum hirsutum* and *H. montanum*), wood sage (*Teucrium scorodonia*), figwort (*Scrophularia nodosa*) and herb bennet (*Geum urbanum*) was a feature of the mixes. On the other hand, conspicuous by their absence were several common woodland herbs (*Mercurialis perennis, Hyacinthoides non-scripta, Oxalis acetosella*, violet (*Viola riviniana*), *Anemone nemorosa, Ranunculus ficaria* and *Adoxa moschatellina*), all with vegetative reproduction or difficult germination. Also absent were the less spectacular, but attractive and readily germinating wood-rushes (*Luzula pilosa* and *L. sylvatica*), sedges (*Carex sylvatica* and *C. remota*) and grasses (*Milium effusum* and *Melica uniflora*) which are common members of stable woodland communities. In contrast, good use was made of several

Table 14.3 Germination success of species in seed trays following summer sowings of seven commercial 'shade' and 'woodland' seed mixes

	Successful species	Unsuccessful species
1. Species associated with ancient woodland sites	*Hypericum hirsutum*	*Allium ursinum*
	Scrophularia nodosa	*Campanula trachelium*
	Aquilegia vulgaris	*Conopodium majus*
		Hyacinthoides non-scripta
		Fragaria vesca
		Primula vulgaris
		Stellaria holostea
2. Fast-colonising woodland herbs	*Geranium robertianum*	
	Geum urbanum	
	Silene dioica	
	Stachys sylvatica	
3. Shade-bearing herbs	*Ajuga reptans*	*Agrimonia eupatoria*
	Betonica officinalis	*Alliaria petiolata*
	Dipsacus fullonum	*Digitalis purpurea*
	Galium mollugo	*Filipendula ulmaria*
	Lychnis flos-cuculi	*Hypericum montanum*
	Myosotis arvensis	*Lilium martagon*
	Prunella vulgaris	*Malva sylvestris*
	Taraxacum officinale	*Mecanopsis cambrica*
	Teucrium scorodonia	*Primula veris*
	Torilis japonica	*Rhinanthus minor*
	Verbascum thapsus	*Succisa pratensis*
	Vicia Sepium	
4. Grasses	*Festuca* spp.	*Anthoxanthum odoratum*
	Poa spp.	*Briza media*
	Agrostis spp.	*Deschampsia cespitosa*
	Cynosurus cristatus	*Deschampsia flexuosa*

rapidly-colonising woodland species, while weedy and invasive species were on the whole studiously avoided, with the exception of the annual (*Torilis japonica*).

To test the robustness of the seven different seed mixes, we sowed them under sub-optimal conditions, using unstratified seed in early summer, at the recommended rate in a replicated seed tray experiment under three light regimes (67 per cent, 16 per cent and 4 per cent of full daylight values), the shade being provided by layers of woven polypropylene mesh (Knight, 1987). The experiment was carried out in an unheated polythene tunnel over a period of ten weeks.

Success of individual species was variable, with about half of the total germinating across all treatments (see Table 14.3). A large proportion of the species associated with continuously wooded sites failed, whereas all four fast-colonising woodland herbs together with the grass, *Poa trivialis*, were successful. An approximately equal number of shade-tolerant species and grasses germinated compared with those that did not.

Table 14.4 Mean percentage cover * of grasses and herbs present in seven commercial seed mixes, recorded after a ten-week growth period in seed trays under three different light regimes

	Light regime (per cent of ambient PAR)		
	67%	16%	4%
% grass cover	163 ± 32	167 ± 11	91 ± 28
% herb cover	43 ± 37	22 ± 3	15 ± 14
% bare ground	1 ± 2	2 ± 3	27 ± 10
grass: herb ratio	3.79	7.59	3.37

* cover recorded using 35-point quadrats on 35 × 21 × 5 cm seed trays; means and standard errors

A majority of grass seed was present in all seven shade mixes, presumably to provide a cover or 'nurse crop' for the herbs. Weight ratios were approximately 4:1 (grass: herbs), giving seed number ratios (grass:herbs) of between 3:1 and 11:1. From Table 14.4 it can be seen that, irrespective of shading level, the grasses dominated the herb layer after ten weeks, in cover ratios similar to those of sowing weight. Both herbs and grasses declined in decreasing light, especially between relative illuminations of 16 per cent and 4 per cent PAR (photosynthetically active radiation), an important threshold in terms of plant performance and productivity. Bare ground increased markedly in the heaviest shade treatment.

The investigation showed no clear effect of light level on the grass:herb ratio, or indeed on the relative responses between woodland species and shade-bearing herbs. This is not really surprising as the majority of the herbs and grasses in the mixtures were not shade-tolerant specialists. Although most of the sown grasses might be expected to decline under severe shade, their high seed densities in sowings of this type are clearly detrimental to accompanying herbs because of their rapid germination and early growth.

Using a sowing rate of about 4g m^{-2}, the cost of an average shade mixture in 1987 was perhaps 15p m^{-2} (4p for grass, 11p for the herbs), about £1,500 ha^{-1}. Omitting grass reduces the sowing rate to 0.8g m^{-2} and the cost to 19p m^{-2} or £1,100 ha^{-1}, a reasonable outlay considering that tree planting is additional.

Container plants

The horticultural alternative to sowing is to set out individual container plants in the field, either in random groups or on some regular planting grid. The technique has distinct advantages for woodland herbs from which it is difficult to obtain quantities of seed, or which have poor germination (see Wells *et al*., in this volume). Container plants are also more likely to be able to withstand weed competition than seedlings, and to flower earlier. Cost, however, is the limiting factor. To represent desirable species such as *Mercurialis perennis, Viola riviniana, Primula vulgaris* or yellow archangel, *Lamiastrum galeobdolon*, adequately in the ground layer of a new wood might require the equivalent of 2,500 plants ha^{-1} at 2m spacing. At 40p per container, this is £1,000 ha^{-1} for each species.

An alternative approach to raising plants in a horticultural operation would be to transfer mature plants from a 'donor' woodland to one which is to be reconstructed or diversified. This has been little tried, although Euarson and Craven (1982) had some success with transplanting wild woodland herbs into a beech-maple woodland in Wisconsin.

Using soil transfer

Transplanting grassland or moorland is feasible by modern earth-moving machinery (or by the manual efforts of conservation volunteers), but for woodlands the task is much more daunting (but see Down and Morton, in this volume). However, despite the presence of tree stumps and roots, it is technically feasible to remove a profile of soil from a 'donor' woodland site and to transport it to a receptor area. Given a reasonable scale of operation, say up to 1 ha, the theoretical cost of this kind of operation might be of the order of £3–5 m^{-3} of soil, provided transport distances are minimal. However it can be quickly calculated that to transport a 30cm layer of soil even a short distance, say 500m, would cost £9,000–£15,000 ha^{-1}.

Soil movement has many theoretical advantages over other methods of restoring the ground vegetation. Not only is the soil profile transferred to the new site then in theory a faithful copy of the original and an appropriate medium for the plants it contains, but it also carries with it the complete vegetation in the form of whole plants, plant fragments (stems, roots, rhizomes and bulbs) and the buried seed bank. Other components of the ecosystem such as mycorrhizae, soil invertebrates, fungi, etc., will be introduced as a bonus. Owing to cost, few examples of soil transfer from semi-natural woodlands have been contemplated or carried out, and the efficiency of the method has yet to be adequately documented.

Investigations into the composition of soil seed banks of different woodlands provide a useful starting point (e.g. Olmstead and Curtis, 1947; Livingstone and Alessio, 1968; Thompson and Grime, 1979; Hill and Stevens, 1981; Brown and Oosterhuis, 1981; Harris and Kent, 1987). These show that: (a) numbers of viable seeds are large, and can germinate rapidly to form a dense vegetation cover; (b) certain species develop considerable buried populations while others have practically no buried seed; and (c) there are remarkable disparities between the buried seed bank, which contains many woodland-edge species and ruderals, and the existing woodland vegetation. Differences between the vegetation and the seed-bank species are especially marked when a wood has remained undisturbed for a long period, allowing shade-tolerant 'woodland' perennials to dominate the ground flora. In contrast, in recently cleared or coppiced woods the vegetation is more closely allied to the soil seed bank which has benefited from disturbance.

The influence of soil transfer on the original understorey vegetation can be best illustrated by a specific example. Biggins Wood, now removed and partly relocated, was originally a five ha fragment of ancient woodland on the site currently being

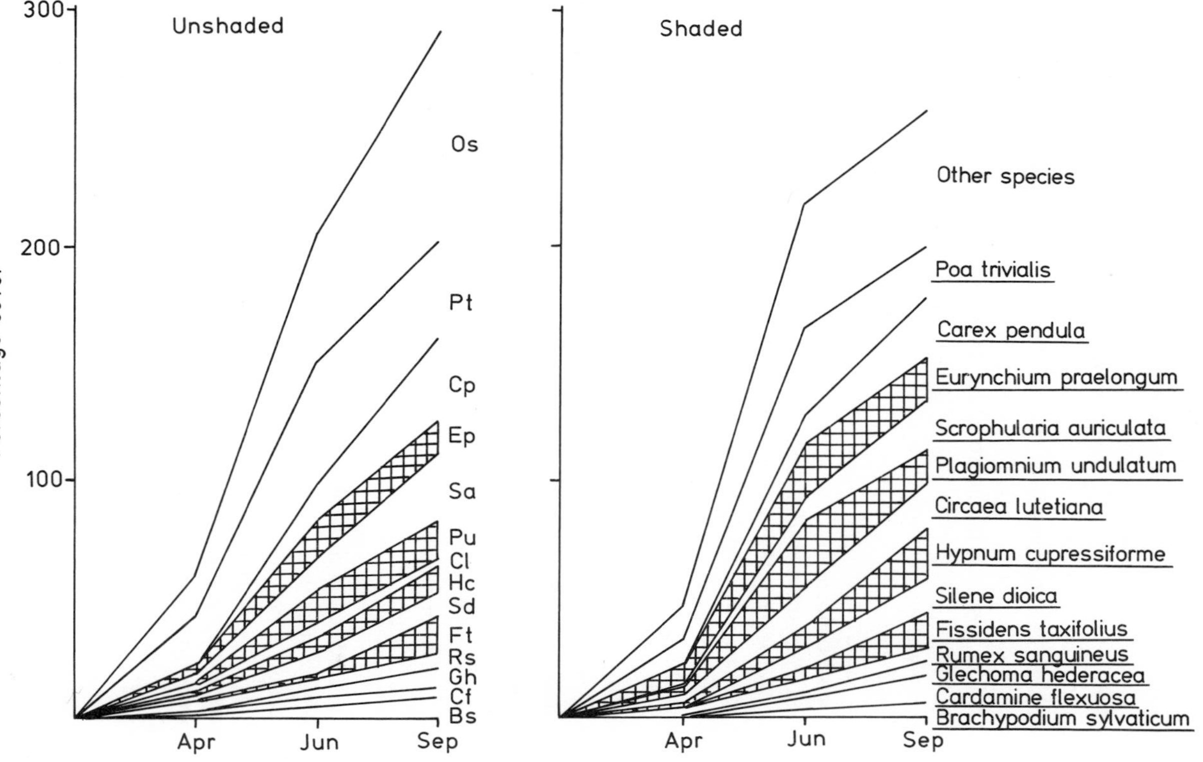

Figure 14.1 Percentage cover of grass and woodland herbs developing in shaded and unshaded soil samples from Biggins Wood, 1986–7.

developed as the Channel Tunnel Terminal at Folkestone, Kent. In order to test in advance whether its ground flora could be successfully relocated at an adjacent site by means of soil transfer, the vegetation arising from soil samples was examined. Random quadrats were set out in the wood in the successive summers of 1986 and 1987, in which the existing vegetation was recorded, while soil samples were taken from these same areas during December 1986 (Buckley, 1988). The soil was placed in seed trays in a polythene tunnel, to allow vegetation development during the 1987 growing season. Two different light regimes, shaded (*c.* 9 per cent of ambient photosynthetically active radiation (PAR) and unshaded, were used. The results are summarised in Figure 14.1.

Vegetation quickly developed on the soil samples, irrespective of shading treatment. This was dominated by the grass *Poa trivialis*, the bryophytes *Eurynchium praelongum* and *Plagiomnium undulatum*, the sedge *Carex riparia* and water figwort (*Scrophularia auriculata*). Other common woodland herbs were enchanter's nightshade (*Circaea lutetiana*), ground ivy (*Glechoma hederacea*) and *Silene dioica*. Overall, 76 vascular species were recorded from an actual sample area within the wood of about 20m^2. Compared with the actual ground flora of the wood, the seed trays contained considerably more cover of *Scrophularia, Poa*, square-stemmed St. John's wort (*Hypericum tetrapterum*), wood bitter-cress (*Cardamine flexuosa*), wood speedwell (*Veronica montana*) and *Geum urbanum*. Their success in the disturbed ground represented by the soil trays was probably not only due to the activation of the soil seed bank (important in the case of *Scrophularia, Poa* and *Hypericum*), but also to the absence of competition.

In the wood itself, *Arum maculatum, Allium ursinum* and *Adoxa moschatellina* were more abundant than in the soil trays. This suggests that their tubers, rhizomes and bulbs were mainly below the sample depth (7cm) used for soil collection. Clearly, in soil transfer operations the depth of soil collection must be sufficient to represent the vegetative fraction (in this case at least 10–15cm), but not so deep as to dilute the soil seed bank present in the upper soil profile. In the Biggin's Wood operation a soil layer about 30 cm thick was moved.

A feature of soil seed bank studies is that they tend to ignore, or actively exclude, the vegetative component. In the seed tray investigation described, plant fragments were removed from some soil samples but not others. Although differences between the two treatments was not large, a higher incidence of *Circaea* and *Arum* was found in control trays, confirming that their rhizomes were important contributors to surface vegetation. Altogether 15 species produced plants from vegetative fragments, including some which also had abundant buried seed. Nearly half of the *Circaea* plant population originated from rhizomes, while significant vegetative proportions of *Glechoma* (29 per cent), *Ajuga* (27 per cent), *Veronica* (12 per cent) and *Urtica* (11 per cent) were also present.

Significant members of the seed tray vegetation were potential weed species which might, under open conditions, compete with and exclude more desirable species. Although their percentage cover in mid-season was small (in the range 10–20 per cent), species such as *Rubus fruticosus* agg., *Urtica dioica*, wood dock (*Rumex sanguineus*) and willowherb (*Epilobium adenocaulon*) were ubiquitous. Also present were large numbers of woody seedlings (for example *Fraxinus* and *Sambucus*, both present at 11–12 seedlings m^{-2}). Management to reduce weed growth and to promote canopy-forming species would clearly be critical.

The problem of the canopy

In order to maintain the shade-tolerant ground flora of a woodland and to prevent species of open habitats from gaining competitive advantage during woodland reconstruction, some form of overhead shade is necessary. Several independent studies have shown that broadleaved canopies can reduce light values at the ground layer to 10 per cent or less of full daylight values in summer (e.g. Coombe, 1966), whereas as little as 15–25 per cent of ambient light will encourage the development of non-woodland flora. In conifer plantations, Hill and Hays (1978) suggested that relative illumination values of 12–13 per cent represented a threshold for most vascular plants, with tolerant bryophytes being eliminated at values of 5 per cent or less. However, deciduous canopies affect the ground flora differently, allowing more opportunity for spring assimilation which may enable plants to persist through the denser shade of mid-summer. In Biggins Wood, the relative illumination levels in August 1987 were within the range five to ten per cent PAR (see Figure 14.2).

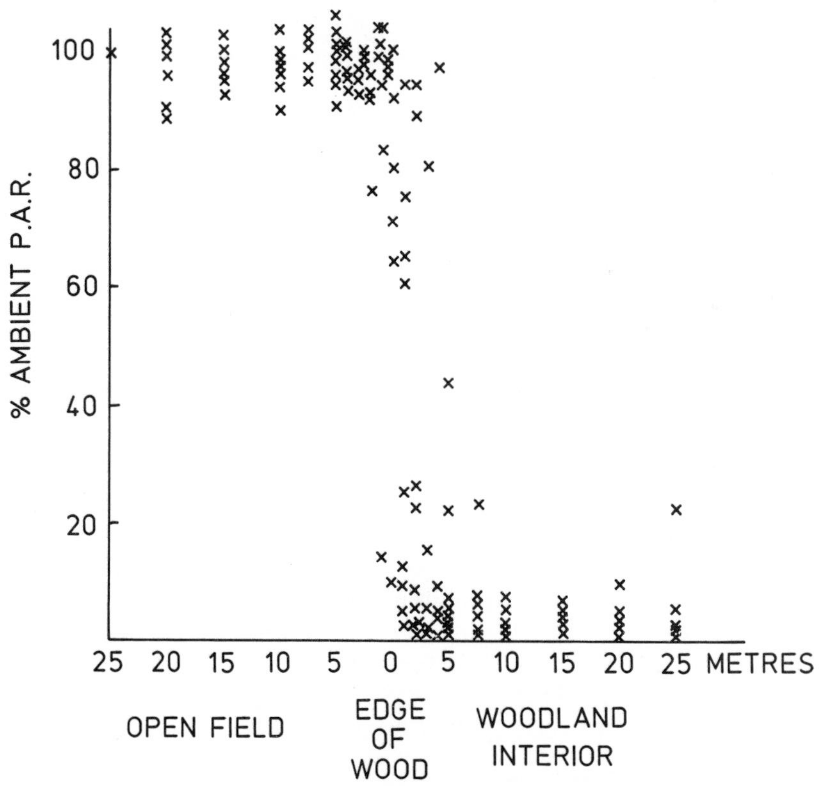

Figure 14.2 Scatter diagram showing light profiles at the edge of Biggins Wood, Folkestone, in August 1987.

These examples are reasonably consistent with our assessment of limiting light values for shade seed mixes, and also with the Biggins Wood soil transfer experiment, which demonstrated that light reductions of eight to ten per cent tended to promote bryophytes and shade-tolerant vascular species, such as *Circaea* compared with controls (see Figure 14.1). Nevertheless it was very noticeable that less desirable species, such as *Rubus* and *Sambucus*, were apparently unaffected by the same treatment. It seems inevitable that, even if low light values can be achieved, additional management will be needed to control non-target species in the short to medium-term. This will be particularly true when establishing new plantings on non-woodland sites which carry a persistent seed bank of arable weeds and wasteground species.

It is often pointed out that small planting areas have high perimeter/area ratios and hence a long 'edge' zone in which light intensity levels may be well above the critical level. Cole (1982) suggested that a true woodland flora was unlikely to survive in an isolated mature tree block of less than 1600m^2 (i.e. 40 × 40m), which, allowing for an 'edge' of 10m, might effectively shade only 20 × 20m = 400m^2. By the same reasoning any woodland block less than 20m in width is unlikely to sustain effectively woodland plants in a relocation exercise. However, the example of Figure 14.2 suggests that the transition in light values from the woodland edge to the woodland interior can be much sharper. The long 'edge' associated with small woods also has the inbuilt advantage of providing a permanent light gradient, sections of which will be optimal for different ground flora species. In a newly planted wood, it will be the gradual transition from open ground to closed-canopy light values which has greater significance for the establishment of the 'correct' flora.

Few remedies are available to overcome the lack of a canopy at the start of a woodland recreation exercise. Ignoring the possibility of mature canopy transplantation, the only viable alternative is to plant closely spaced trees. In Britain, close planting to achieve woodland conditions has been pioneered by the New Town Development Corporations, especially at Warrington and Milton Keynes, where experimental work on woodland herb establishment has been carried out. At Oakwood, Warrington New Town, broadleaved whips and transplants were planted at 0.9m in ground kept weed-free for five years subsequently. Thinning was carried out at the end of the fifth year and again in the eighth year. At this time shade had reduced the cover of open-grown plants in the field layer to <10 per cent, providing suitable conditions for the introduction of spring and summer-flowering woodland plants (Tregay, 1985).

Although in new plantings the introduction of woodland herbs has to be delayed until light levels are suitable, canopy closure of trees and shrubs planted at 1m centres or less will be rapid and considerably accelerated by the initial blanket herbicide treatments necessary to reduce weed growth. Indeed in some cases thinning has been thought necessary after only three years (Scott *et al.*, 1986). The level of herbicide use during these pre-canopy closure periods may be vital in reducing the buried seed bank of non-target species to acceptable levels.

Managing the canopy

Nowadays many tree-planting schemes are ecologically motivated, and may even be perceived as habitat creation exercises in their own right. The mass plantings described at Warrington are typical of those pioneered in the Netherlands consciously to promote nature conservation in cities. Such 'ecological' plantings are loosely based on classical models of woodland succession, providing for a scrub development phase, with or without 'pioneer' or 'nurse' tree species, and including sufficient 'climax' tree species to form the eventual canopy (although the latter can also be planted retrospectively). Additional light-demanding 'edge species', often attractive flowering shrubs, were sometimes planted at the perimeter to add further understorey interest. Early plantings even specified invasive and troublesome species such as *Rubus fruticosus* agg. (e.g. Eardley, 1973), although ironically, as Table 14.1 shows, it is a major species of semi-natural woodlands.

As already pointed out, mass plantings are an excellent way to achieve the rapid reductions in light level which are a prerequisite for the introduction of woodland herbs. However, early 'ecological' plantings specified roughly equal densities of 'pioneer' species (e.g. alder, birch, poplar and willow, but these often proved too vigorous for the 'climax' species (e.g. oak, beech, hornbeam and lime and had to be coppiced or removed prematurely. Such treatments would be expected to cause considerable disturbance, increasing light levels and activating the soil seed bank, to the detriment of shade-tolerant woodland herbs already in the ground flora. The same would apply where the 'climax' species were retrospectively introduced into 'pioneer' plantings.

The other major drawback, which applies to almost all new landscape plantings, is that they are essentially even-aged. They will therefore form closed canopies, perhaps becoming too dark to sustain some woodland herbs, and over a long period may even exhaust their buried seed banks (Hill and Stevens, 1981). However, unlike monocultures, mixed broadleaved canopies are unlikely to accumulate thick litter layers which might otherwise limit some ground flora species (Anderson, 1979) so that, although diversity in the ground layer may be reduced, the vernal flora can be expected to remain prominent.

Management ultimately has a casting role which decides the composition of the understorey species. Theoretically, thinnings could be carried out with the sole objective of maintaining an optimum light regime conducive to maximum diversity in the ground layer. These would need to be somewhat more frequent than in commercial forests, giving lighter conditions, if the ground flora is to thrive. Alternatively wide rides or extraction racks could be cut through the woodland where the less shade-tolerant species can survive (see Anderson, in this volume). Coppicing will also encourage a rapid turnover of these species in the soil seed bank, but at the same time would tend to discourage a ground flora tolerant of deep shade. Uneven-aged forestry, using silvicultural techniques such as group felling, appears to offer a compromise where both the shade-tolerant and ephemeral components of the woodland flora can coexist (Harris and Kent, 1987).

Discussion

Although the ground layer of an ancient, semi-natural woodland might not always fit in with our visual or ecological stereotypes, only a rash optimist would seriously attempt the reconstruction of something so complex. A good illustration of the gap between theory and practice is the phenomenon of the 'woodland' seed mix. Such mixtures contain not only relatively few of what might be described as the 'correct' species, but also a large number of potential competitors. This is not so much the fault of the horticultural industry as an illustration of the fact that 'shade' mixes are still garden-based products, while their wider use is eschewed by professional conservationists who consider them unethical. With greater recognition, the product could be refined and could contain, for instance, several more of the reliable fast-colonising species of recent woodland. Although this would improve the image of seed mixtures, it may still be necessary to introduce some of the 'difficult' species vegetatively in order to satisfy the stereotype of semi-natural, ancient woodland. A major problem remaining is how to achieve a diverse herb layer by orchestrating together sowing, planting and management techniques.

Sowing and transplanting woodland herbs is a technique available for new land prepared for the purpose, but it can equally be done to 'repair' or diversify existing low-grade woodlands which have suffered neglect, overgrazing or disturbance. Small farm woods containing an understorey of *Sambucus* and *Rubus* with cow parsley (*Anthriscus sylvestris*), cleavers (*Galium aparine*) and *Urtica dioica* are a case in point. The problem here is not so much to control light levels, but to eliminate the invasive species and to replace them with other, more desirable plants. In such situations the extensive use of herbicide, accompanied by the sowing or planting of fast-colonising woodland species, may be appropriate.

The prospects for woodland reconstruction by transplantation are best of all, provided a 'donor' wood can be found and the original soil profile can be used. The species list for the herb layer will be much more extensive than in woodlands reconstructed horticulturally by seeding and planting, while a higher proportion of vegetative perennials can be expected. But soil movement is bound to cause distortions in species balance within the herb layer through several agencies, namely the activation of the buried seed bank, the arrival of competitive species and fundamental changes in site conditions, notably an increase in the light regime. Although drastic, it is perhaps useful to remember that soil movement is only a more extreme form of the disturbance which normally accompanies felling and extraction, from which the flora regularly recovers after the canopy has been re-established. The problem here is that there may not be an adjacent area of suitable woodland from which shade-tolerant plants can colonise.

It has been assumed that reconstructing a structured woodland canopy from trees and shrubs is of minor consequence compared with the ground flora, since nursery stock, although often impoverished or inappropriate in its genetic base, is freely available. As in any woodland, the silvicultural system has an overriding effect on the ground flora. Mass 'ecological' plantings, although of native species, have the potential to become as dark as commercial forests, and must be kept structurally diverse if the ground flora is to remain prominent and interesting in the habitat. A variety of techniques is available to foresters to achieve this structural diversity (see Anderson, in this volume), but the effects of different silvicultural systems on the herb layer are poorly researched.

Further diversity in woodlands is possible, from the simple putting up of bird-nesting or bat-boxes, to the importation of logs to create a habitat for wood-boring insects and fungi, as has been done in some Berlin parks (Wittkugel, 1988). Ponds and glades are further refinements already used by conservation organisations to 'improve' existing woods, although these are rarely perceived as habitat creation *per se*. But providing for every single feature of the woodland ecosystem is impossible, while the ecological implications of each new addition became ever more complicated. Ultimately the extent to which woodland reconstruction is done will have more to do with visual and aesthetic appeal, and, as always, cost.

References

Anderson, M.A., 1979, 'The development of plant habitats under exotic forest crops', in S.E. Wright and G.P. Buckley (eds.), *Ecology and Design in Amenity Land Management*, Wye College, University of London, Kent, pp. 87–109.

Brown, A.H., Oosterhuis, L., 1981, 'The role of buried seed in coppice woods', *Biological Conservation,* **21**, 19–38.

Buckley, G.P., 1988, 'Soil transfer as a means of relocating woodland', Channel Link Studies report No. 4, Wye College, University of London.

Cole, L., 1982, 'Does size matter?' in A. Ruff and R. Tregay (eds.), *An Ecological Approach to Urban Landscape Design*, occasional paper No. 8, Department of Town and Country Planning, University of Manchester, 70–82.

Coombe, D.E., 1966, 'The seasonal light climate and plant growth in a Cambridgeshire wood', in R. Bainbridge, R.C. Evans and O. Rackham (eds.), *Light as an Ecological Factor*, British Ecological Society Symposium 6, Blackwell Scientific Publications, Oxford,. pp. 148–66.

Donelan, M., Thompson, K., 1980, 'Distribution of seeds along a successional gradient', *Biological Conservation,* **17**, 297–311.

Eardley, J.P., 1973, 'Forestry: a tool for creating mixed woodland', *Landscape Design,* **103**, 34–7.

Euarson, R.S., Craven, S.R., 1982, 'Herb re-establishment in a beech-maple woodlot, *Restoration and Management Notes*, Vol **1** no. 2, 15–16.

Harris, M.J., Kent, M., 1987, 'Ecological benefits of the Bradford-Hutt system of commercial forestry: II. The seed bank and the ground flora species phenology', *Quarterly Journal of Forestry*, Oct 1987, 213–24.

Hill, M.O., Hays, J.A., 1978, 'Ground flora illumination under differing crop species in a Forestry Commission experiment in North Wales', CST Report No. 171, Nature Conservancy Council, Peterborough.

Hill, M.O., Stevens, P.A., 1981, 'The density of viable seed in soils of forest plantations in upland Britain', *Journal of Ecology,* **69**, 693–709.

Knight, D.G., 1987, 'Woodland wildflower mixes: an evaluation', unpublished M.Sc. thesis, Wye College, University of London, Kent.

Livingstone, R.B., Alessio, M.L., 1968, 'Buried viable seed in successional field and forest stands, Harvard Forest, Mass.', *Bulletin of the Torrey Botanical Club,* **95**, 58–69.

Olmstead, N.W., Curtis, J.D., 1947, 'Seeds of the forest floor', *Ecology*, **28**, 49–52.

Peterken, G.F., 1981, *Woodland Conservation and Management*, Chapman & Hall, London.

Rodwell, J., 1986, 'National Vegetation Classification: woodland communities, a preliminary conspectus', (unpublished) University of Lancaster.

Scott, D., Greenwood, R.D., Moffat, J.D., Tregay, R.J., 1986, 'Warrington New Town: an ecological approach to landscape design and management', in A.D. Bradshaw, D.A. Goode and E.H.P. Thorp (eds), *Ecology and Design in Landscape*, British Ecological Society Symposium 24, Blackwell Scientific Publications, Oxford, pp. 143–60.

Sowter, F.A., 1949, 'Biological flora of the British Isles. *Arum maculatum* L.', *Journal of Ecology,* **37**, 207–19.

Stirreffs, D., 1985, 'Biological flora of the British Isles. *Anemone nemorosa* L.', *Journal of Ecology,* **73**, 1006–20.

Thompson, K., Grime, J.P., 1979, 'Seasonal variation in the seed banks of herbaceous species in ten contrasting habitats', *Journal of Ecology,* **67**, 898–921.

Tregay, R.J., 1985, 'A sense of nature', *Landscape Design,* **156**, 34–8.

Wells, T.C.E., Bell, S., Frost, A., 1981, *Creating Attractive Grasslands Using Native Species*, Nature Conservancy Council, Shrewsbury.

Wittkugel, U., 1988, 'Urban park management in Berlin', *Landscape Design*, **171**, 23–6.

15

Reconstructing freshwater habitats in development schemes

C.J. Bickmore and P.J. Larard

Introduction

This paper reviews some of the practical problems encountered in the creation of freshwater habitats, especially those constructed within development schemes. Here, provision for nature conservation interests is often only a chance by-product and may be overridden by commercial desires or functional needs. The scope for purposely incorporating habitat reconstruction projects into such schemes is illustrated and the problems encountered in achieving successful implementation discussed.

The first part of the paper briefly reviews the nature of water-related development projects in relation to the potential for active habitat reconstruction. This is followed by an appraisal of the key factors which should be considered at the planning and design stage of schemes. Finally, a more detailed discussion of the practical aspects of creating schemes on the ground is presented.

The nature of commercial freshwater schemes

The continuing loss and degradation of natural wetland habitats in Britain and many other parts of the world has increasingly focused attention on developing reconstruction techniques for these habitats. Over the last hundred years or so, artificially created lakes and marshes have become, often by chance, common components of many nature reserves, local community wildlife projects and conservation schemes. Great potential exists to design features of nature conservation interest into a variety of projects where the primary aim is development for functional or recreation needs. Water bodies may be included in a diverse range of projects associated with highways, housing and retail complexes, leisure facilities, mineral extraction, power generation and water supply schemes. Such developments vary considerably in the size of their water component, ranging from small (say 1.0ha) flood-storage reservoirs, to medium-sized water sports lakes (say 7.0ha), designed to accommodate water-ski runs, rowing courses or yachting facilities, to major reservoirs for storage or energy purposes (up to 200ha plus). The size and nature of a project may influence the scope for habitat reconstruction, and the degree of conflict arising in site use and management.

The water aspects of development schemes usually fall into the following functional categories:

1. drainage/flood storage – commonly associated with road schemes and building (e.g. housing) development in the form of balancing ponds;
2. water supply – as a storage reservoir for public abstraction and irrigation purposes;
3. power generation and industrial purposes – a water feature created to supply hydroelectric stations or to provide a station cooling system;
4. after-use following mineral extraction or excavation.

Other objectives include the provision of recreational facilities involving active water sports (e.g. bathing, fishing, boating, sail-boarding or water-skiing) or simply the creation of a visually attractive feature (as a backdrop or focal point of a new development). While, in some instances, these may be primary objectives in their own right, they are often only of secondary consideration. Nature conservation is usually low on the list of priorities and often occurs as an incidental consequence of the provision of a new water feature.

Although one development objective usually takes priority in any scheme, a combination of objectives is often possible. If opportunities to include a positive habitat reconstruction input are to be realised, contributions from ecologists, landscape scientists or managers must be made early on, at the planning and design stage. A water-based scheme which attempts to combine a variety of features, including specific habitat reconstruction techniques, benefits from a multi-disciplinary approach. For instance, it requires co-ordination and understanding between engineers (civil and drainage), architects, biologists, planners, statutory authorities, local government, land owners and local wildlife organisations. This is particularly important when potentially conflicting site functions are envisaged.

Using an iterative design process, a cohesive development strategy and site design can be achieved. Often site factors considered 'problems' by one discipline can be used to the benefit of another. For example, surplus 'unsuitable' fill from one part of a site may be used to advantage around a water body to create islands or graded profiles at the water's edge.

The following section highlights a variety of factors which should be identified at the feasibility and planning stages of a water-based development. These factors which will directly affect development proposals and the synthesis of the site design are summarised in Figure 15.1.

Feasibility planning and design

The influence of scheme function

The overall objective of a development scheme will define the need for and the nature of its water component, often determining the basic layout and form. Where new water bodies are created primarily to fulfil engineering requirements, modifi-

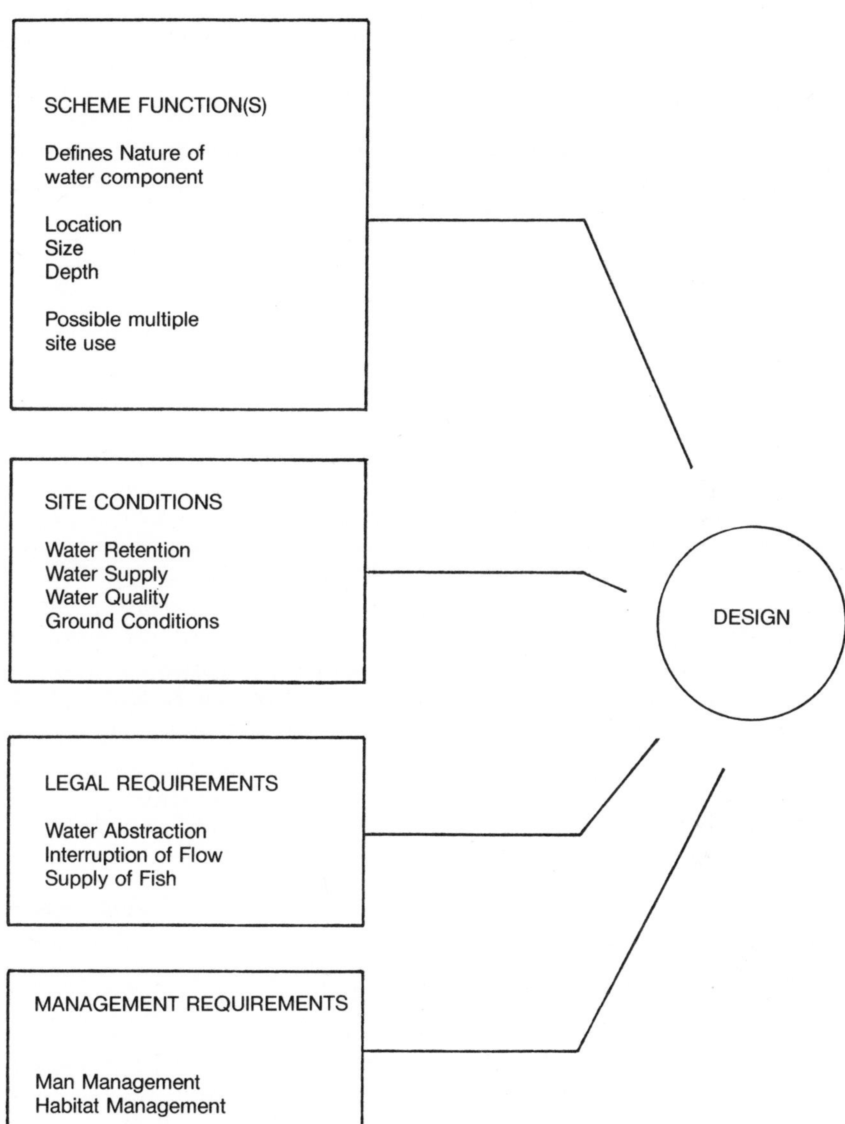

Figure 15.1 Key factors which influence site planning and design.

cation of such schemes may be possible, broadening initial objectives to include a nature conservation input. The latter should have defined objectives e.g. whether to attract particular 'target' species, or to provide an area of general local interest or educational benefit.

Less specialised requirements are usually more easily achieved. For instance, many balancing reservoirs required for the drainage of new road schemes are designed as relatively small, uniform structures often with a trapezoidal cross-section and hard surfacing. Some relatively small changes in their design make them of greater benefit for wildlife, generally providing a site of local interest. Such changes could involve the omission of a hard surface lining, and the over-deepening and extension of the pond to create a variety of water depths and slope profiles while still maintaining the required water storage capacity. In most cases such provisions are relatively easy to achieve provided they are considered at an early stage.

Site functions which affect factors such as size and depth of water body therefore need to be established at the planning stage, since they will influence the design approach (planting, site use) and long-term management requirements. On sites where mineral extraction (e.g. of gravel, clay) is the primary development objective, achieving sufficient overall water depth and variation of topography for a new wetland feature is not usually a problem if considered at an early stage. On many existing worked-out extraction sites the resulting deep and steep-sided excavations are often safety hazards where public access is considered. Conversely, shallow mineral extraction sites, where the cost of excavation, ground modelling and disposal of unwanted spoil may restrict water to depths of less than 2.0m, can limit the ultimate habitat variety achievable. This can also pose long-term site management problems in relation to the need for regular clearance of vegetation and silt.

Water features associated with housing, retail or leisure developments, mineral extraction restoration schemes and many water supply reservoirs, commonly aim to provide multiple interest and to maximise high recreation usage. Here conflicts arise not only in the construction design but in the long-term use of the area; for example when public activity and the provision of 'wildlife' areas coincide.

Problems arising from site conditions

The potential of a site for freshwater habitat reconstruction depends on the prevailing physical and chemical conditions such as topography, ground substrate, hydrological, geotechnical and water quality conditions, and the existing and/or proposed surrounding land-use and vegetation. Preliminary site investigations to determine these factors need to be undertaken at the planning stage, followed by more detailed appraisal prior to final design. Local topography and hydrological considerations influence the initial engineering problem of how and where to site a new water feature and how to ensure that it will receive and retain water or be assured of regular inundation. In turn the choice of siting may affect existing wetland habitats and influence future habitat reconstruction proposals.

Some sites naturally hold water if excavated to sufficient depth, e.g. to the ground water or an impermeable stratum. Others require a liner to ensure year round water retention. Ideally this should be a natural lining such as puddled clay, but this may not be feasible if none is locally available. Alternative artificial liners such as polythene, butyl rubber or bentonite matting have varying cost implications. Some artificial liners can be punctured by plant species with thick piercing roots, and can restrict planting design around a water body. The cost of an artificial liner may also limit the flexibility of the design; for example, if large water habitats are wanted over and above the basic functional needs, it might be difficult to justify a more extensive liner.

The source of water for both the initial filling of a water body, subsequent 'top-up' (to combat surface evaporation losses) and regular run-off from the surrounding catchment should be established at the start. Similarly, potential fluctuations in water level and periodicity of flooding may determine the potential for such habitats as water meadows, and the location of plant species sensitive to inundation or drying out. Bird-nesting platforms and boat storage areas will also need to be sited where they will not get washed away. Conversely, advantage can be taken of flood channels to create areas of reed beds, or fluctuations in water levels can be used to positive benefit to provide a muddy shore area for waders (see Hill, in this volume).

Local water quality should also be investigated to establish any potential sources of chemical pollution, suspended solids and nutrient loading. Such information may be available from the Water Authority, particularly when the expected source of water is a nearby river. The nearest monitoring station may, however, be several miles from a new pond location and supplementary sampling will be necessary.

Schemes are often subject to nutrient-rich run-off from the surroundings. This applies to both rural and urban catchments, with agricultural run-off often being a major contributor in the countryside. Regular influxes of run-off high in phosphates and nitrates pose continual site management problems, and may also necessitate reconsideration of the water source. Problems of pollution from industrial chemicals, sewage, silage, suspended solids or road run-off may also exist despite attempts to curtail these by legislation (e.g. Control of Pollution Act 1974). Some can be alleviated at the design stage by incorporating filter systems (e.g. oil interceptors, petrol traps), settling-out lagoons, installing weirs and restricting the number and location of run-off inputs. Plants such as the common reed (*Phragmites communis*) can also provide a valuable living filter system to remove excess nutrients and heavy metals from water. This is achieved in part by promoting microbial activity in the reed root zone which helps to break down nutrient compounds. The encouragement of wildfowl and the addition of fish and water snails will also help to maintain water quality by controlling troublesome insects (e.g. midge fly larvae) and algal growth. A reed bed will also trap silt and restrict silt accumulation to a localised area. These functions may prove useful in a variety of schemes enabling wildlife habitat reconstruction to provide both functional and nature conservation benefits.

Future site management

No new habitat reconstruction scheme should be designed without a view to both the establishment and long-term maintenance requirements. Where practical management is limited it may influence the initial design, possibly restricting the range of habitats proposed and plant species chosen. Thus, easily established and potentially invasive species may need to be restricted to areas where they are naturally limited by ground formations, water levels or other environmental conditions, or omitted entirely. Some schemes may require special management provisions for particular site uses. For example, if multiple site use is envisaged, wardening might be required to discourage public access to sensitive wildlife areas.

Preparation of an outline site management plan, identifying the major tasks required and estimating long-term inputs and costs, should form an important part of the early planning. The landowner or developer should be made aware of the future management commitments for the site, particularly those involving capital expenditure and the purchase (or hire) of specialist equipment (e.g. boat-mounted weed cutters, or use of aquatic herbicides). It may be possible to hand over the long-term management of a site to a local wildlife group who may then develop the site further for wildlife benefit.

Synthesis of the design

Once scheme function(s), site conditions, legal and management requirements have been investigated, the design can be drawn together. This is usually a two-stage process, first establishing the layout of the basic site elements, and secondly the superimposing detailed designs of habitat types or site features.

Where a scheme attempts to fulfil several objectives, the layout should be designed to minimise potential conflicts. One way of achieving this is to zone different activities after drawing up a table or matrix to consider the compatibility of various site uses. For example, provision can be made for areas where recreation pressure is permissibly high, while reserving other parts of the site as undisturbed refuges where habitat reconstruction techniques can be maximised. However, this does not exclude measures to encourage wildlife throughout the site.

Although zoning is possible on a small site, generally the technique is much more successful on a larger scale. Figure 15.2 illustrates a water-theme development for a post-extraction mineral restoration scheme. Here, uses have been zoned in order to isolate sensitive wildlife habitats from the more active recreation pursuits (e.g. sailing, water-skiing). Sports such as fishing can be controlled by the provision of special fishing platforms, or restricting access to prevent bank degradation and the destruction of waterside vegetation.

On sites where regular public use is envisaged, safety factors may limit scope for habitat variety. The range of water depths needed to maximise plant and animal colonisation, or conversely to prevent the spread of invasive, emergent plant species such as reed-mace (*Typha latifolia*) and *Phragmites communis*, may be dictated by safety requirements. Sudden changes in water depth and overall depths greater than 1.5m are often a problem where children have access to the water. This further supports the case for separation of activities where possible. Control of sporting and waterside activities may also help to reduce the impact of potential minor water pollutants (e.g. boat engine petrol, run-off from car parks, litter accumulation).

As is the case in other types of habitat reconstruction, a comprehensive knowledge of wetland communities, and the autecology of their dominant species is a prerequisite. Where possible, a complete framework of plant material should be planned, corresponding to the natural gradation from dry land and marginal species, through emergent vegetation to floating-leaved and free-floating plants of open water. Their zones correspond principally to water depth variation, with certain species being characteristic to definable depths.

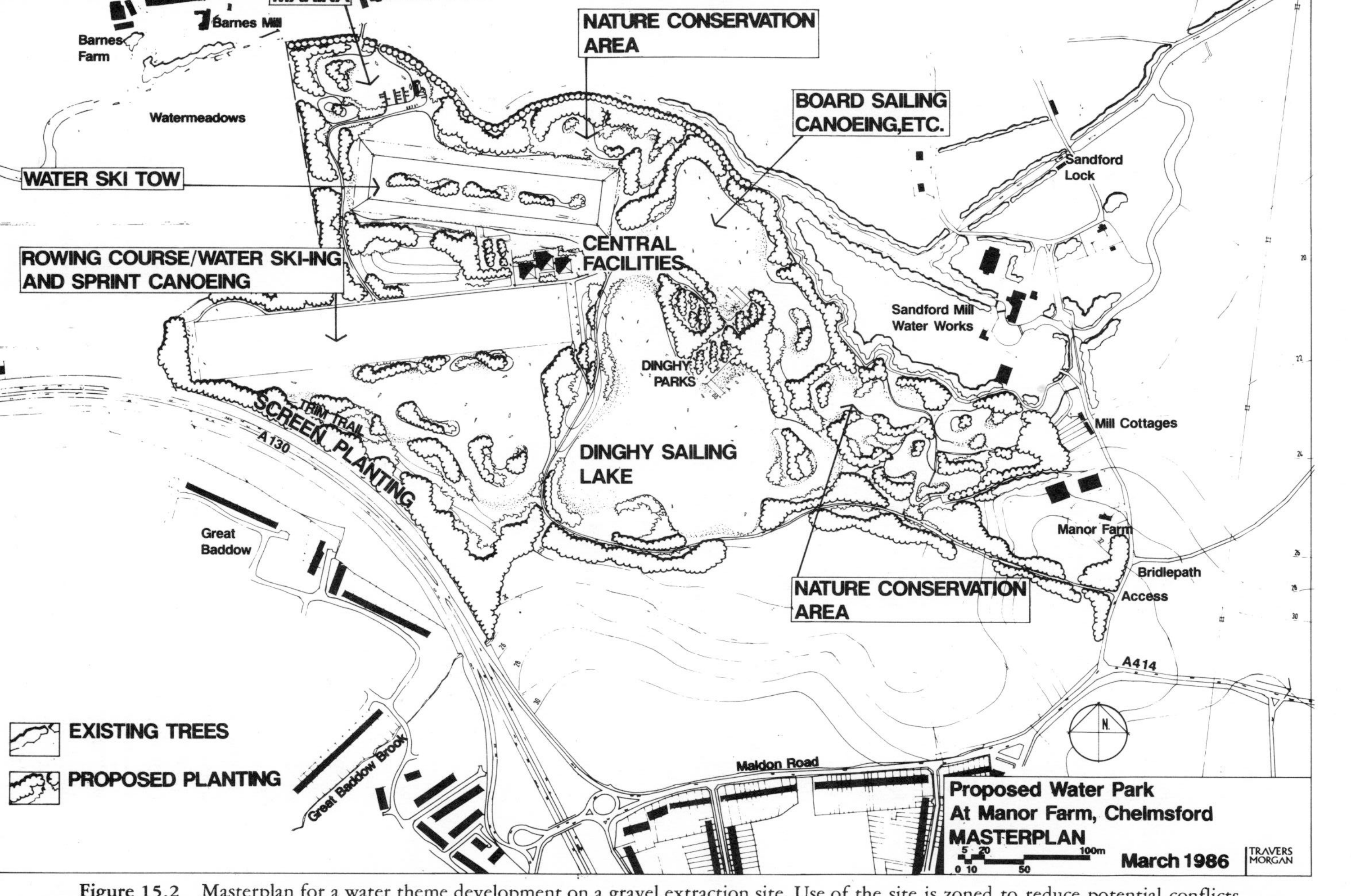

Figure 15.2 Masterplan for a water theme development on a gravel extraction site. Use of the site is zoned to reduce potential conflicts.

The scope for topographic variation in and around a water body will, therefore, influence the variety of habitats available and the palette of usable plant species. Where variation in water depth is limited and overall depths are shallow (e.g. less than 2.0m), plant species which are naturally invasive and competitive (e.g. *Phragmites communis, Typha latifolia*) must be positioned with care. These species are usually rapid colonisers and often dominate the water's edge down to 1.0–1.5m. Where maximum depths are not much greater than this, open water features may eventually become clogged with emergent plants. These species provide valuable wildlife habitats, but areas of open water are desirable aesthetically as well as to increase overall habitat variety (e.g. in providing wildfowl flight paths). This problem is often resolved by limiting the planting of potentially invasive species to restricted areas and planning a programme of long-term control and maintenance.

Once the basic layout and form have been established, the availability of plants and other materials for habitat reconstruction purposes should be investigated. Natural colonisation is one solution which can be designed for, but the time-scale for vegation development may be too long and there is little control over the species which come in. Thus, although consideration should be given to the natural colonisation, the majority of schemes will use introduced plant material. The commercial availability of native aquatic plants is somewhat restricted, so removing them from other water bodies is often the best way to stock a new water feature. Local Water Authorities, if contacted early in the design stage, may be able to advise on availability and can sometimes supply plant material from their routine river maintenance operations.

Local wildlife organisations may also be able to advise on sources, but care must be taken not to violate statutory legislation relating to the removal of wild plants (e.g. Section 13, Wildlife and Countryside Act, 1981). Some Water Authorities possess wetland nurseries from which material can be purchased. As local availability may limit the choice of plant species, habitat reconstruction can only aim to provide an 'average' rather than 'special' site of nature conservation interest. It offers an initial framework on which nature can build with time.

Areas within a water body identified as subject to possible future erosion (e.g. from natural wave action or recreation impact) will require special protection. Sometimes there is little scope for creating a 'natural' edge, and to combat this a 'hard' solution (e.g. brick, concrete or a suitable geotextile) may be selected. Often, however, plant materials can be used to provide shore protection and also create a valuable wildlife habitat: suitable species on banksides are soft rush (*Juncus effusus*), reed canary-grass (*Phalaris arundinacea*), alder (*Alnus glutinosa*) or willows (*Salix* spp.). Stands of such vegetation will also divert users from vulnerable areas but provide an attractive water's edge for them to enjoy. Vegetated geotextiles or gabion baskets may also be considered.

Providing the objective is not to encourage invertebrates, e.g. Odonata, many freshwater schemes benefit from stocking with fish, either as a basis for angling or to provide a framework of animal life which will aid in controlling invasive vegetation and maintaining water quality. Consultation with the relevant Water Authority and local angling societies will identify local availability and procedure for fish supply. Where this is done early in the design and specification stages, it is usually possible to ensure both varied and adequate fish stocks.

Getting the scheme on the ground

The work-force

For most commercial schemes, works are undertaken by specialist building, earth-moving or landscape contractors. In many cases these specialists are not sufficiently experienced to ensure successful reconstruction of wetland habitats, particularly in the handling, planting and maintenance of aquatic plants, and the requirements need to be stated precisely (see *Specification of operations*).

On-site assistance can be obtained from organisations with experience of habitat reconstruction work (e.g. Water Authority, local authority conservation officers, or local conservation/wildlife organisations). In many cases, such help may be voluntary and can help to reduce overall budget costs. However, this may cause problems where contract labour is already being used, for example to implement engineering works or ornamental planting, and can lead to confusion of responsibilities.

Choice of contract

Most works to create freshwater habitats which are associated with commercial development schemes will require standard contract documentation. This provides a binding and legal agreement which helps to clarify responsibilities and to co-ordinate the phasing of the works.

In the United Kingdom most external works projects are conducted under engineering or building works contracts. Often, creating a new water habitat is only part of a scheme which incorporates wider civil engineering or building aspects, and therefore determines the form of contract used. For predominantly landscape works (ground preparation and planting), a special standard form of contract already exists in the UK. This is the Joint Council for Landscape Industries (JCLI) Standard Form of Agreement for Landscape Works which is particularly tailored to the special requirements of dealing with plant materials.

Conversely, projects which include a large civil engineering input (road schemes, dams, bridges), or large amounts of earth moving (often the case where new water bodies are created), are often executed under the Institute of Civil Engineering (ICE) Form of Contract. This generally gives the 'engineer' wider ranging and more strictly defined powers of control than the JCLI Form. In this situation landscape works associated with habitat reconstruction may be carried out by the main (engineering) contractor or a subcontractor. A standard form of subcontract exists for these circumstances.

Suitable contracts for habitat reconstruction works ultimately depend on the overall size of a scheme and its various components. Various facets of a project might be split into separate contracts which may be under different forms, so that site preparation, including major earth moving, extraction and filling of a water feature is carried out under an engineering contract, while minor ground modelling and planting works is completed in a separate landscape contract. This has important consequences as regards phasing and execution of operations on the ground.

Specification of operations

The scheme layout, materials required, methods and quantities are conventionally set out in Specification and Bill of Quantity documents and working drawings. Ideally, the specification should be a 'performance' specification detailing the type and standard of materials to be used and the required end result. Methods of working can be detailed but should be clear, testable and flexible enough to suit unforeseen circumstances. Over or under-specifying an item can lead to unnecessarily high prices being quoted for the works. Specification writing is thus a skilled discipline, requiring a thorough knowledge of how a job will be phased and the tolerances which are permitted. It may be possible to develop tried and tested specification clauses for works on wetland sites, but each site has its own particular requirements and 'standard' instructions for certain operations may be difficult to apply.

In practice, many specialist tasks such as handling and planting wetland plants can be adequately described in the documents and reinforced by careful supervision in the field. If an experienced ecologist or landscape scientist is not able to be present, detailed background notes and briefings should be prepared for the supervisor on site.

The Bill of Quantities is a measurement of the extent of works to be carried out. It usually itemises the required tasks and materials which the tendering contractor has to price. For example, where extensive areas of marginal or emergent vegetation are envisaged it is useful to obtain a rate for planting per m^2.

SITE PREPARATION

Engineering contractors who are involved in major earth moving and ground modelling often have limited experience of creating water features for nature conservation. In practice it is a fairly simple task to specify the contours of a pond and to provide varied depths and side slopes. However, this requires an appreciation of the large-scale nature of such works and the limited degree of accuracy achieved by large earth-moving machinery. A useful way of specifying the required levels in and around a water body is to use plus and minus tolerances. This ensures an irregular finish rather than uniform grades (see Figure 15.3). A skilled machine operator can make smaller-scale local regrading adjustments as required. The cost of this may be estimated on a time basis, however, and can be expensive.

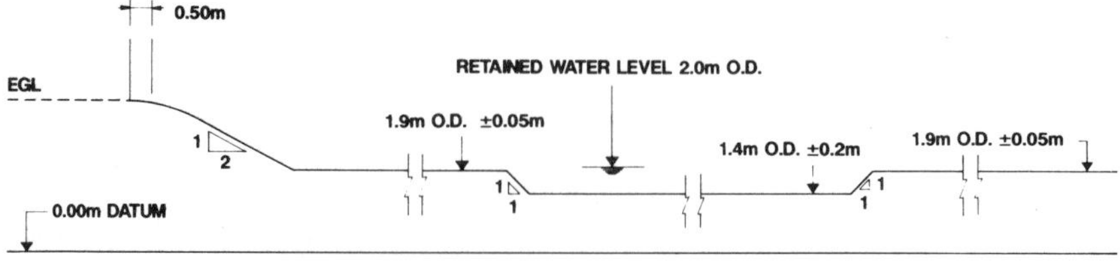

Figure 15.3 Detailing of lake profile specification. Plus and minus tolerances allow flexibility for machine operators and ensure an irregular surface finish.

PLANT MATERIAL

Contract specifications may group plants by species name and size range. Where plants are to be obtained from a nursery, precise numbers can be quoted and their proposed locations indicated on a planting plan. If plants are to be obtained from local river-dredging operations, the source may be nominated and the contractor instructed to make prior arrangements with the relevant supply authority (e.g. Water Authority). If the source is not nominated, details of the plants' location and prevailing soil type should be provided by the contractor for approval by the Supervising Officer. Since the exact composition of dredgings may be uncertain, some flexibility of specification should be allowed for.

When species are to be established by transplanting clumps, the sod size required will need to be defined. This may be determined in practice by the tolerance of sod size to damage from mechanical or manual handling, as plants in the centre of larger sods are less likely to be damaged. For preference, delivered material should be sorted into species groups, labelled and adequately protected until planting takes place. It may be necessary to put a time-limit on the storage period to avoid deterioration of the plant material.

Apart from local river dredgings, a source of plant material might be a nearby stream, river or pond which may be destined to be rerouted or relocated. Dependent on the phasing of works, material can be transferred from old to new sites by bucket excavator. This can be carried out by direct on-site supervision with an estimate of the machine time required included in the contract documents.

PLANTING OPERATIONS

Specifications should include a definition of the planting season. The autecology of particular species will determine the best period for transplanting to take place, (e.g. February and March for emergent plants, when the rhizomes are dormant but are about to shoot). Documents and plans should clearly describe or illustrate required plant positions. Final positioning of plants can be supervised on site, to ensure placement in the correct plant zones. This is particularly important if the contractor is unfamiliar with the plant material concerned.

Planting techniques will depend on the type of plant considered. These are not considered in detail here, but generally the introduction of free-floating, non-rhizomatous submerged and floating rooted plants is relatively easy. Free-floating plants can simply be added to the water surface while submerged, and floating-leaved plants can be weighted and dropped into the water at the desired place: some nurseries even supply plants ready weighted for this purpose. Rhizomatous plants (e.g. emergent species) can be more difficult to establish. One method of propagation is by transplanting bundles of rhizomes or sods of rhizome mat. Species such as *Phragmites communis* can be established by placing rhizomes in long trenches dug below the normal water level, allowing plants to spread up and down the margin. Clumps of emergent plants can also be pit planted.

Emergent plants can also be established by placing bundles of rhizomes in weighted hessian sacks placed just below the water level. Alternatively *Phragmites communis* can be established from cuttings rather than rhizomes. These are placed in a narrow trench above the water line with about a third of the cutting remaining exposed so that the plant roots from the leaf nodes. New planting, particularly at the water's edge, may be easily dislodged by wave action. This is especially so in large lakes or where there is likely to be early recreation pressure (e.g. boating, water-skiing). It may be necessary therefore temporarily to 'de-water' a planting area to encourage plant establishment, or to provide temporary protective 'booms' or rows of sand bags to reduce wave pressure. In other cases it may be desirable to build a temporary dam in order to raise water levels and encourage the establishment of some emergent/marginal species.

Conclusion

There is much scope for the positive incorporation of habitat reconstruction projects within freshwater development schemes. This can be achieved by providing a basic framework of suitable site conditions throughout the scheme, which will attract an influx of water-associated plants and animals. Provision for the creation and management of specific habitats and the encouragement of particular species is also possible.

For the majority of schemes a multi-disciplinary approach is invaluable, with ecologists and landscape scientists or managers being involved in early decision-making and site planning. The synthesis of a well thought out, cohesive design is essential and a thorough understanding of development functions, site conditions, potential site use conflicts and future management requirements will provide the basis for this.

Successful practical implementation of habitat reconstruction projects within a development scheme may often rely on work carried out by non-ecologists who have little knowledge of wildlife needs (e.g. engineering contractors, landscape contractors). The value of a comprehensive Specification for the works carried out under contract is critical, although flexibility in implementation methods and the need for simple instructions is however essential. Once a basic framework has been established it may be desirable to involve specialist groups (e.g. local wildlife groups) in more detailed site manipulations for wildlife needs, with possible wardening of the site to co-ordinate management and intepretation.

16

Wild flower seed mixtures: supply and demand in the horticultural industry

Richard J. Brown

Introduction

A variety of methods are employed today in the enhancement of the environment for the benefit of both people and wildlife. Although at one time habitat reconstruction was mainly undertaken by voluntary wildlife groups, many of the initiatives are now taken by professionals within local authorities and within the horticultural industry. The creation of wild flower meadows is an increasingly popular option, because results can be achieved within a relatively short time, and the seeding and subsequent maintenance of such areas can be based on traditional grassland management techniques. For these reasons grassland reconstruction can be considered on a wider scale than, for example, the construction of pond or woodland habitats.

The potential for using native wild flower species in amenity landscaping was first brought to the attention of the UK horticultural industry in the early 1970s. Changes in public attitudes towards the environment, and pressure within public authorities to reduce the costs of maintaining grassland, have stimulated interest to the point where 'wild flower meadows' (or floristically diverse grassland) are now an accepted extension to the range of landscape options for areas of amenity grass. As four per cent (8,500km^2) of the UK land surface is covered by amenity grass, i.e. 'all grass with recreational, functional, or aesthetic value, and of which agricultural productivity was not the primary aim' (NERC 1977), the scope is large. One major category making up 55 per cent of the total consists of less intensively managed untrampled and semi-natural grassland with potential for wildlife management, including urban parks, golf 'rough', road verges, country parks and picnic sites and building surrounds. Areas which are newly created (e.g. new road verges, golf courses etc.) offer the best opportunity for habitat creation. Approximately one per cent (9,000ha) of the area of amenity grassland is seeded or reseeded each year. The total cost of maintaining amenity grass runs to hundreds of millions pounds, of which just over one half is the responsibility of public authorities.

The first commercial response to the concept of creating new flower-rich meadows was the marketing of 'Nature Conservation Blends' for use as alternatives to the usual grass, or grass/clover mixtures. These early mixtures were unsatisfactory as seed of appropriate native plants was not available at the time. Agricultural and foreign varieties were used which were highly competitive, produced large amounts of herbage and tended to be short lived, with species such as *Chicorium intybus* (chicory) and *Papaver somniferum* (opium poppy) often being included. Fortunately, during this time a study to review the feasibility of producing seed of native grass and wild flower species was carried out by the Institute of Terrestrial Ecology for the Nature Conservancy Council (NCC). This culminated in the publication of a booklet *Creating Attractive Grasslands Using Native Plant Species* (Wells *et al*., 1981) which has served as a blueprint for further developments of the concept over the last seven years. Having established that many of the most appropriate native species could be successfully grown as seed crops using established horticultural methods, the development of better and larger-scale techniques of seed production has now moved into the hands of commercial seedsmen.

Several hundred wild flower and grass species are now offered commercially (albeit some in only very small quantities) in ready-made seed mixtures, or as individual items. Although the available range represents only a proportion of the total number of species occurring in British grasslands, the landscape industry has an extensive and complex range of possible choices when deciding upon the appropriate species combination for any given site. As seedsmen we face this daunting range of new crops, all needing simultaneous development. The traditional amenity grass market by comparison uses eight species and 80 per cent of sales are of two species: ryegrass (*Lolium perenne*) and red fescue (*Festuca rubra*).

If we are to define the role of the horticultural industry in this area of habitat creation we must assess the requirements of users involved in habitat creation, both in terms of the type of products needed and the scale of the demand, and combine this information with our knowledge of the limitations of seed supply and the techniques available for establishing wild flowers.

Consumer demand for wild flower mixtures

In the spring of 1987 we mailed 2,000 questionnaires to customers of our amenity products. We received 135 replies (*c*. 7 per cent), which, given the wide target group, was the level of response we had expected. Approximately 85 per cent of the response came from customers who had experience of using wild flower seed mixtures, and were thus motivated to participate. The small response from customers with no experience of wild flower mixtures was disappointing, and the results should be interpreted with this imbalance in mind. Figure 16.1 shows the origin of replies classified into the main customer groups included in this survey.

The questionnaire was divided into two parts: the first asked respondents about criteria important to the specification of wild flower seed mixtures, while the second asked those familiar with wild flower seed mixtures about their experiences (see Appendix 16.1). In addition to the questionnaire, we collected together details of wild flower seed orders received in 1985–6 in an attempt to answer questions such as: who buys wild flower seed, on what scale is it being used and in what form is the seed ordered?

Two aspects of demand for wild flowers are relevant to this discussion: the size of current demand (i.e. the size of the market, the size of individual wild flower orders/schemes and the sources of these orders) and the nature of the demand (i.e.

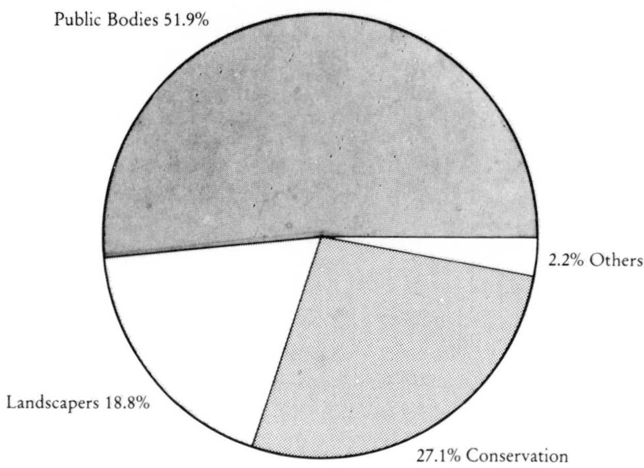

Figure 16.1 The origin of questionnaire replies.

the type of products required – their intended purpose, content, complexity and cost).

Market size

The wild flower seed market in 1982–3 was estimated at about three tonnes of mixture, containing 15–20 per cent wild flower seed (Wells, 1984) and sufficient to seed about 85ha. Demand since 1982 has expanded at a rate of 20–50 per cent each season; but this growth has to some extent been limited by the seed availability, so that sales figures may not represent the true size of the market. Accurate figures for the current size of the market are not available, but must be in the region of 10 tonnes (1987–8), sufficient seed for up to 300ha with a value of about £350,000. To put this in context, wild flower seed accounts for only a very small proportion of the amenity grass market in which around 8,000 tonnes of seed is sold annually in Britain, with a value of about £15–20 million. The potential for growth within this market, given the level of interest in developing wild flower seed use and changes in agricultural policies, is enormous.

The size of individual seed orders

The majority of wild flower schemes/orders are small. The general pattern of orders has not changed significantly from that reported for 1982–3 (Wells, 1984), when 75 per cent of orders/schemes involved less than eight kg of seed, sufficient for only 0.2ha (see Figure 16.2). There is, however, an underlying trend for an increase in the number of more ambitious schemes as users, particularly in local authorities, become more confident of the techniques involved. Larger schemes, while small in

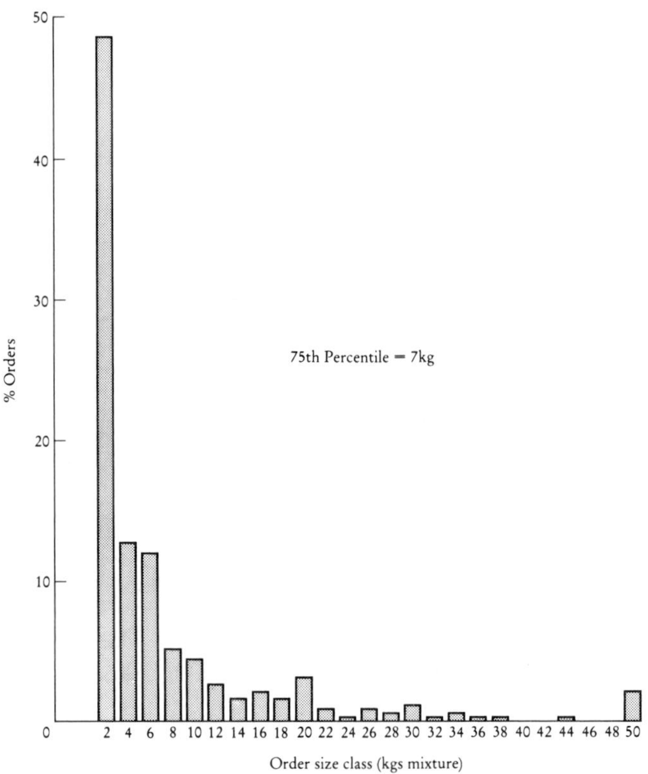

Figure 16.2 Frequency distribution of wild flower seed orders, 1985–6.

number, can account for a significant proportion of the seed used; in 1986 for example, although only 20 per cent of orders we received were for more than 8kg, they accounted for 75 per cent of the wild flower seed sold.

Sources of demand

The main customer groups ordering wild flower seed in 1985 – 6 were as follows: public authorities, landscapers (landscape architects, landscape contractors etc.), conservation groups (wildlife trusts, Farming and Wildlife Advisory Groups, etc.) and others (private individuals, farmers, etc.). Figure 16.3(a) shows the proportion of our orders originating from each of these groups, indicating where the interest lies, whereas Figure 16.3(b) shows the destination of orders by quantity to each of the customer groups. Comparison of the two pie diagrams shows that whilst on an

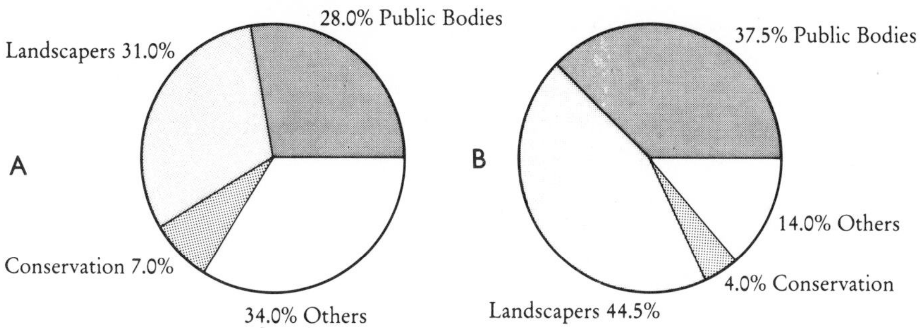

Figure 16.3 The origin (a), by number and destination (b), by weight of wild flower seed orders in 1985–6.

order basis sales originate in approximately equal proportions, public authorities and landscapers are by far the most important groups of customers, accounting for 82 per cent of wild flower seed use. On average these two groups undertook schemes requiring two to three times the quantity of seed used by those in the 'others' group. Conservation groups accounted for only a small percentage of direct sales, but may have been involved in other sales in an advisory capacity. The information acquired through sales figures should be interpreted with caution as it is only truly a measure of buying patterns and to some extent purchasing power. Seed bought by a landscape contractor, for example, may be bought following a specification from any of the other customer groups, and the seed ordered is not always used in a single scheme.

Motives of consumers

Some knowledge of the motives of customers using wild flowers is useful in predicting current and future trends in the type of wild flower seed products required. In our questionnaire we asked customers to rate a number of set reasons and benefits for using wild flower seed. The highest ratings were given to the desire to create environmental interest/variety and to creating wildlife interest (see Table 16.1). Other factors were of secondary importance, or given greater importance by certain customer groups (e.g. the need to create educational interest). The potential for reduced maintenance (mowing) on areas sown with wild flower seed when compared with conventionally mown grass areas was rated as being unimportant by a significant number of respondents (see also Funnell, in this volume).

Seed specification

In our questionnaire we asked customers to select from four wild flower seed mixtures the one most likely to meet their future requirements:

1. A complex mixture of more than 20 native species costing £35/kg (£1,050/ha).
2. A mixture containing 15–20 native species costing £25/kg (£750/ha).

Table 16.1 Responses given to a question relating to the importance of a number of common motives for deciding to use wild flowers

Response:	Very important	Important	Don't know	Unimportant	Total	mean weighted score
weighted score:	(3)	(2)	(0)	(1)		
Creation of environmental interest	86	45	2	0	133	2.62
Reduced maintenance	21	66	9	23	119	1.83
Wildlife interest	86	41	3	1	131	2.60
Education interest	40	60	14	16	130	1.96
Public demand	12	56	25	20	113	1.48
Ability to sow on subsoil	19	8	68	17	112	0.80
Total no. responses:	264	276	121	77	738	(1.9)

Note: The main body of the table gives the number of responses to each structured question. Weighted means are also given for comparison between questions.

3. A simple mixture of 10–15 native species costing £20/kg (£600/ha).
4. A mixture with some agricultural flower species costing £15/kg (£450/ha).

The most popular choice was for a relatively simple mixture of species containing only native wild flowers at a relatively low cost (see Figure 16.4). Cheap mixtures containing agricultural flower species and at the other extreme, more complex, and thus more expensive mixtures, were only selected for a minority of situations. A proportion of the respondents gave more than one selection (mean = 1.37 per respondent), on the basis that different situations would require different mixtures. These demand patterns appear to suggest that both simplicity and cost are major factors influencing mixture specification, but that quality is also important and should not be compromised. Further insights into consumer demand can be obtained by looking at the sales pattern of the main wild flower seed product groups in the Johnson's range (see Appendix 16.2). The mixture groups are as follows (where the sales by weight are given as a percentage):

1. Meadow mixtures suitable for most situations: general purpose mixture and standard meadow mixtures: 60 per cent.
2. Mixtures for more specialised situations (e.g. shade, chalk/limestone): 20 per cent.
3. Supplementary mixtures only recommended for very specialised situations (e.g. pond margins, woodland): < 1 per cent.
4. Custom mixtures, made to individual specifications: 5 per cent.
5. 'Straight' individual wild flower items: 15 per cent.

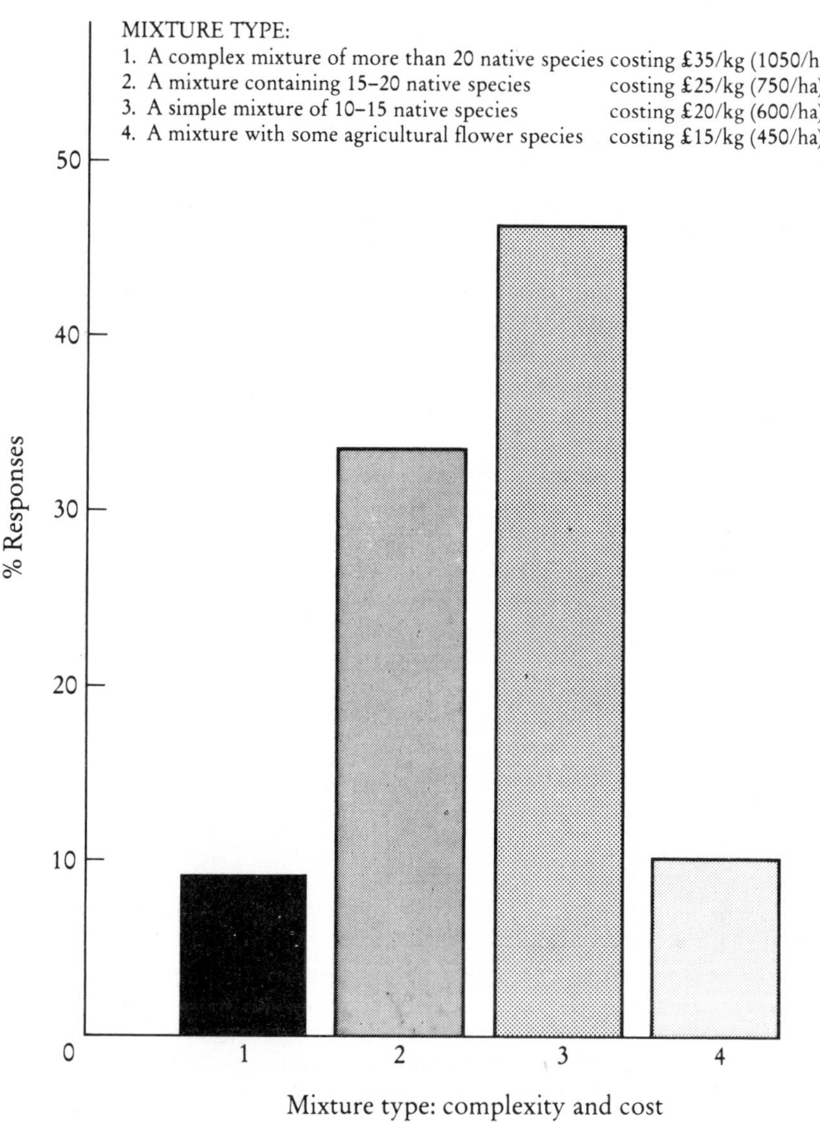

Figure 16.4 Types of mixture chosen as most suitable for wild flower schemes.

Standard wild flower seed mixtures account for approximately 80 per cent of our sales of wild flower seed, showing that the majority of customers are guided by the advice provided by our literature and technical service. This is clearly of great importance, as many customers have limited access to expert guidance within their own organisation, and little, if any, practical experience of using wild flowers upon which they can base their specifications. Our experience from marketing wild flower seed mixtures over the last eight years emphasises the importance of clear and simple mixture recommendations linked with equally clear directives as to their use.

Our 'general purpose' wild flower seed mixture is by far the best selling type. This mixture falls into category '4' of question 3 in our questionnaire (Appendix 16.1), i.e. a cheaper mixture containing some agricultural flower species. This suggests that cost and simplicity (this mixture is suitable for many types of sites) are much more important than the questionnaire response would suggest.

The general purpose mixture currently contains commercial strains of yarrow (*Achillea millefolium*), birdsfoot trefoil (*Lotus corniculatus*) and ribwort plantain (*Plantago lanceolata*), for reasons of both cost and availability, but excludes problematic species like red and white clover (*Trifolium pratense* and *T. repens*) which are common in broadly similar mixtures marketed by other suppliers. We have a policy of replacing foreign and agricultural constituents with more appropriate native species as and when this becomes feasible.

Agricultural clovers are frequently cited as reasons for poor establishment (e.g. Wathern and Gilbert, 1978) because of their smothering effect on other seedlings which are slow in establishing. The undesirable effects of clovers are most pronounced on soils of lower fertility as the nitrogen-fixing ability of these plants exaggerates their competitive advantage. Clovers also have unwanted side effects in the long term by building soil fertility which ultimately favours more vigorous species (e.g. grasses), thereby increasing the mowing requirement.

Product quality

In conventional seed markets, quality is a multiple concept involving both the seed itself (e.g. species purity and germination capacity) and of the varieties used (Thomson, 1979). Both aspects of quality are covered by EEC seeds legislation, which sets out statutory minimum standards for each species. In the absence of specific legislation governing the sale of wild flower seed or mixtures, the need for seed of high germination and purity is obvious, but the concept of varietal quality is less apparent.

Most wild flower seed is sold without any assessment of its purity or germination capacity. Seed quality under such circumstances is governed by the competence of the grower in correctly harvesting and storing seed, supplemented by visual checks on colour and boldness of seed samples. Figures 16.5 and 16.6 summarise data collected for all wild flower seed lots tested in a single year (1985), originating from both our own seed production and from seed bought in from other suppliers. The general quality of seed is quite variable. Certain subjects present particular problems, such as the genera *Succisa, Scabiosa* and *Knautia* of the *Dipsacaceae*, which rarely give good germination even when selectively harvested by hand. Where our seed testing picks out seed with particular quality problems we are sometimes able to improve the quality of the seed lot by re-cleaning and grading. The general trend has been for a steady improvement in standards of purity and germination, but the industry will never be able to match the high standards expected with cultivated species.

Wild flower seed growers are saddled with a number of inescapable restrictions.

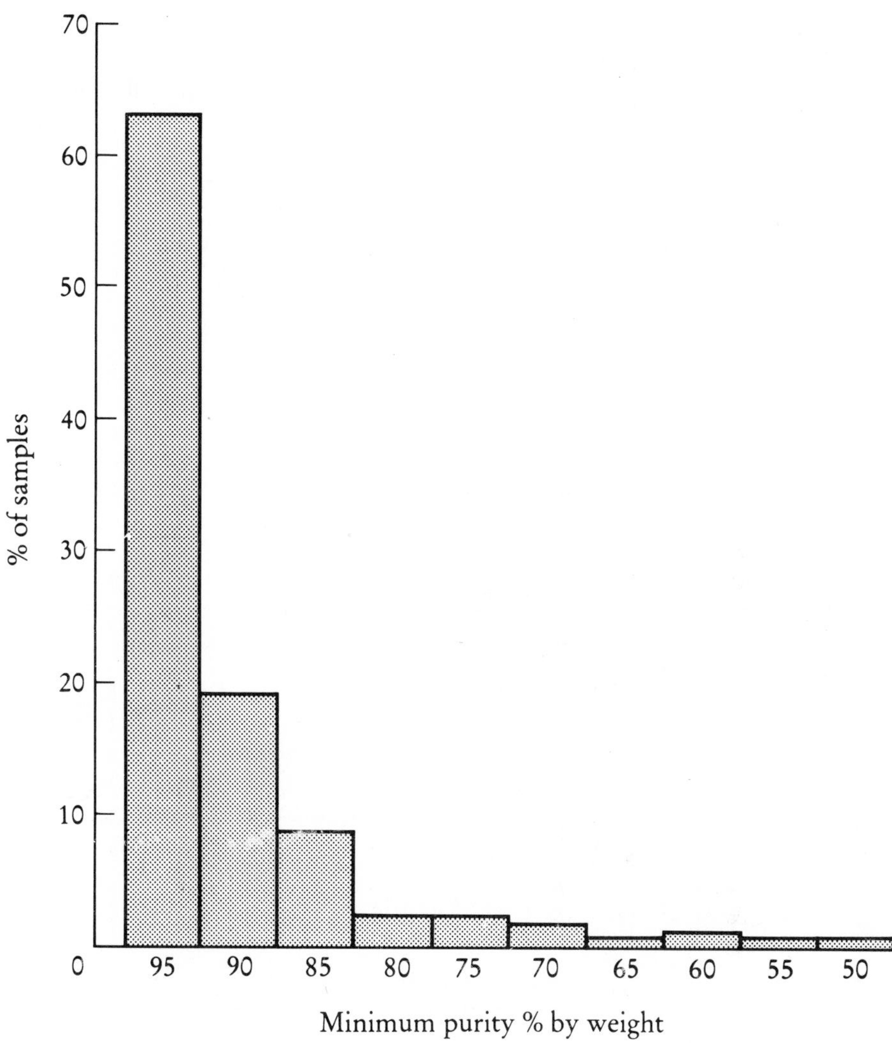

Figure 16.5 The purity of wild flower seed lots tested in 1985.

Firstly, seed production must take place in Britain and thereby suffers harvesting difficulties associated with an uncertain climate (seed production of comparable cultivated species now almost exclusively takes place overseas in regions with more equitable conditions for ripening and harvesting). Secondly, growers must retain the wild characteristics of seed stocks and cannot select for seed production characteristics such as uniformity, seed yield and non-shattering seed heads. Most wild flowers are naturally indeterminate, produce seed over a period of time and usually possess efficient mechanisms for seed dispersal soon after the seed reaches maturity. Seed growers as a consequence are presented with a variable crop of mixed maturity

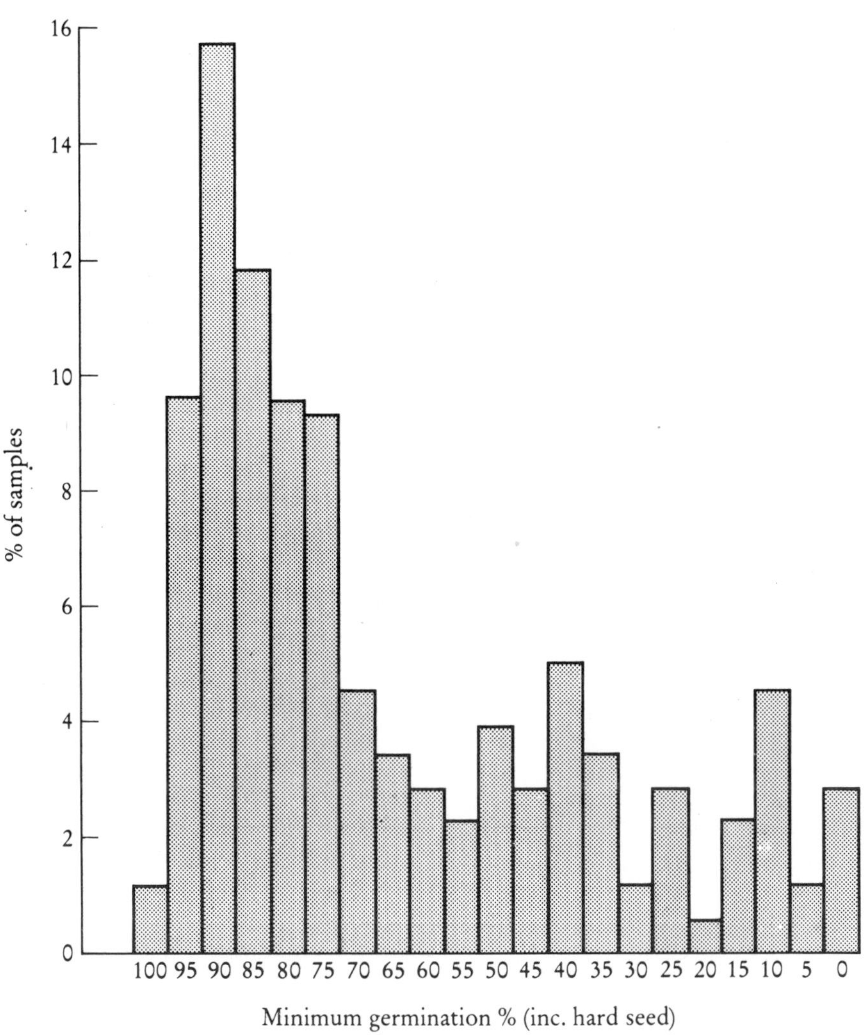

Figure 16.6 The germination of wild flower seed lots tested in 1985.

and can only expect to harvest a percentage of the plants' seed production at full maturity in a given year.

The concepts and regulations covering varietal quality for most crops are divided into two components: cultivars must be (genetically) distinct, uniform and stable; and secondly, they must be shown to have 'value for cultivation and use' when compared with other varieties available to the user. The first of these concepts is clearly inappropriate in this instance, as the wild flower seed user requires diversity and the genetic flexibility to succeed in a variety of circumstances. In practice this means the use of the widest range of ecotypes feasible, and taking care in the multiplication of seed stocks to avoid the narrowing or shifting of the genetic base through the unintentional selection which can result from cultivation practices.

The second concept, 'value for cultivation and use', has some relevance to wild flower seed. For agricultural crops and amenity turf grasses, 'value' can be measured in terms of yield, wear tolerance, all year green colour, less frequent mowing, speed of establishment, etc. (NERC, 1977). However, with wild flowers and mixtures the objectives are less precise, and different users have differing ideas on this subject. In practice we adopt a mixture of aesthetic, ecological and practical ideals. For individual species we aim to use appropriate native ecotypes, preferably originating from a typical locality and habitat, avoiding conscious selection. Wells *et al.* (1981) also provide some useful guidelines on the subject of rarity, distribution and competitiveness of sown species.

In our questionnaire we asked customers to rate the relative importance of specifications affecting the quality of wild flower seed mixtures. The criteria were that wild flower mixtures should:

1. state clearly the percentage species composition and sowing rates of mixtures;
2. contain seed that has been tested for its purity and germination;
3. contain well-known species which reliably establish;
4. contain only native wild flowers originating and grown in the UK;
5. contain mainly native wild grasses originating and grown in the UK;
6. *not* contain aggressive species such as cheap agricultural clovers; and
7. *not* contain rare or difficult species which are often expensive and/or hard to establish.

In general most respondents agreed with the importance of all of these objectives relating to wild flower seed mixture composition (see Table 16.2). However, as we have already seen, the extent to which this agreement is translated into buying patterns is variable; the relative importance given to each of the points is therefore of some interest. Within the five most highly rated statements there were no statistically significant differences, the respondents giving more or less equal importance to all aspects of seed quality in wild flower seed mixtures and to the need to use appropriate species of proven reliability (points 1, 2, 3, 4 and 6). However, significantly less importance ($p<0.001$) was attached to the desirability of using native grass seed (rather than the commercially available cultivars used at present) and the advisability of excluding rare or difficult species which are often expensive and/or hard to establish.

Seed supply

Costs

All the evidence of the previous discussion points to the not very surprising conclusion that price has a very strong influence on buying patterns, irrespective of theoretical ideals. An average wild flower seed mixture at current prices costs

Table 16.2 Responses given to a question relating to the importance of a number of objectives in wild flower seed mixture specification

Response:	Very important	Important	Don't know	Unimportant	Total	mean weighted score
weighted score:	(3)	(2)	(0)	(1)		
Stated % composition	90	34	3	3	130	2.62
Quality tested seed	75	50	3	2	130	2.52
Reliable species used	72	48	4	5	129	2.46
Only native wild flowers used	70	44	9	4	127	2.38
Mainly native grasses used	50	42	14	10	116	2.10
Aggressive species excluded	86	32	6	5	129	2.53
Rare/unusual species avoided	54	38	15	11	118	2.11
Total no. responses	264	276	121	77	738	(2.40)

Note: The main body of the table gives the number of responses to each structured question. Weighted means are also given for comparison between questions.

around £35 per kg, equivalent to £1,000 per ha or 10 pence per square metre. By comparison traditional grass seeding would cost about half, even though the seeding rates are around ten times that recommended for wild flower mixtures. Advancements in seed production techniques should, however, bring the seeding costs per unit area of wild flower seed mixtures down to that of amenity grass mixtures or less. In contrast, wild flowers supplied as pot plants cost from 5 to 50 pence each, equivalent (based upon mixture costs) to 0.2–2 plants per square metre. Plants do not compare on a cost basis with seeding, but this method may be more suitable for introducing difficult subjects, or those with seeds in short supply (see Wells *et al.*; Hodgson, in this volume).

Seed availability

The availability of native wild flower seed has been a major limiting factor in the growth in the use of wild flower seed mixtures. Until recently seed production was exclusively based upon relatively small-scale techniques in wild flower seed nurseries, supplemented by some (less than five per cent) collection from the wild of species producing abundant seed (e.g. *Rhinanthus minor*). The scale of future demand dictates that we increase the scale of seed production significantly by means of the full use of modern techniques such as mechanisation and herbicides. Not all species are suitable for such treatment and development cannot proceed for all species at the same rate. As was mentioned earlier, wild flowers are a recalcitrant group because they are generally indeterminate, are often unpredictable and are difficult to harvest efficiently. Plants which possess efficient wind dispersal mechanisms, such as the compositae (e.g. *Leontodon hispidus*), and others with explosive capsules (e.g. *Cardamine pratensis*), present particular problems which may be solvable on a small scale, but defy attempts at mass production.

Ironically, perhaps, weed control is a major problem facing growers of wild flower seed crops. Even if herbicides were available which could be safely applied to these crops, their development will be slow and expensive, and limited within the new pesticide legislation. Most wild flower seed crops are perennial, and weed problems tend to accumulate over time. From both the practical and economic point of view some compromises in the range of species used are inevitable, particularly in the short term.

Conclusions

Most habitat reconstruction projects involving the use of wild flower seed involve members of the horticultural industry either as seed growers and suppliers, or as specifiers and users of mixtures. Whilst conservationists and researchers will continue in their efforts to refine and expand the possibilities of habitat creation, the responsibility for the execution of these ideas, and ultimately the continued growth and success of the concept, lies squarely with the horticultural industry.

This review clearly demonstrates the role of seed suppliers in providing initial guidance and direction in the specification and use of wild flower seed mixtures, given that the majority of customers follow advice in seedsmens' literature. A purely ecological model for wild flower seed use would be both highly complex and impractical for a variety of reasons, and could not be successful in meeting the needs and demands of the consumers. It is important that we make sure that seed mixtures are easy to specify, reliable across a range of site conditions and represent value for money. However, while some compromises are inevitable in the content of commercially available seed mixtures, care is essential to ensure that the principal objectives are not compromised as this leads to failure and disappointment. The quality and value of the seed and advice which results from the balancing of these conflicting requirements will be variable, and will reflect the commitment and expertise possessed by individual seedsmen.

The selection of an appropriate wild flower seed mixture by the user is only the first step in successful habitat creation using this method. Specifiers and users of wild flower seed mixtures have the most important role to play, as only by attending to the correct management of newly created wild flower meadows in the years after seeding can we hope genuinely to see an increase in the number and quality of such habitats around us.

References

NERC, 1977, *Amenity Grasslands – the Need for Research*, Publications series 'C', No. 19, Natural Environment Research Council, Swindon.

Thomson, J.R., 1979, *An Introduction to Seed Technology*, Leonard Hill, London.

Wathern, P., Gilbert, O.L., 1978, 'Artificial diversification of grassland with native herbs', *Journal of Environmental Management* 7, 29–42.

Wells, T.C.E., 1984, 'Creating attractive grasslands', National Turfgrass Council workshop report No. 5, 8–9.

Wells, T.C.E., Bell, S.A., Frost, A., 1981, *Creating Attractive Grasslands Using Native Plant Species*, Nature Conservancy Council, Shrewsbury.

Appendix 16.1 Part of a questionnaire survey of the demand for wild flower seed mixtures. Questions 1 and 2 have been completed to show the modal response and question 3 to show the overall order of preference.

A survey into the demand for wild flower seed mixtures

The aim of this survey is to obtain information about the types of wild flower mixtures you would like to use. Your participation will help us respond to the ever increasing demand for our wild flower and grass seeds, and will enable us to maintain the standard of our mixtures and service in the future.

Please complete as much of the questionnaire as you can, most of the questions require you to simply tick the appropriate box(es), please feel free to add any further comments that you think may be useful.

	V. Important	Important	Don't know	Unimportant
Creation of environmental interest/variety	✓			
Reduced maintenance		✓		
Wildlife interest	✓			
Educational interest		✓		
Public interest/demand		✓		
Ability to sow on infertile soil			✓	
eg. subsoil avoiding cost of topsoil				
None of the above: eg. Seed specified by other person(s)				
Other reasons (please state)				

2. *Mixture composition:*

We feel that the following statements should be important objectives when specifying wild flower seed mixtures; please rate each statement from your own point of view using the scale provided:

Wild flower mixtures should:

State clearly the % species composition and sowing rates of mixtures.	✓			
Contain seed that has been tested for its purity and germination	✓			
Contain well known species which reliably establish	✓			
Contain only native wild flowers originating and grown in the UK	✓			
Contain mainly native wild grasses originating and grown in the UK	✓			
NOT contain aggressive species such as cheap agricultural clovers	✓			
NOT contain rare or difficult species which are often expensive and/or hard to establish	✓			

3. Which of the following types of wild flower seed mixture would be most likely to meet your requirements for most situations? (tick appropriate box(es):

1. A complex mixture of more than 20 native species	costing £35/kg (£1050/ha)	4
2. A mixture containing 15–20 native species	costing £25/kg (£750/ha)	2
3. A simple mixture of 10–15 native species	costing £20/kg (£600/ha)	1
4. A mixture with some agricultural flower species	costing £15/kg (£450/ha)	3

Appendix 16.2 Examples of typical wild flower seed mixtures, each fulfilling different roles in terms of cost and situation.

General Purpose mixtures

A: *Simple mixture of 10–15 native species* (JF40)

Species in this mixture are common to most grasslands and soil types in Britain.

Application: All areas except those with extreme pH or moisture stress where a flowery alternative to the all grass sward is required, from urban highways verges to rural applications eg set-aside schemes.

% by weight	SPECIES Botanical name	Common name	weight	Botanical name	Common name
	WILD FLOWERS: 1 PART (20%)			GRASSES: 4 PARTS (80%)	
10	Centaurea nigra	KNAPWEED	10	Agrostis capillaris	BROWNTOP
20	Leucanthemum vulgare	OXEYE DAISY	20	Cynosurus cristatus	CRESTED DOGSTAIL
25	Malva moschata	MUSK MALLOW	40	Festuca ovina	SHEEPS FESCUE
25	Prunella vulgaris	SELFHEAL	20	Festuca rubra	RED FESCUE
20	Ranunculus acris	BUTTERCUP	10	Poa pratensis	SMOOTH S.M. GRASS
100%			100%		

B: *Mixture with some agricultural species* (JF41)

Species in this mixture are frequently found in a diverse range of habitats and soil types.
Application: Most situations including Highway Verges for example, except very acid or saturated soils. (F) = Cultivated strains

% by weight	SPECIES Botanical name	Common name	% by weight	Botanical name	Common name
	WILD FLOWERS: 1 PART (20%)			GRASSES: 4 PARTS (80%)	
6.0	Achillea millefolium	YARROW (F)	10.0	Agrostis capillaris	BROWNTOP
2.0	Daucus carota	WILD CARROT	20.0	Cynosurus cristatus	CRESTED DOGSTAIL
3.0	Digitalis purpurea	WILD FOXGLOVE	40.0	Festuca ovina	SHEEP'S FESCUE
4.0	Filipendula ulmaria	MEADOW SWEET	20.0	F. rubra ssp pruinosa	RED FESCUE
3.0	Knautia arvensis	FIELD SCABIOUS	10.0	Poa pratensis	SMOOTH S.M. GRASS
9.0	Prunella vulgaris	SELF HEAL			
10.0	Leucanthemum vulgare	OXEYE DAISY	100%		
10.0	Lotus corniculatus	BIRDSFOOT TREFOIL (F)			
2.5	Malva moschata	MUSK MALLOW			
11.0	Plantago lanceolata	RIBWORT PLANTAIN (F)			
4.0	P. media	HOARY PLANTAIN			
1.0	Hypericum perforatum	C. ST. JOHN'S WORT			
4.0	Ranunculus acris	MEADOW BUTTERCUP			
10.0	Rhinanthus minor	YELLOW RATTLE			
3.0	Silene alba	WHITE CAMPION			
12.0	S. dioica	RED CAMPION			
3.0	Tragopogon pratensis	GOAT'S BEARD			
2.5	Vicia sativa	COMMON VETCH			
100%					

Standard meadow mixtures
Mixture containing 15–20 native species: e.g. wet, loamy soils (JF43)

Species in this mixture are frequently found in heavier, wetter soils subject to occasional (e.g. seasonal) water logging.
Application: Low lying areas on heavier clays and silts, water meadows and other areas with poor drainage.

% by weight	SPECIES Botanical name	Common name	% by weight	Botanical name	Common name
	WILD FLOWERS: 1 PART (20%)			GRASSES: 4 PARTS (80%)	
5.0	Achillea millefolium	YARROW	8.0	Agrostis capillaris	BROWNTOP
6.0	Centaurea nigra	BLACK KNAPWEED	5.0	Alopecurus pratensis	MEADOW FOXTAIL
3.0	Leontodon hispidus	ROUGH HAWKBIT	16.0	Cynosurus cristatus	CRESTED DOGSTAIL
7.0	Leucanthemum vulgare	OXEYE DAISY	1.0	Deschampsia caespitosa	TUFTED HAIRGRASS
5.0	Lynchis flos-cuculi	RAGGED ROBIN	32.0	Festuca ovina	SHEEP'S FESCUE
8.0	Plantago lanceolata	RIBWORT PLANTAIN	16.0	F. rubra ssp pruinosa	S. C. RED FESCUE
7.0	Prunella vulgaris	SELF HEAL	9.0	Hordeum secalinum	MEADOW BARLEY
3.0	Pulicaria dysenterica	COMMON FLEABANE	8.0	Poa pratensis	SMOOTH S.M. GRASS
7.0	Ranunculus acris	MEADOW BUTTERCUP	5.0	Trisetum flavescens	YELLOW OATGRASS
10.0	Rhinanthus minor	YELLOW RATTLE			
6.0	Rumex acetosa	SORREL	100%		
7.0	Sanguisorba officinalis	GREATER BURNET			
5.0	Silaum silaus	PEPPER SAXIFRAGE			
5.0	Stachys officinalis	BETONY			
3.0	Succisa pratensis	DEVILSBIT SCABIOUS			
3.0	Thalictrum flavum	MEADOW RUE			
10.0	Vicia sativa	COMMON VETCH			
100%					

Shaded area mixture

Mixture for specialised situations: e.g. shaded area mixture (JF46)

Species in this mixture are frequently found in partial shade.
Application: Open woods, tree plantings, hedge bottoms, woodland edges, clearings

% by weight	SPECIES Botanical name	Common name	% by weight	Botanical name	Common name
	WILD FLOWERS: 1 PART (20%)			GRASSES: 4 PARTS (80%)	
10.0	Agrimonia eupatoria	COMMON AGRIMONY	9.0	Agrostis capillaris	BROWNTOP
9.0	Alliaria petiolata	HEDGE GARLIC	18.0	Cynosurus cristatus	CRESTED DOGSTAIL
2.0	Conopodium majus	PIGNUT	3.0	Deschampsia caespitosa	TUFTED HAIRGRASS
6.0	Digitalis purpurea	FOXGLOVE	36.0	Festuca ovina	SHEEP'S FESCUE
6.0	Dipsacus fullonum	TEASEL	18.0	F. rubra ssp pruinosa	SL. C. RED FESCUE
8.0	Geum urbanum	HERB BENNET	7.0	Poa nemoralis	W. MEADOW GRASS
5.0	Hypericum hirsutum	HAIRY ST. JOHN'S W.	9.0	P. pratensis	SMOOTH S.M. GRASS
3.0	Lamiastrum galeobdolon	YELLOW ARCHANGEL			
8.0	Prunella vulgaris	SELF HEAL	100%		
3.0	Scrophularia nodosa	FIGWORT			
12.0	Silene dioica	RED CAMPION			
7.0	Stachys sylvatica	HEDGE WOUNDWORT			
5.0	Teucrium scorodonia	WOOD SAGE			
8.0	Torilis japonica	U. HEDGE PARSLEY			
4.0	Verbascum thapsus	GREAT MULLEIN			
4.0	Vicia sylvatica	WOOD VETCH			
100%					

Supplementary mixture
For very specialised situations: e.g. wetland supplement (JF432)

Species in this mixture are found in soils saturated for most of the year.
Application: Marshy areas: margins of streams, rivers, lakes, ponds, ditches, etc.

% by weight	SPECIES Botanical name	Common name
	WILD FLOWERS	
6.0	Althaea officinalis	MARSH MALLOW
10.0	Dipsacus fullonum	TEASEL
25.0	Filipendula ulmaria	MEADOW SWEET
12.0	Geum rivale	WATER AVENS
10.0	Iris pseudacorus	YELLOW FLAG
8.0	Lotus ulginosus	GRTR. BRDT. TREFOIL
5.0	Lychnis flos-cuculi	RAGGED ROBIN
7.0	Lycopus europaeus	GIPSY WORT
5.0	Lythrum salicaria	PURPLE LOOSESTRIFE
3.0	Potentilla palustris	MARSH CINQUEFOIL
9.0	Thalictrum flavum	MEADOW RUE
100%		

17

Soil conditions and grassland establishment for amenity and wildlife on a restored landfill site

B.N.K. Davis and *R.P. Coppeard*

Introduction

Local authorities responsible for waste disposal are often also involved with the final restoration of landfill sites to an agreed after-use. In many cases the planned after-use has been agriculture since that was the original land-use. Restoring land to agriculture generally requires a covering of 1.0 m of subsoil followed by 0.3 m of topsoil and installing sub-surface drainage. It may also be necessary to use a thick clay layer or impermeable membrane over the refuse to prevent the emergence of methane gas (Crawford and Smith, 1985). This whole operation can cost up to £100,000 per ha if it is necessary to purchase large quantities of soil for sites where the original cover material has been removed. It is also questionable whether such restoration is appropriate at a time when agricultural land may be going out of production.

In 1983, Essex County Council commissioned studies by the Institute of Terrestrial Ecology, University of Essex, North East London Polytechnic and the Colchester and Essex Museum on the ecology of landfill restoration in order that certain sites could be developed as county parks. The particular remit for ITE was to develop grassland habitats which would attract butterflies. This paper describes how such an objective can be met given the constraints imposed by waste disposal, low cost restoration, site landscaping and management.

Experimental site: landform and soil conditions

The landscaping of landfill sites must fulfil several criteria. Sites must generally blend in with the surrounding landscape but they are often somewhat mounded to accommodate more waste and to allow for settlement. A sloping surface also helps to shed surface water; too much water entering the waste can create problems of leachate disposal.

The Martin's Farm experimental site lies two km north of St Osyth, Essex (grid reference: TM 117175). It is roughly 24 ha in extent and bounded by coastal marshes on the western edge. It was worked for gravel between 1945 and 1965, and then used for landfill until 1983. The cover material was subsoil derived from various local excavation and building works. The salient feature of this experimental site was that only 0.6 m of soil covering over the waste was required in the planning consent. Site contours were made to blend with the natural coastal landform. The eastern two-thirds were therefore finished almost level while the western third had a slope of about 5.7° (1:10).

It was decided that large plots would be laid out along this western slope for trials on creating grassland and for monitoring colonisation by butterflies.

Auger borings along this slope gave values of 45 to 60+ cm depth of soil cover generally, but some samples gave shallower values of 13–23 cm. Particle-size analysis of soil (Bascomb, 1974) again gave variable results but most samples were in the sandy clay loam category. The stone content (>2 mm) was often very high (mean and standard error 25.8 ± 4.2). Bulk density measurements (Smith and Thomasson, 1974) gave values of 1.64 ± 0.08 for the top 10 cm and 1.72 ± 0.04 for 10–20 cm depth. These high values were probably due to the strong tendency for soil with this texture to compact easily when spread by bulldozer.

Analysis of nitrogen, phosphorus and potassium (Allen *et al.*, 1974) gave variable but generally rather low values compared with typical agricultural soils (see Table 17.1). Calcium levels tended to be rather high and pH levels were around 7.5.

Objectives and methods

In order to create conditions suitable for grassland butterflies their particular requirements need to be understood. These needs are essentially threefold: the presence of appropriate larval food plants, suitable flowers to provide nectar for the adults and the right habitat structure. The browns (*Satyridae*) and most skippers (*Hesperidae*) are grass feeders in Britain. Particular herbs are needed for most species of blues (*Lycaenidae*), whites (*Pieridae*) and fritillaries and vanessids (*Vanessidae*). Some butterflies need a degree of shade and shelter, some favour tall open grassland while others favour short or sparse vegetation for oviposition with bare ground for basking.

The experimental programme therefore involved three stages:

1. the layout and initial recording of plots in order to establish baseline data on natural vegetation and butterfly colonisation;
2. sowing of selected grass/herb mixtures; and
3. monitoring vegetation establishment and butterfly response.

This paper deals with the vegetation aspects. The effects on butterflies are summarised here and given in more detail by Davis (1989).

Four treatments were proposed. These consisted of two sown mixtures of grasses and wild flowers, plots which were cultivated but not sown and undisturbed controls. The seed mixtures were selected from the tall and the short herb mixtures given in Wells *et al.* (1981) for clay and alluvial soils, with a few modifications to meet the particular objectives and site conditions (see Table 17.3). The four treatments were called Tall, Short, Mound and Control. (The Mound plots each contained a bank of fine limestone for growing additional lime-loving plants but this

Table 17.1 Analyses of soil cover used for Martin's Farm landfill site.

pH	Organic matter %	Total N %	Extractable ppm P	K	Ca
7.45	3.10	0.09	35	88	3320
±0.03	±0.23	±0.02	±14	±6	±734

Note: Means and standard errors (6 samples)

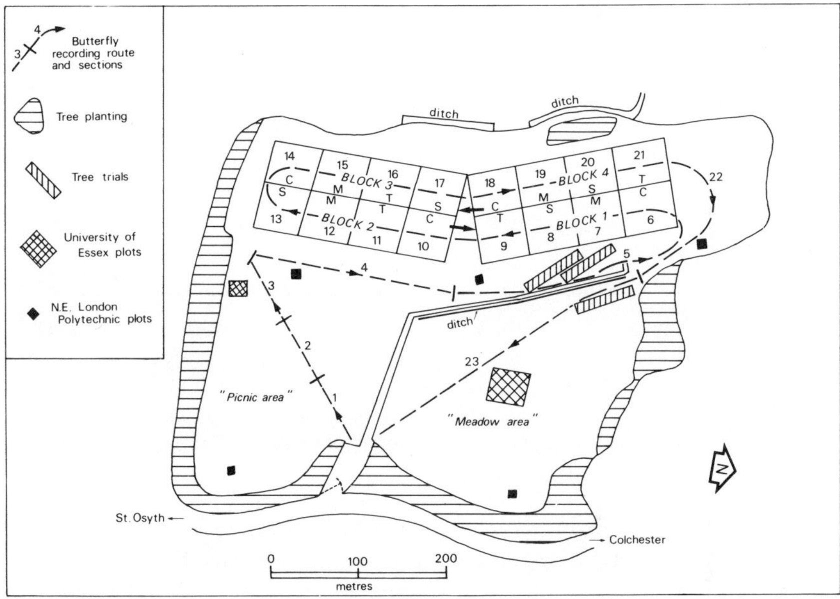

Figure 17.1 Plan of Martin's Farm landfill site showing experimental areas and butterfly recording route. Treatments: C = Control, M = Mound, S = Short, T = Tall.

aspect is not elaborated here). The treatments were replicated in four randomised blocks (see Figure 17.1), each plot being about 63 × 53 m (0.33 ha). This large plot size was considered the minimum necessary to measure butterfly response.

Total vegetation cover and presence of all species were recorded in eight stratified random quadrats (0.25 m^2) in each plot. Vegetation height was assessed by dropping a light card or plastic disc (30 cm diameter) in each quadrat and measuring height above the ground at its centre. Records were made in July 1983, and in June 1984–7 supplemented by records from four permanent quadrats in each plot in July, with additional site records usually in May, August and September.

The two grass/herb mixtures were sown in September 1983 after cultivation to kill existing vegetation and to prepare a seed bed. A nurse crop of Westerwolds

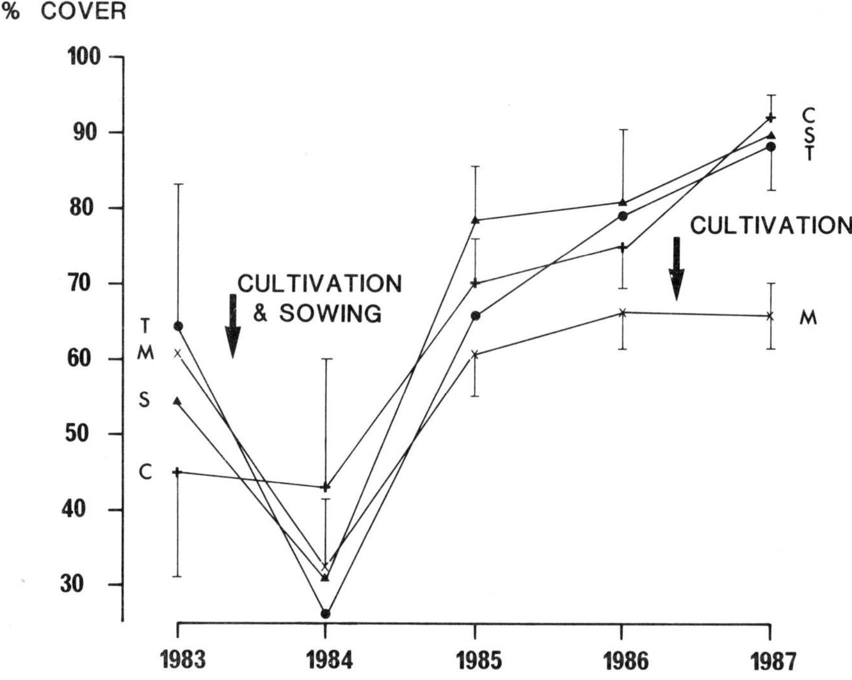

Figure 17.2 Changes in percentage cover July 1983, and June 1984–7 after cultivation of all but control plots and sowing of S and T plots. M plots recultivated in 1986. Treatment means and standard error bars. C = Control, M = Mound, S = Short, T = Tall.

ryegrass (*Lolium multiflorum*) (recommended by Wells *et al.*) was not used. A granular fertiliser (Nitrophoska) was applied in April 1987 to encourage flowering and taller growth. The Tall plots received 250 kg/ha and the Short plots 125 kg/ha of Nitrophaska (containing N:P:K at 12:12:17).

Results

Cover and height of vegetation

The reconnaissance survey in 1983 showed that the oldest part of the Martin's Farm site, in the south-east corner, already had a dense sward of couch grass (*Elymus repens*). This was almost certainly introduced as rhizomes with soil. The western fringe of this area and the south end of the experimental area were dominated by creeping bent (*Agrostis stolonifera*), while the more recently finished areas were mainly bare ground.

The cultivations destroyed the existing cover in all but the Control plots. However, plots which had had more vegetation in 1983 started again with a richer seed bank than the less vegetated plots, and therefore recovered more rapidly (see Figure

Table 17.2 Vegetation height (cm) by dropping disc method.

	Control	Mound	Short	Tall	LSD (P<0.05)
1986	7.1	5.4	8.0	8.5	6.3
1987	9.3	5.3	10.3	13.3	4.2

Note: Means for the 4 treatments in 1986 and 1987.
Least significant difference by Tukey test.

17.2). This factor largely masked the effect of the sowing treatments so that differences between treatments were not significant until 1987, following a second light cultivation of the Mound plots in October 1986.

Vegetation height remained generally very low with no significant difference between treatments until 1987 after the application of fertiliser (see Table 17.2).

Establishment of sown vegetation

All ten of the sown grass species were established in the plots from sown seed (see Table 17.3). Crested dog's-tail (*Cynosurus cristatus*) and yellow oat-grass (*Trisetum flavescens*) became the dominant tall grasses where they were sown, while red fescue (*Festuca rubra*) provided a more or less continuous understorey. On the other hand, quaking grass (*Briza media*) and smooth-stalked meadow-grass (*Poa pratensis*) were recorded for the first time in 1987, probably as a result of the fertiliser application.

The 21 sown species of wild flowers could be divided into four main categories:

1. species appearing in the first year (1984) and becoming well established e.g. ox-eye daisy (*Leucanthemum vulgare*), bird's-foot trefoil (*Lotus corniculatus*) and wild carrot (*Daucus carota*);
2. species which appeared only in the second year but then established strong plants at various densities e.g. lady's bedstraw (*Galium verum*) and hoary plantain (*Plantago media*);
3. annuals and short-lived perennials, such as yellow rattle (*Rhinanthus minor*) and kidney vetch (*Anthyllis vulneraria*), which were established only in some plots and then very patchily;
4. four species which either died out or never appeared, including lady's smock (*Cardamine pratensis*).

The application of fertilisers stimulated flowering in several species in 1987, notably in *Leucanthemum vulgare*, which produced extensive drifts of white in June, and in common knapweed (*Centaurea nigra*).

Until 1987, there were very few records of sown species having colonised adjacent unsown plots.

Table 17.3 Establishment frequency of sown grasses and wild flowers, 1985–7.

Short	Category	1984	1985	1986	1987	Tall	Category	1984	1985	1986	1987
			#Grasses								
Alopecurus pratensis	II	–	✓*	18*	16*	*Alopecurus pratensis*	II	–	✓*	16	21*
Anthoxanthum odoratum	III	1	5*	9*	3*	*Cynosurus cristatus*	I	19*	29*	20*	31*
Briza media	III	–	–	–	3*	*Dactylis glomerata*	III	✓	5	5*	4*
Festuca rubra	I	5	23*	26*	29*	*Festuca rubra*	I	17	26*	24*	29*
Trisetum flavescens	I	1	19*	26*	29*	*Holcus lanatus*	I	26*	29*	21*	21*
						Phleum pratense	II	1	3	10	9*
						Poa pratensis	III	–	–	–	4*
			Wild flowers								
Anthyllis vulneraria	III	4	4*	–	✓*	*Centaurea nigra*	II	4	8	3*	8*
# *Cardamine pratensis*	–	–	–	–	–	*C. scabiosa*	–	–	–	–	–
Galium verum	I	–	7*	16*	10*	*Daucus carota*	I	11	14	11*	8*
Leontodon hispidus	II	–	5	9	2*	*Hypochaeris radicata*	II	7	15*	8*	2*
# *Medicago lupulina*	I	14*	14*	18*	9*	*Knautia arvensis*	III	1	2	–	–
Plantago media	I	–	11*	14*	14*	*Leucanthemum vulgare*	I	15	23*	22*	26*
Primula veris	III	✓	✓	2*	1*	# *Lotus corniculatus*	I	14	13*	17*	19*
Prunella vulgaris	II	1	5*	6	4*	*Plantago lanceolata*	I	6	15	12*	10*
Rhinanthus minor	III	3*	1*	✓*	✓*	*Ranunculus acris*	III	–	✓	5*	5*
						# *Rumex acetosa*	III	–	1	2*	1*
						Sanguisorba minor	III	5	5	3	2
						Silene alba	–	–	(*)	–	(✓*)
Total species		8	12	11	13			13	16–17	15	16–17

Note: Occurrence in 32 random quadrats in 'short' and 'tall' plots.
Categories I good, II moderate, III poor.
✓ indicates presence in plots but not in quadrats.
* indicates flowering.
( ) indicates possible natural seed source.
# indicates larval food plants for potential butterfly colonists.

The unsown vegetation

Fifty-three species of plants were recorded in the whole experimental area (5.28 ha) in 1983, before the plots were sown. A cumulative total of 143 naturally colonising species were recorded up to September 1987 of which 90 per cent occurred in the unsown plots (Control and Mound) at one time or another (see Appendix 17.1). These included a few of the sown species mostly as very occasional sightings, e.g. four plants of native *Lotus corniculatus* in 1987, but large numbers of locally derived *Holcus lanatus* and *Medicago lupulina* from 1984 onwards. Eighty-four species (59 per cent) were annuals or biennials, some of which are widespread but of distinctly local occurrence e.g. *Apera spica-venti, Myosurus minimus, Ranunculus parviflorus* and *Trifolium subterraneum*. Others are frequent mainly in coastal areas such as *Carduus tenuiflorus, Juncus maritimus, Parapholis incurva* and *Ranunculus sardous*. (Ten species were recorded only on the mounds. They were almost certainly introduced with the material and are not included here).

Agrostis stolonifera was the most common species over most of the site; after an initial set-back in 1984, the result of plot cultivation in 1983, it was recorded in 90 out of the total 128 quadrats in 1985–7. Among the other common species, three main trends could be discerned in the control plots: (1) species which remained more or less consistently frequent throughout the five-year period e.g. *Epilobium tetragonum, Picris echioides, Ranunculus repens*; (2) species which increased steadily in frequency e.g. *Holcus lanatus, Poa trivialis, Vulpia bromoides*; (3) species which were common in 1983, or increased during the first few years, and then declined e.g. *Medicago lupulina, Plantago major, Sagina apetala, Tripleurospermum inodorum*.

The trend for overall species-richness in quadrats reflected the trend of this last group in an initial increase and subsequent decline. The decline followed the reduction in bare ground as vegetation cover increased between 1985 and 1987 from 70 per cent to 92 per cent. The light harrowing in the Mound plots in autumn 1986 produced an upturn again in floristic richness in 1987 (see Figure 17.3). As the sown vegetation became established in the Tall and Short plots, the naturally colonising species were further displaced. The number of unsown species in the Tall plots, for example, was significantly lower ($P < 0.05$) than in the Controls in 1985 and 1986, and lower than in the Mound plots in 1987 ($P \sim 0.05$) (see Figure 17.3).

Butterflies

Eighteen species of butterflies were recorded on the site during 1983–7. Of these, the common blue *Polyommatus icarus* and five to six species of browns (*Satyridae*) and skippers (*Hesperidae*) established strong breeding populations. The sown *Lotus corniculatus* formed a major attraction to the adult common blue whose numbers were significantly higher in plots containing this species. The skippers and browns were all grass feeders. They showed some preference for denser and taller grassy areas rather than for any particular sowing treatment. The use of fertilisers in the Tall plots produced a positive response in Essex skippers (*Thymelicus lineolus*).

The most important nectar-producing plants for butterflies were thistles (*Cirsium arvense* and *C. vulgare*), which occurred throughout the site and densely on parts of

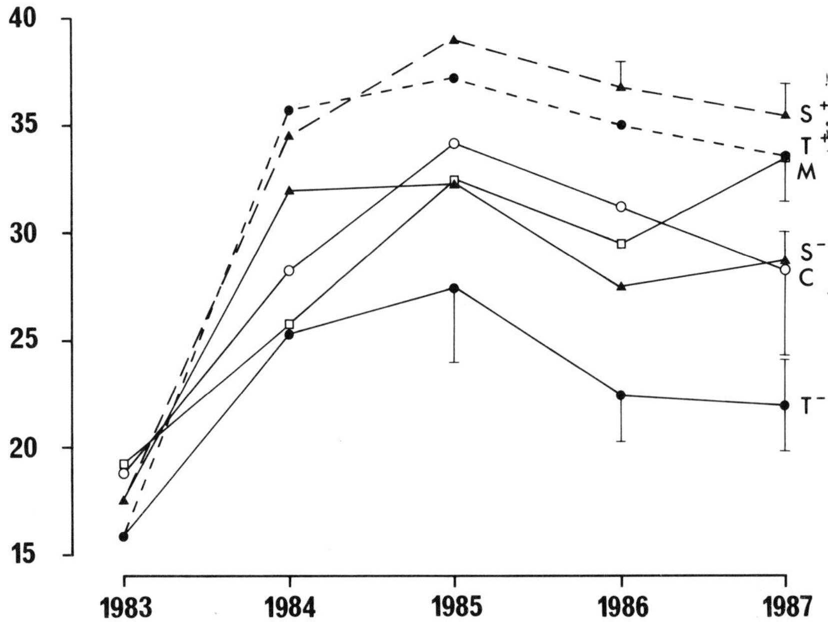

Figure 17.3 Trends in floristic richness in 1983–7. Mean numbers of species in each treatment (per 8 random quadrats of 0.25 m²) with standard error bars. C = Control, M = Mound, S = Short, T = Tall, −/+ = excluding/including sown species.

the limestone mounds, bristly ox-tongue (*Picris echioides*) and, more locally, teasel (*Dipsacus fullonum*). The sown common knapweed (*Centaurea nigra*) flowered well in 1987 and other sown species were probably used. Floristic diversity is thus indirectly useful to butterflies in maintaining a series of nectar sources through the season.

Discussion

Nutrient deficiency, compacted soils and poor structure are common features on many types of reclamation sites (Bradshaw and Chadwick, 1980). The Martin's Farm site demonstrated these and several additional problems which can limit the establishment of vegetation.

1. The compacted refuse beneath the soil cover prevented good drainage. This caused waterlogged conditions after heavy rain, especially in winter and where hollows developed as a result of subsidence, and so prevented surface run-off. Soil pits dug in April 1985 exposed a dark anaerobic zone at the interface between soil cover and refuse. This feature tended to limit rooting depth, and was probably the cause of failure for *Centaurea scabiosa.*
2. The steeper slopes led to a certain amount of sheet erosion and incipient gulley erosion in the early years before vegetation cover became well established. Many young plants were seen in 1985 growing proud of the soil surface as miniature

tussocks where soil had been washed away around them, leading to some loss of seed and death of seedlings.

3. The combination of slope, south-west facing aspect and poor soil structure for water retention created drought conditions in the summer. Water stress on plants was exacerbated by the shallow rooting depth caused by the winter waterlogging. It was usually most evident in the late summer but was observed even in June 1986 after a dry period. Drought was almost certainly the cause of failure in *Cardamine pratensis*.
4. The texture of the subsoil cover rendered it particularly prone to slumping. This meant that attempts to reduce compaction by cultivation would be frustrated when wet weather caused the soil to resettle. It also prevented the use of shallow mole drains. Improved winter drainage would have encouraged deeper rooting by plants rendering them less vulnerable to summer drought.
5. Methane production from landfills can cause stunted growth and bare patches through anaerobism. The policy at Martin's Farm was to allow methane to diffuse out over the whole area rather than to become concentrated. Gas smells were noted locally and intermittently, but the presence of some persistent bare patches was more consistent with initial surface compaction by vehicles.
6. A few small surface emissions of leachate on the lower slopes were treated with lime. A larger emission point was tapped and leachate pumped back into the middle of the site in 1985 and 1986.
7. The shallow soil cover meant that some solid items of waste (concrete and metal) lay close to the surface and could interfere with cultivation.

Techniques for creating attractive grasslands using native species of plants have been developed on relatively fertile and easily worked soils. The only modifications made at Martin's Farm were in the modification of seeding rate, the inclusion of *Cardamine pratensis* and the omission of Westerwolds ryegrass (*Lolium multiflorum*) as a nurse crop. These modifications were made primarily for the benefit of butterflies and to minimise the need for any management (mowing) in the early years.

Large arable weeds such as charlock (*Sinapis arvensis*) and mayweeds (*Tripleurospermum maritimum* and *Matricaria* spp.) were not a problem on these nutrient-poor soils. A certain degree of 'weediness' and open ground was considered to be actually beneficial for butterflies and other insects until the sown plants were big enough to flower and provide nectar. The cultivation of the Mound plots in 1986 was a deliberate attempt to retain open ground conditions for certain butterflies and for some of the interesting wild flowers that had colonised the site naturally. Different types of management would be required on different soils or to meet different objectives.

Most of the problems for grassland establishment listed above were a necessary consequence of the engineering requirements for waste disposal and the decision to minimise the depth of soil cover to reduce costs. The Control plots showed that, even if no vegetation had been sown, the site would have acquired a short grassy sward of 90 per cent cover after four years (and probably a moderate butterfly population). This sward, however, was dominated by *Agrostis stolonifera* whose light seed, surface rooting and vegetative reproduction makes it a particularly well adapted species for rapid colonisation of such sites. Outside the experimental plots at Martin's Farm it formed extensive swards suppressing most other species and producing a low floristic diversity after a few years. *Elymus repens* can become even more dominant if it gets into a site. It had started to invade the plots from the adjacent areas in the south-east corner. It was controlled by spot treatment within the plots with a selective grass herbicide (alloxydim-sodium) in the plots, and by cultivating the main area, resowing with *Lolium perenne* and mowing.

The best way to combat these aggressive grass species is to sow down the required grassland vegetation before they become established, in stages if possible, as parts of the site become completed. It would then be safe to use a light fertiliser dressing to encourage growth without fear that the established wild grasses would suppress the young seedlings.

The sown plots demonstrated that many attractive meadow grasses and wild flowers can be established, even under quite harsh conditions. However, seed is much more expensive than for standard grassland mixtures without wild flowers, and growth is slow. Every care should therefore be taken to ensure that conditions are as favourable as possible for germination and growth. These sown plots are more likely to retain a more diverse vegetation under a mowing or grazing regime than the unsown plots whose species depend largely on periodic disturbance.

All reclamation projects should have well defined objectives. This is particularly true of experimental work so that successes and failures can be recorded and lessons learned. Monitoring is an essential part of this process. The establishment of many wild plants is slow, even on good soils, and changes in sward composition takes place for several years. The results after only one or two years can therefore be disappointing and misleading. This is even more true for insects which need to build up populations gradually as the habitat improves.

This was the first large-scale attempt to create habitats to attract butterflies. The results are quite encouraging. Such open, 'unimproved' grassland has become scarce and is favoured by many insects as well as butterflies. Lack of shelter and failure to establish certain food plants (see Table 17.3) are limiting factors which will take longer to overcome on this and most reclamation sites.

Acknowledgements

This work was done under contract to Essex County Council. The first author thanks his colleagues who have helped with this study, especially Mrs C. Adlard (née Brown), Mr A. Holland, Mrs S. Lahav, Mr G. Prior, Mr A.J. Frost and Mr T.C.E. Wells.

References

Allen, S.E., Grimshaw, H.M., Parkinson, J.A., Quarmby, C., 1974, *Chemical Analysis of Organic Materials*, Blackwell Scientific Publications, Oxford.

Bascomb, C.L., 1974, 'Physical and chemical analyses of < 2 mm samples' in B.W. Avery and C.L. Bascomb (eds.), *Soil Survey Laboratory Methods*, Soil Survey technical monograph No. 6, Harpenden.

Bradshaw, A.D., Chadwick, M.J., 1980, *The Restoration of Land*, Blackwell Scientific Publications, Oxford.

Crawford, J.F., Smith, P.G., 1985, *Landfill Technology*, Butterworth, London.

Davis, B.N.K., 1989, 'Habitat creation for butterflies on a landfill site', *The Entomologist,* **108**, 109–22.

Smith, P., Thomasson, A.J., 1974, 'Density and water-release characteristics' in B.W. Avery and C.L. Bascomb (eds.), *Soil Survey Laboratory Methods*, Soil Survey technical monograph No. 6, Harpenden.

Wells, T., Bell, S., Frost, A., 1981, *Creating Attractive Grasslands Using Native Plant Species*, Nature Conservancy Council, Shrewsbury.

Appendix 17.1. Frequency of occurrence of naturally colonising species in experimental plots at Martin's Farm landfill site, St Osyth 1983–7.

R = rare (< 4 out of 128 quadrats in best year), O = occasional (5–16 quadrats), F = frequent (17–64 quadrats), A = abundant (> 64 quadrats in best year), L = locally, * = in unsown plots.

Species	Years 198–	Occurrence
Achillea millefolium	5	R
Agrostis stolonifera	3 4 5 6 7	A
Aira praecox	3 5 6	R
Alopecurus geniculatus	4 5 7	R
Alopecurus myosuroides	3 4 5 6 7	R
Anagallis arvensis	3 4 5 6 7	F
Anthemis cotula	4 5 6 7	O
Anthriscus caucalis	5 7	R
Apera spica-venti	3 4 5	R
Aphanes arvensis	4 5 6 7	F
Arabidopsis thaliana	7	R
Arctium sp.	4	R
Arenaria leptoclados	3 4 5 6 7	F
Arrhenatherum elatius	3 4 5 6 7	R
Artemesia vulgaris	6	R
Atriplex hastata	3 4 6	R
Barbarea vulgaris	6	R
Bellis perennis	4 5 6 7	R
Brassica nigra	7	R
Bromus hordeaceus	3 4 5 6 7	O
Bromus sterilis	4 5 6 7	R
Capsella bursa-pastoris	3 4 7	R
Cardamine hirsuta	6 7	R
Carduus tenuiflorus	5 7	R
Carex otrubae	4 7	R
Centaurium erythraea	3 4 5 6 7	FL
Cerastium diffusum	4	R
Cerastium fontanum	3 4 5 6 7	F
Cerastium glomeratum	4 5 6 7	F
Chenopodium album	3 4 5 6 7	R
Chenopodium ficifolium	3	R
Chenopodium rubrum	3	R
Cirsium arvense	3 4 5 6 7	F
Cirsium vulgare	3 4 5 6 7	F
Convolvulus arvensis	5	R
Crataegus monogyna	5 6 7	R
Crepis capillaris	4 6 7	R
Crepis vesicaria	4 5 7	R
Dactylis glomerata	3 6 7	R*
Desmazeria rigida	4 5 6 7	R
Dipsacus fullonum	4 5 6 7	O
Elymus repens	3 4 5 6 7	O
Epilobium hirsutum	3 4 5 6 7	R

Species	Years 198–	Occurrence
Epilobium tetragonum	3 4 5 6 7	F
Equisetum arvense	5 6	R
Erigeron canadensis	4 5	R
Erodium cicutarium	4 6 7	R
Erophila verna	5 6 7	R
Euphorbia exigua	4	R
Euphorbia helioscopia	4	R
Filago vulgaris	3 4 5 6 7	O
Galium aparine	4 5 6 7	R
Geranium dissectum	4 5 6 7	F
Geranium molle	4	R
Holcus lanatus	3 4 5 6 7	F*
Hordeum murinum	4 5 7	R
Juncus bufonius	3 4 5 7	R
Juncus effusus	4 6	R
Juncus inflexus	7	R
Juncus maritimus	7	R
Kickxia elatine	3	R
Kickxia spuria	3 4	R
Lathyrus nissolia	4 5 6	R
Lolium perenne	3 4 5 6 7	F
Lotus corniculatus	6 7	R*
Malva sylvestris	3 4 7	R
Matricaria recutita	4 5 6 7	O
Medicago arabica	4 5 6 7	FL
Medicago lupulina	3 4 5 6 7	A*
Melilotus alba	4	R
Melilotus ?officinalis	7	R
Mentha arvensis	6 7	R
Myosotis arvensis	3 4 5 6 7	R
Myosurus minimus	6	R
Odontites verna	4 5 6 7	R
Papaver rhoeas	3 4	R
Parapholis incurva	5 7	R
Parapholis strigosa	4	R
Phleum bertolonii	5 6 7	R
Phragmites australis	5 6	R
Picris echioides	3 4 5 6 7	A
Plantago coronopus	4 5 6 7	R
Plantago lanceolata	3 4 5 6 7	R*
Plantago major	3 4 5 6 7	F
Poa annua	3 4 5 6 7	F
Poa pratensis	4 5 6 7	R*
Poa trivialis	5 6 7	A
Polygonum aviculare	3 4 5 6 7	F
Prunella vulgaris	5	R*
Pulicaria dysenterica	5 6 7	R
Quercus robur	4	R
Ranunculus parviflorus	5	R
Ranunculus repens	3 4 5 6 7	F

Species	Years 198–	Occurrence
Ranunculus sardous	4 7	R
Ranunculus sceleratus	4	R
Reseda luteola	4 5 7	R
Rubus fruticosus	3 4 5 6 7	R
Rumex acetosella	5 6 7	R
Rumex conglomeratus	5 7	R
Rumex crispus	3 4 5 6 7	F
Rumex obtusifolius	4	R
Rumex sanguineus	4 7	R
Sagina apetala	3 4 5 6 7	A
Sagina procumbens	5 6 7	O
Salix cinerea	6 7	R
Scirpus maritimus	5 6	R
Sedum acre	5 7	R
Senecio jacobaea	3 4 5 6 7	F
Senecio viscosus	4	R
Senecio vulgaris	4 5 6 7	R
Sherardia arvensis	4 5 6 7	R
Silene alba	5 7	R*
Sinapis arvensis	4 7	R
Sisymbrium officinale	7	R
Sisymbrium orientale	4	R
Solanum nigrum	3	R
Sonchus arvensis	3 4 5 6	R
Sonchus asper	3 4 5 6 7	O
Sonchus oleraceus	4 7	R
Spergularia marina	4	R
Stellaria media	4 5 6 7	R
Taraxacum officinale	5 6 7	R
Torilis nodosa	5	R
Trifolium campestre	3 5 6 7	R
Trifolium dubium	4 5 6 7	F
Trifolium fragiferum	5	R
Trifolium hybridum	5 7	R
Trifolium micranthum	6	R
Trifolium pratense	4 7	R
Trifolium repens	3 4 5 6 7	O
Trifolium striatum	4 5 7	R
Trifolium subterraneum	4	R
Tripleurospermum inodorum	3 4 5 6 7	A
Urtica dioica	4 5 7	R
Veronica arvensis	4 5 6 7	O
Veronica persica	3 4 5 6 7	O
Vicia cracca	7	R
Vicia hirsuta	4 5 6 7	R
Vicia sativa	3 4 5 6 7	F
Vicia tetrasperma	3 4 5 6 7	F
Viola arvensis	3 4 5	R
Vulpia bromoides	3 4 5 6 7	A
Vulpia myuros	3 4 5 6 7	A

18

Modelling and shaping new habitats in landscaping works

Penny Anderson

Much lip-service is currently paid to the desirability of incorporating habitat creation into landscaping schemes. Yet there is a stark contrast between the average amenity grass mix with scattered patches of trees or shrubs and the successful integration of habitat creation into landscape design. This is particularly frustrating when the potential for sound habitat creation is so great. For good habitat design, close co-operation is needed between ecologists, landscape designers and sympathetic engineers. There are several distinct phases in the development of the composite landscape scheme: habitat survey and evaluation, habitat retention, habitat creation and habitat design.

Habitat survey and evaluation

First, the existing site needs to be surveyed to ascertain the range and types of plant and animal communities present, along with the environmental characteristics. Observations on the adjacent land-uses and the existence of any interconnecting habitats will be important elements at the design stage (see Bunce and Jenkins, in this volume).

The habitats present then need to be evaluated, and this is probably the most difficult element in the project. High quality habitats stand out, and are relatively easily assessed using accepted conservation criteria (e.g. Ratcliffe, 1977; Usher 1986) and by comparing them with Sites of Special Scientific Interest (SSSIs). Sites of lower scientific value are more problematic, and may only be important as vestiges of the habitat mosaic of an area.

The relative rarity of individual species can be readily assessed by reference not only to Red Data Books (Perring and Farrell, 1983; Shirt, 1987) but also to national and county distribution maps for plants and animals, but other criteria are more difficult to evaluate. As it is unusual to be able to set local community types into the national context, the final evaluation has to be, to some extent, subjective.

Habitat retention

Habitat evaluation is not just an ecological nicety, but is essential to the development plan. The key is to be able to carry out the preliminary ecological work as part of an initial sieve analysis that any potential developer should undertake. Just as steps are taken to avoid archaeological monuments, etc., the more important habitats of a site need to be kept intact and incorporated wherever possible *in situ* into the design and layout. This has obvious advantages. Part of the landscape for the development is thus not only in place, but also mature, before development commences. This provides an instant, attractive work environment, can retain local views and might screen the development. Ecologically, apart from the protection of declining semi-natural habitats, it provides a core of native species, some of which may be able to colonise new adjoining habitats at a later stage. In economic terms, there is kudos to be gained from a sensitive approach to such environmental issues, and house prices can be higher, and sell more readily if they overlook community wildlife areas or public open space.

This approach has been carried out with some success on a number of schemes. For example, one housing project in the City of Manchester is to incorporate a 1.4ha marsh in which there is a fine show of northern marsh and common spotted orchids (*Dactylorhiza purpurella* and *D. fuchsii*) and other locally scarce species. On a larger scale, the same approach has been adopted at Stansted Airport, Essex, where a preliminary survey for the British Airports Authority (BAA) revealed small fragments of ancient woodland, a network of hedges and some well-established grassland and fen-type communities, many of which have been retained *in situ* during the airports expansion. Such site protection is not necessarily straightforward. Secure fencing of areas to be protected is critical, accompanied by close liaison with site managers, clear instructions in the contract preliminaries and penalty clauses requiring any damage to be repaired. It must be stressed to contractors that whereas replacement of a built structure may be straightforward, repair to a damaged habitat may be impossible, or at best, mediocre.

Habitat creation

Selecting appropriate habitats

Having protected the best habitats, the next stage in the design process is to identify the potential for habitat creation. This is not a substitute for properly conserving the better habitats, but rather an opportunity to forge corridors and links between them and to provide a measure of compensation for the extensive habitat losses over the last forty years (Nature Conservancy Council, 1984). The aim is to build new habitats on the disturbed ground of a development site which are characteristic of the local environmental conditions and geographical location. This is likely to be far more successful and cheaper if it is tailored to the site conditions. Thus, if the soils are naturally wet, base-rich boulder-clays, appropriate habitats would be marshes, ponds, damp grassland or woodland. If the soils are sandy or acidic, heathland creation may be more appropriate.

Species selection

There is an ecological case for limiting the sown species selected for new habitats to regions in which they already occur, even though the seed or plant source may not necessarily be local in origin. Many species are cosmopolitan within Britain and north-west Europe and can be widely used, provided their soil and habitat preferences are observed. However, others have distinctive geographical distributions limited by often unknown factors or combinations of conditions. Various investigations have revealed a complexity of limiting mechanisms: for example, Pigott (1974) found that low seed viability limits the northern spread of stemless thistle (*Cirsium acaulon*). Jarvis (1960) in contrast, found that summer drought dictated the southern limit of bird cherry (*Prunus padus*). Since habitat creation should be aiming to develop long-term, self-sustaining communities, it is logical to use species suited to the local conditions on the assumption that they are adapted to them, and will be able to reproduce and maintain themselves over many generations given a suitable management regime.

It is clear from even the earliest comprehensive ecological studies (e.g. Tansley, 1965), that there is a striking community variation throughout the country. To the naturalist and ecologist these differences give a region character and distinctiveness, which should provide the raw material for landscape design. Twenty years ago, Nan Fairbrother (1970) made an impassioned plea for planting on new roadsides to match the propensities of the countryside: 'We should be conscious of the difference between the Midland clays, or the chalk, or the farming loams, or the sand and gravel country, or the sterner rocks of the uplands . . . If developed as local landscapes, motorways could be a fascinating cross-section of the countryside they traverse'. Her sentiment is equally relevant today, and the more poignant since so little seems to have been achieved in the intervening years. There is a strong case to concoct our own seed mixes, selecting compatible species suited to the soil and other conditions. Unsympathetic landscaping works, which add locally inappropriate species, confuse the ecological patterns and could lead to hybridisation and potential loss of native taxa.

The next stage is to decide whether it would be feasible just to facilitate natural colonisation or to add species (Bradshaw, in this volume). Given suitably infertile soil conditions, and adjacent seed sources from semi-natural habitats, an alternative to natural colonisation is to use a thin seed mix of non-vigorous grasses, such as sheep's fescue (*Festuca ovina*) or common bent (*Agrostis capillaris*) to act as initial soil stabilisers and nurse species. Subsequent management could be critical in ensuring the long-term availability of colonisation gaps, and experience from other sites, such as some quarries (Davis, 1979), suggests that it could take 40 or more years for a reasonably diverse plant and animal community to colonise.

Ecological knowledge of a range of species suggests that many will not be able to disperse across alien or hostile terrain when there are no adjacent semi-natural habitats. This may apply equally to woodland plants (Peterken, 1981), or to animals (e.g. some butterflies, Heath *et al.*, 1984). Individual colonising ability may also vary geographically; (for example the greater woodrush (*Luzula sylvatica*) is an ancient woodland indicator-species in Lincolnshire (Peterken, 1974), but disperses freely in the south Pennines (Anderson and Shimwell, 1981)). There often is a case, therefore, for introducing locally native and compatible species which are unlikely to reach a site of their own accord. Pressure to produce immediate results may also mean planting rather than relying on natural colonisation.

The selection of species represents a critical point at which the opportunism and comparative costs of landscape works can compromise ecological principles. Many of the desirable plants may not be available commercially, and when the optimum seed mix is costed a cheaper alternative may be requested. Even if the optimum mix requires only a minimal and low-cost maintenance regime, this may be irrelevant when the initial landscaping and subsequent management are funded from different budgets.

At this point there are a variety of options. First, the ecological survey should have highlighted lower-grade habitats containing species which are unavailable commercially or which represent a useful, free resource. Depending on their nature, and the type of development, it may be feasible to 'rescue' them for use in landscaping. There is scope for much greater opportunism, not only within a site, but also elsewhere where habitats are still being destroyed. Great care must be taken though to ensure such operations are not seen as an alternative to *in situ* habitat conservation. Secondly, seed mixes can be modified by reducing the sowing rates to an absolute minimum, allowing some natural colonisation to help diversify the sward, or by omitting expensive seeds, providing that this does not lose the distinctiveness of the resulting community.

Having selected the appropriate species, the relative quantities and patterns of each have to be determined. These can be gauged by collecting quantitative data from nearby equivalent habitats (where available), or by searching for clues in floras or the literature. These approaches were adopted at Stansted Airport where the woodland and hedgerow specifications in the landscaping scheme have been derived from ecological survey data supported by reference to the flora of Essex (Jermyn, 1974). The mixtures thus defined may then need modification to suit the development. For example, a particular visual effect may be required, or, as at Stansted Airport, bushes producing prolific berry crops could only be used sparingly to reduce the risk of bird strike.

Habitat design

The next phase in habitat creation within landscape design is to pass the community prescriptions and list of habitat types selected to the landscape architect, but there are still several ecological principles which it is important to observe. Consideration of the extent of each new habitat, and the way they link together and to the adjacent land-uses are important. The minimum size of a habitat will vary for each animal species, and although average ranges may be known for some of these, it is probably more important to aim for viable populations of as great a variety of characteristic species as possible. Although some aspects of the theory of island biogeography may be relevant here, they do not provide practical guidance for a further problem, the determination of an optimum proportion of different habitats, such as grassland to woodland or scrub. There is a tendency in some landscape schemes to incorporate too many different habitats in too small a site in the belief that this creates diversity.

In theory, the more catholic mobile species capable of utilising all or most of the habitats may thrive at the expense of specialists. Since it is the latter which are scarce and in need of extra habitat provision, habitat creation should logically concentrate on these. In deciding the optimum balance of habitat types in a scheme a combination of ecological experience, environmental sensitivity and opportunism is probably the most effective and pragmatic approach.

Perhaps a more obvious connection between theory and practice arises from the need to link communities together. Habitats abutting a site can be continued across the landscaped area, to form not only the corridors and stepping-stones promoted in urban areas (e.g. West Midlands County Council, undated), but also equivalent linkages in intensively farmed regions. At the same time, the optimum design conditions for each habitat type need to be considered. For example, ponds need to be deep enough to avoid total freezing in winter, and wavy edges, with irregular water depths and an open sunny aspect are among standard recommendations. (e.g. British Trust for Conservation Volunteers, 1976; Baines and Smart, 1984). By combining good ecologically based design with the needs of management, parallel guidelines for other habitats can also be prepared.

Implementation

Landscape contractors may be familiar with the techniques of modern woodland planting, even when trees and shrubs are grouped randomly rather than in rows, but pond excavation, land modelling to create diversity and meadow or marsh establishment pose different problems. Communication with machinery drivers and site contractors is often best achieved by personally directing excavations. At Stansted Airport for example, it was more effective to mark the new pond with leafy twigs which the excavator driver could see, than to produce detailed drawings. In the same way, the diver's urge to smooth the pond sides could be restrained, and a lumpy texture created which produced an appropriate micro-topography for the aquatic and emergent plants which were subsequently introduced.

For other types of habitat creation, it may be better to use agricultural contractors since cultivation and seed establishment is more often part of their normal repertoire. Whatever the project, a sympathetic site manager is crucial. This has been the key to the successful habitat creation at Stansted Airport.

Habitat design in practice: Stansted Airport

The foregoing may seem to present an ideal approach to obtain a utopian dream. That this need not be so is exemplified on nearly all counts by the landscape design and habitat creation being incorporated into the expansion of Stansted Airport.

The site was surveyed in 1981 prior to the public inquiry into its expansion. No high value wildlife sites were found, but several of medium interest, and others of more local importance were identified. They consisted mostly of small fragments of ancient woodland, a network of hedges, a small fen and small patches of flower-rich grassland on abandoned field headlands or overlying rubble. The BAA committed themselves at the inquiry to incorporating all the better habitats as far as was possible *in situ* into the development (mostly woodlands, some hedges and grassland); to rescuing a mixture of the smaller fragments (mostly of grassland); and to implementing a substantial landscaping scheme.

Once the government approved the scheme, development began in earnest in 1986. Although construction of this massive project was divided into phases, each tends to be run by a different contractor and many are synchronous rather than sequential. Successful habitat rescue without the communities being inadvertently damaged has exercised skills ecologists may not habitually use! Early in the planning process it was decided to move all the rescued habitats to one area to facilitate subsequent management, to increase the potential for viable populations of as many species as possible and, because of the nature of the development, to have all the areas of value safely in one place.

A narrow field and adjoining area totalling 4.5ha was selected as a convenient area (see Figure 18.1). This 'wildlife area' is partly surrounded by thick, diverse hedges, is easily accessible and has a ditch passing through it. The field, which forms a gentle valley, had been used for cereals, and soil tests showed enhanced phosphorus levels compared with soils under the grasslands to be rescued. The first operation, therefore, was to strip the topsoil and remove it from the site for use elsewhere in the development. The habitat design in the Wildlife Area sought to replace the cereals with flower-rich grassland, with the grasslands containing ancient woodland edge species, like greater burnet saxifrage (*Pimpinella major*) (Rackham, 1980), placed next to the old hedges. Because of its rarity in Essex, grassland was to fill most of the area, but top-soil clearance broke most of the field drains and a small, temporary stream developed down the shallow valley. This facilitated the construction of a pond at the lower end, which was also fed by diverting the ditch. The pond was designed specifically for amphibia and invertebrates rather than birds, since bird-strikes are an ever-present risk. It is therefore quite small, with a centre over 1m deep, and shallow, irregular edges. Aquatic and marsh plants were collected by hand in buckets from other ponds which will disappear, and introduced into the new pond, along with some 500 newts and numerous dragonfly larvae, all rescued from a condemned concrete tank.

Since not all the thick, old hedges containing ancient woodland shrubs and herbs could be retained in the development, the opportunity was taken to move some of them to the wildlife area. These form the northern fringe of the site which is contiguous with the roadside woodland landscape planting. The admixture of woodland planting, and an irregular scrub boundary fringing grassland to the south, was felt to promise the optimum habitat conditions for a variety of plants and animals.

The rescue operation

Because of the exceptionally wet winter in early 1986, which left the sticky subsoil in the wildlife area nearly unworkable, the first phase of translocation was not undertaken in the winter as planned, but in April and May. An empirical approach was adopted with regard to machines and methods. Shrubs and small trees were most successfully removed by first coppicing the selected specimens, then using a

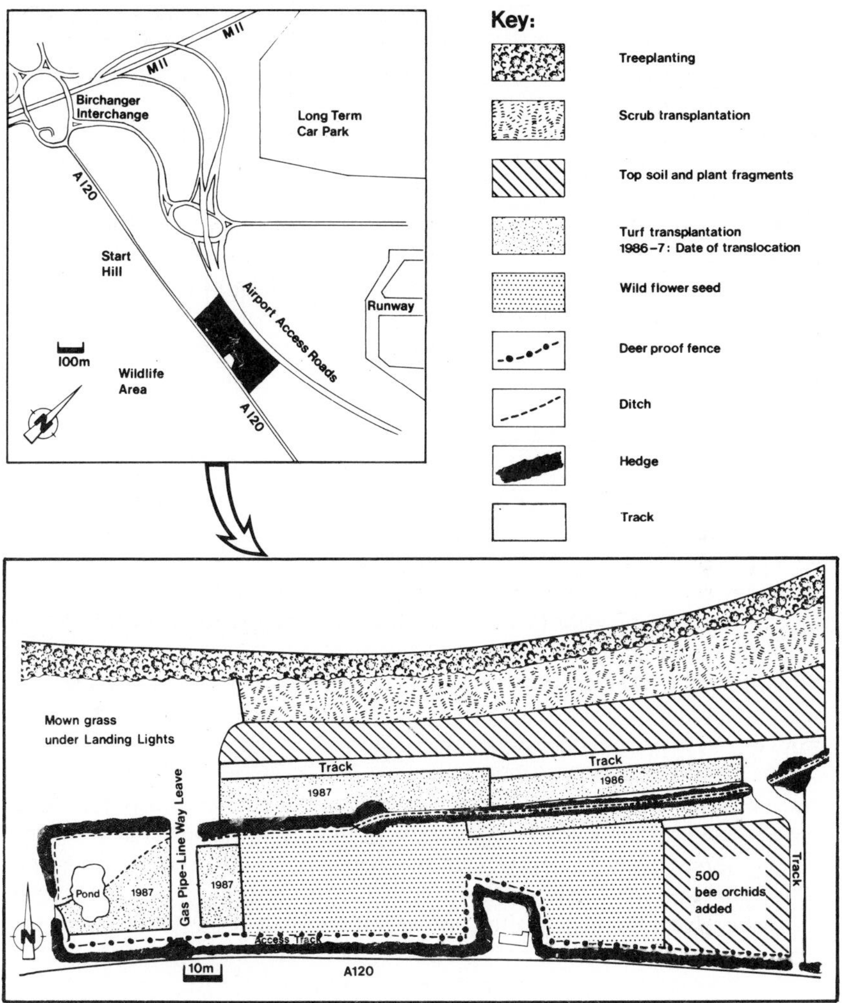

Figure 18.1 The wildlife area at Stansted Airport.

360° tracked excavator with a medium-sized bucket to dig up individuals or groups of specimens. These were transferred to the Wildlife Area on a trailer, placed on the subsoil, and then the spaces between them infilled with soil taken from the old hedges. Since this first trial, topsoil from a small portion of a woodland which was cleared to house a new road has been spread, and the rest of the shrubs removed in March 1988 planted into this soil.

The first areas of grassland turves were transplanted by lifing in a 90cm × 45cm bucket with a specially fitted guillotine attached to a wheeled excavator, and placed two at a time onto pallets on a trailer. Initially they were laid by hand, but the operation was later accelerated by using a fork-lift truck to tip the turves gently into place off the pallets. In 1987, turves were cut using a larger bucket (2m × 1.5m) on a Volvo loading shovel, carried individually to the wildlife area and tipped carefully into place. This last method was more cost effective, partly because the distance from the translocation site was only some 700m. The earlier turves were moved about 2.75km. The turves were cut 20–25cm thick, because it was found that the roots or tubers of plants like bee orchids (*Ophrys apifera*) extended to nearly this depth. Most of the turves were relaid in roughly the order they were excavated so that community types were kept intact.

Turf transfer concentrated on the oldest, longer-established grasslands, but one younger area, which had colonised after abandonment with such species as bristly ox-tongue (*Picris echioides*) and common centaury (*Centaurium erythraea*), was scraped up, re-spread and then power-harrowed to provide a good seed bed. Into this some 500 turf cores containing bee orchids were inserted. These orchids were found to have survived broadleaved herbicide application on the runway grasslands which were otherwise strikingly species-poor. A corer 20cm diameter × 25cm deep was made which fitted onto a Hymac.

The turves and rotovated material did not cover the whole site allocated for grassland, so a wild flower seed mix was specified for the remaining area. However, this was judged by the client as being too costly and a cheaper alternative was found by purchasing a hay field mix from Dr Miriam Rothschild's Sudbury meadow which lies on similar soils to those at Stansted. Seed analysis of this mixture revealed a very similar composition to that required. Dr Rothschild also added some characteristic annual species to the mix including corncockle (*Agrostemma githago*) and cornflower (*Centaurium cyanus*) (See Figure 18.2).

Results

The grassland turves have transplanted very well despite initial concern due to drought. The main area moved contains bee, pyramidal and common spotted orchids (*Ophrys apifera*, *Anacamptis pyramidalis* and *Dactylorhiza fuchsii* respectively), all of which have flowered healthily annually since transplantation. Very few plant species have been lost and most of these seem to be related to the introduction of a mowing regime rather than to the translocation process. Some species have also appeared, due possibly to the creation of small colonisation gaps by the turf removal process, or by mowing. Table 18.1 lists the species recorded in one small area of grassland before and after the operation.

There is no evidence that the transplantation did not also transfer a variety of invertebrates. Unfortunately no monitoring of this aspect was done, but since the grasslands are isolated from others, it is likely that some of the species seen were translocated along with the plants. The degree of efficiency of this operation for invertebrates needs careful scrutiny in future research programmes (see Park in this volume). Butterflies were monitored unsystematically in 1987 and 1988 by members of the Bishops Stortford Natural History Society, and futher invertebrate monitoring is envisaged.

An early-successional community has redeveloped on the rotovated material, with many of the species which had been present previously. The sward after three growing seasons is still sufficiently open for new species to colonise, including small

Figure 18.2 Hay flower mix from Miriam Rothschild's Sudbury meadow.

quantities of common meadow species which were added as seed as a safety measure. The survival rate for the bee orchid plants from the cores is quite high (around 70 per cent), and flowering was prolific in 1988 (226 flowers were counted on a sample of 116 cores). The hay meadow seed mix produced a flush of colour from the annual plants in 1987, which was replaced by the longer-lived species in 1988. The grasslands are being cut in September–October whenever the ground is dry enough for access. The cut material has been raked off and laid over bare areas to provide seeds, or left on the access tracks as a protective mulch. Regular monitoring (which is essential) will determine the future management prescription. Mowing in September will be the most appropriate measure, but if excessive grass growth develops, a May cut may be applied as Wells (1971) showed this to be most effective in reducing vigour of some grasses.

Of the first cohort of transplanted trees and shrubs, only 52 per cent survived two seasons due to rabbit grazing and deer browsing, which removed the new shoots and killed a significant number of regrowing specimens in the first season after relocation. The much higher survival rate (86%) between 1987 and 1988 when grazing was excluded is more encouraging (see Table 18.2). The later phase of transplantation should provide more useful data on survival rates since deer and rabbits are now fenced out.

Monitoring of the ground flora of the first shrub removal trial revealed a mixture of woodland species introduced with the soil, such as dog's mercury (*Mercurialis perennis*), ground ivy (*Glechoma hederacea*), bird's-eye speedwell (*Veronica chamaedrys*)

Table 18.1 The flowering plants in an area of translocated grassland at Stansted Airport

		Before Moving	After Moving		
		1985	1986	1987	1988
Achillea millefolium	Yarrow	+		o	
Alliaria petiolata	Hedge garlic	+			
Alopecurus pratensis	Meadow foxtail	+	r		0
Angelica sylvestris	Angelica		r		r
Anthriscus sylvestris	Cow parsley			o	
Aquilegia sp.	Columbine	+	r	r	o
Arrhenatherum elatius	False oat-grass	a	a	a	a
Brachypodium sylvaticum	Slender false-brome	+	o	o	o
Bromus hordeaceus	Soft brome	+			o
Carex flacca	Carnation-grass	+	o		
Carex otrubae	False fox-sedge	+			r
Carex pendula	Pendulous sedge	+	r	o	
Carex sylvatica	Wood sedge	+		+	r
Centaurea nigra	Knapweed	+	o	o	f
Cirsium arvense	Creeping thistle		r		r
Cirsium vulgare	Spear thistle	+			
Clematis vitalba	Old man's beard	+			
Clinopodium vulgare	Wild basil		lf	o	f
Dactylis glomerata	Cock's foot			f	
Dipsacus fullonum	Teasel	+	o	o	o
Elymus repens	Couch grass		r		
Epilobium hirsutum	Great hairy willowherb	+	o		r
Equisetum arvense	Field horsetail	+		vo	r
Festuca arundinacea	Tall fescue	+	o	f	
Festuca rubra	Red fescue	o	+	o	a
Filipendula ulmaria	Meadow-sweet	+	o	vo	o
Galium mollugo	Hedge bedstraw	+	o	o–lf	f
Geranium dissectum	Cut-leaved cranesbill	+	r		r
Geranium robertianum	Herb Robert	+	r	o	r
Glechoma hederacea	Ground ivy	+	f	f	o
Heracleum sphondylium	Hogweed	+		f	o
Holcus lanatus	Yorkshire fog	+	o	o–lf	a
Hypericum hirsutum	Hairy St. John's wort	+	o	+	
Hypericum perforatum	Perforate St. John's wort	+			
Lathyrus pratensis	Meadow vetchling	+			o
Leontodon hispidus	Rough hawkbit	+			
Leucanthemum vulgare	Ox-eye daisy	+	f	o	r
Malva moschata	Musk mallow	+	o	o	r
Malva sylvestris	Common mallow	+			
Medicago lupulina	Black medick	+		vo	r
Mercurialis perennis	Dog's mercury	+	f	f	f
Myosotis arvensis	Field forget-me-not	+	o	o–lf	r
Origanum vulgare	Wild marjoram	+			
Pimpinella major	Greater burnet saxifrage	+	lf	o–lf	f

		Before Moving		After Moving	
		1985	1986	1987	1988
Poa pratensis	Smooth-stalked meadow-grass	+	+	r	
Potentilla reptans	Creeping cinquefoil	+		vo	o
Primula veris	Cowslip	+	lf	o	f
Prunella vulgaris	Self-heal	+	+		f
Pulicaria dysenterica	Fleabane			r	r
Ranunculus repens	Creeping buttercup	+	o	vo	o
Rubus fruticosus agg.	Blackberry, bramble	+		+	
Rumex conglomeratus	Sharp dock	+	o	o	o
Rumex crispus	Curled dock	r		vo	r
Rumex obtusifolius	Broadleaved dock	+			r
Scrophularia auriculata	Water betony	+	lo		
Solanum dulcamara	Bittersweet	+			
Stachys sylvatica	Hedge woundwort	+	o	lf	o
Taraxacum officinale agg.	Dandelion	+	o		r
Tragopogon pratensis	Goat's beard	+			
Trisetum flavescens	Yellow-oat grass		+		o
Tussilago farfara	Coltsfoot	+			
Urtica dioica	Stinging nettle	+	r	vo	o
Veronica chamaedrys	Germander speedwell		o	o	o
Vicia hirsuta	Hairy tare			vo	
Viola arvensis	Field pansy	+			
Viola hirta	Hairy violet	+	r	vo	f

Key:
a – abundant
f – frequent
o – occasional
r – rare
l – local
v – very
\+ – present

NB. Not all of this grassland area was moved, some remains *in situ*, therefore comparisons of species can only be made with a knowledge of the site.

and bugle (*Ajuga reptans*). Individual oxlips (*Primula elatior*) were planted manually having been rescued from a blackthorn thicket elsewhere on the airport. Many of the other herbaceous species are ruderals, including teasel and welted thistle (*Dipsacus fullonum* and *Carduus acanthoides*). The mixture is reminiscent of the flush of species appearing after coppicing (Rackham, 1980) or those known to be dormant as seeds in ancient woodland soils (Ball and Stephens, 1981). It is anticipated that once the tree canopy emerges, light-demanders will decline, and the woodland herbs will reassert themselves.

Little management is envisaged for the scrub-woodland areas, although a coppic-

Table 18.2 Transplanted wood, scrub and hedges at Stansted Airport

English name	Latin name	No. transplanted	No. surviving 1986	1987	1988
Trees					
Field maple	*Acer campestre*	4	3	2	0
Hornbeam	*Carpinus betulus*	3	3	1	0
Ash	*Fraxinus excelsior*	4	3	2	2
Elm species	*Ulmus* spp.	4	3	2	2
Shrubs					
Dogwood	*Cornus sanguinea*	23	22	13	11
Hazel	*Corylus avellana*	14	9	5	5
Blackthorn	*Prunus spinosa*	183	147	85	78
Rose species	*Rosa* spp.	35	32	30	27
Hawthorn	*Crataegus monogyna*	40	39	28	20
Willow species	*Salix* spp.	2	2	2	1
Elder	*Sambucus nigra*	1	1	1	1
No. unidentifiable when dead			19		
	Total alive:	332	264	172	148
*Tree seedlings**					
Field maple	*Acer campestre*		7		
Hornbeam	*Carpinus betulus*		2		
Hawthorn	*Crataegus monogyna*			2	
Ash	*Fraxinus excelsior*			1	
Sessile oak	*Quercus petraea*			1	
Field maple	*Acer campestre*			2	

* These were not counted in 1988

ing programme may eventually be applied. However, some of the scrub has been planted more spaciously into a dewberry (*Rubus caesius*)-grassland mixture, and this scrub will be maintained by occasional, cyclical cutting to prevent a dense canopy from developing.

The pond vegetation spread so rapidly that it had to be reduced in its second year. The aquatic life will be monitored and it is hoped that the plants will provide sufficient cover for the newts to breed and to attract more dragonflies as well as other species (see Figure 18.3).

Conclusions

The close co-operation between the landscape designers and ecologists has resulted in an exciting venture. Important habitats have been retained, fragments rescued and a valuable wildlife area created as an integral part of the landscaping scheme.

Figure 18.3 New pond within the development, Stansted, Essex.

Furthermore, the landscape design has developed on ecological principles, in an attempt to satisfy not only visual requirements but also to create new habitats. Wildlife conservation has benefited from a new awareness by a major developer who now approaches many land-use issues with a wider perspective.

Acknowledgements

Adrian Lisney & Partners (Landscape Consultants) and Brian Salmon (Engineering Department, BAA Stansted) provided invaluable information and comments for this chapter. I would like to thank them for their assistance over the past few years, but also the BAA for permission to present this chapter. The views expressed are not necessarily shared by the BAA but are those of the author.

References

Anderson, P., Shimwell, D., 1981, *Wild Flowers and Other Plants of the Peak District, an Ecological Study*, Moorland Publishers, Ashbourne.

Baines, C., Smart, J., 1984, *A Guide to Habitat Creation*, GLC Ecology Handbook No. 2, Greater London Council.

Ball, D.F., Stephens, P.A., 1981, 'The role of "ancient" woodlands in conserving "undisturbed" soils in Britain', *Biological Conservation*, **19**, 163–76.

British Trust for Conservation Volunteers, 1976, *Waterways and Wetlands*, BTCV, London.

Davis, B.N.K., 1979, 'Chalk and limestone quarries as wildlife habitats', *Minerals and the Environment*, **1**, 48–56.

Fairbrother, Nan, 1970, *New Lives, New Landscapes*, The Architectural Press, London.

Heath, J. *et al.*, 1984, *Atlas of Butterflies in Britain and Ireland*, Viking, Harmondsworth.

Jarvis, M.S., 1960, *The Influence of Climatic Factors on the Distribution of Some Derbyshire Plants.*, Ph.D. Thesis, University of Sheffield.

Jermyn, S.T., 1974, *Flora of Essex*, Essex Naturalists' Trust Ltd., Fingringhoe, Colchester, Essex.

Nature Conservancy Council, 1984, *Nature Conservation in Great Britain*, Nature Conservancy Council, Peterborough.

Perring, F.H., Farrell, L., 1983, *British Red Data Books: 1. Vascular Plants*, 2nd Edn., Royal Society for Nature Conservation, Nettleham.

Peterken, G.F., 1974, 'A method for assessing woodland flora for conservation using indicator species', *Biological Conservation*, **6**, 239–45.

Peterken, G.F., 1981, *Woodland Conservation and Management*, Chapman & Hall, London.

Piggott, C.D., 1974, 'The responses of plants to climate and climatic changes', in F. Perring (ed.) *The Flora of a Changing Britain*, Classey, Faringdon, pp. 32–4.

Rackham, O., 1980, *Ancient Woodland – its History, Vegetation and Uses in England*, Edward Arnold, London.

Ratcliffe, D.A., 1977, *A Nature Conservation Review*, Vol. I, Cambridge University Press, Cambridge.

Shirt, D.B. (ed.), 1987, *British Red Data Books: 2 Insects*, Nature Conservancy Council, Peterborough.

Tansley, A.G., 1965, *The British Isles and Their Vegetation*, Cambridge University Press, Cambridge.

Usher, M.B. (ed.), 1986, *Wildlife Conservation Evaluation*, Chapman & Hall, London.

Wells, T.C.E., 1971, 'A comparison of the effects of sheep grazing and mechanical cutting on the structure and botanical composition of chalk grassland', in E. Duffey and A.S. Watt (eds.), *The Scientific Management of Animal and Plant Communities for Conservation*, 11th Symposium of the British Ecological Society, Blackwell Scientific Publications, Oxford, pp. 497–515.

West Midlands County Council (no date), *The Nature Conservation Strategy for the County of the West Midlands*, County Planning Dept., West Midlands County Council, Birmingham.

SECTION 6

Transplanting and transferring whole habitats

As discussed in Section 5, 'creating' habitats requires careful site preparation, followed by appropriate seeding or planting. However, the process has severe limitations which militate against a passable habitat stereotype eventually being achieved. Site and soil conditions will inevitably be sub-optimal, plant lists shortened and over-simplified, and refinements such as the introduction of soil organisms or animal populations will rarely be considered an integral part of the scheme. Too much depends upon introducing suitable community members (usually only the correct plants), when other factors may be at least as important. It follows that, in order to improve the accuracy of the reconstruction, the importation of more and more authentic habitat components, culminating in the transference of whole habitats from one place to another, will be expected to give more reliable results.

Several types of habitat transfer have been attempted. Apart from the use of seed and horticulturally propagated plants, it has been shown to be feasible to import turves (widely spaced to act as nucleii for colonisation or laid as a carpet), thin layers of seed-rich soil, and even whole soil profiles from suitable 'donor' sites. Such transplantation schemes are becoming increasingly common, and it has been estimated that in the past seven years at least thirty habitat transfers have been carried out in England alone (Prigmore, 1987), many of them 'rescue' operations to compensate for the destruction of semi-natural vegetation on nearby development sites. The scale of these habitat transfers varies enormously according to the manpower and machinery available, but those involving manual transplanting of turves from one site to another by conservation volunteers are by their very nature extremely limited in area.

It is perhaps significant, therefore, that the three cases of habitat transfer discussed in this section are all development schemes involving mineral extraction industries, where heavy earth-moving equipment is freely available and can be adapted to move species-rich or scientifically interesting communities. In the case of limestone grassland described by Park, the process of habitat transfer is made a particularly viable proposition by a relatively flat site, a shallow rendzina soil profile which is easily defined and handled, and short haul distances between respective 'donor' and 'receptor' sites. The value of the mineral, taken against the scientific value of the grassland itself, makes it worthwhile for industry to invest in the special machinery and the additional soil-moving operations demanded by habitat transfer.

As well as moving soil as intact turves or profiles, another technique is to carry out soil-stripping as part of a conventional earth-moving exercise, directly transferring soil layers containing vegetation as seed, storage organs and plant fragments. Halliwell shows how a number of such layers can be moved in order to re-create the complete soil profile of wetland site, including the engineering of the water table of the 'receptor' area so that it resembles the 'donor' site.

Given the availability of large machines, there is no reason why habitat transfer should be restricted to herbaceous vegetation and to soil. In this volume several references are made to transferring scrub or hedgerows, while Down and Morton show that it is even possible to move quite large trees considerable distances, complete with soil and ground flora plants, in order to form an instant woodland canopy on a new site.

The results of habitat transfer schemes will take time to assess, as subtle changes will occur as the transferred community adapts both to the initial disturbance and to new environmental conditions. Fortunately some of the larger-scale examples of habitat transfer are being carefully monitored by the industries responsible. The onus is on them to demonstrate the feasibility of such operations, and to be able to justify them to the satisfaction of planners, conservationists and ecologists.

References

Prigmore, D.S., 1987, *The Role of Habitat Transplanting in the Conservation of Semi-natural Communities*. (Unpublished M.Sc. thesis), Imperial College of Science and Technology, University of London.

19

A case study of whole woodland transplanting

G. S. Down and *A.J. Morton*

Introduction

Between 4 April and 19 May 1978 a Public Inquiry was held at Dartford, Kent, United Kingdom, into an appeal by Associated Portland Cement Manufacturers Ltd. (now Blue Circle Industries PLC) against the refusal by Kent County Council to permit the extension of chalk quarries supplying Northfleet and Swanscombe cement works. The proposed extension, for convenience termed the Darenth Quarry, lay immediately to the south of the existing Western Quarry. Within the 154 ha extension area the land uses were: a hospital (85 ha); two areas of agricultural land (37 ha); 5 ha of housing; and Darenth Wood, a Site of Special Scientific Interest (SSSI), some 28 ha in extent. Quarrying proposals would have resulted in the excavation of the majority (23 ha) of the wood.

At the Inquiry the quarry company undertook, in the event of permission for quarrying being granted, to create by transplanting or other means a new woodland (on a 6 ha adjoining site) containing many, perhaps most, of the Darenth Wood species growing in similar soil types. After the Inquiry closed, we were asked by the quarry company to conduct trials into ways of achieving a reasonable replica of Darenth Wood. However, our research and trials did not continue to completion because when planning permission was refused the investigations were terminated. This paper therefore includes discussion of what was intended as well as what was actually done.

Darenth Wood SSSI

This SSSI was designated by the Nature Conservancy in 1968 as 'A woodland on two rock types although much disturbed is still of value particularly entomologically'. The total area was 121.24 ha, but the wood had subsequently been severed by construction of the new A2 road, the northern 28 ha portion being part of the area proposed for quarrying.

The basic woodland comprised sessile oak standards with hornbeam and sweet chestnut coppice. Silver birch was also common. Evidence suggested that it was an ancient semi-natural woodland which had been actively coppiced up to 1939–45 and subsequently neglected. Soils were derived from three rock types (Upper Chalk, Thanet Sand, Plateau Drift) which had given rise to five main soil types. It had been found that each soil was anomalous, with characteristically acidic soils being, in their upper horizons, alkaline and containing high levels of exchangeable potassium. This observation was consistent with a long history of atmospheric pollution in the area caused by alkaline dusts from cement manufacture.

The problem

The basic task was considered to consist of three elements: (1) re-establishment of coppice and standard trees; (2) re-establishment of ground flora; and (3) replication of soil types. The latter, which would have entailed soil transfer from Darenth Wood to the new woodland site, was not considered in detail.

A study of the literature revealed that although there existed ample information on coppice management, there was virtually no written experience on establishing coppice woodland from scratch. However, in all coppices there would arise the need to establish new stools to replace dead individuals, and the methods employed had included planting seedlings or cuttings, direct seeding, selecting existing seedlings for growing-on and layering.

Unfortunately, the need in this case included the rapid re-establishment of ground flora. Although it would have been possible to plant seedling trees and wait 15–20 years before initiating coppice management, this would not have permitted ground flora to be planted early, since woodland species in open ground would be likely to die or be crowded out with weeds. This led to the conclusion that early provision of shade was required. The costs and logistics of artifical shading were considered unattractive, and the decision was made to base shade provision on the transplanting of existing woodland trees.

No matter how well the transplanted trees might grow, it was recognised that there would be a period (perhaps 2–4 years) before they provided useful shade, and thus interim provision for maintaining ground flora would be required. This consideration impinged upon the choice of method for re-establishing ground flora. While the transfer of woodland soils would have automatically included a variety of propagules, it was thought that the random species mix likely to result from this could be undesirable, and that a more positive approach was needed. Seed collection would have been possible but germination reliability of wild seed is not always good (Brown, in this volume). Thus it was decided that transplanting vegetative stock into a nursery, and propagating the required species, would be the most positive way of achieving the desired species composition.

The final matter of concern was to provide some insurance against failure of the mature tree transplants. Even in commercial landscape work, using semi-mature or mature trees transplanted by British Standard methods of preparation and movement, failure rates can be high, while considerable after-care is required. Accordingly it was decided to supplement transplanting with wild seedling trees and commercial nursery stock.

Biological survey and design

As a precursor to any full-scale scheme, a survey of Darenth Wood was essential to provide a model upon which to construct the new wood. In particular, it was important to answer four questions:

1. What are the main components of the woodland canopy and how are they distributed?
2. What are the main components of the ground flora and how are they distributed?
3. What is the distribution of any species of particular rarity or aesthetic value?
4. What soil: vegetation relationships exist?

A survey was undertaken in July 1979 to answer these points. Vegetation was surveyed in 102 5m × 5m quadrats placed at the intersections of a 50m grid, for percentage cover (ground flora) and species present (all flora). Soils were surveyed on a 100m grid, as were measurements of light intensity. Additionally, certain areas were mapped in greater detail, to yield spacing and dimensional data useful in laying out the new woodland.

The survey information gave an overall description of the wood in terms of canopy and ground flora types and their relative proportions and distributions within the wood. Information on relationships with soil and light conditions was also gained. If the ultimate aim was to produce a replica of the wood, then this detailed information could have been used to define a model of the whole wood. In the circumstances it seemed desirable and more practical, however, to produce a simplified model of the wood, using those elements which were of highest conservation value. We therefore selected for our model the areas of hornbeam coppice only, and used this to define the ground flora types, soil conditions and light conditions which should be created in the new wood.

An ordination technique was used to structure the ground flora data, and it was found that the hornbeam coppice contained three main ground flora types:

1. Dominated by bramble, (*Rubus fruticosus* agg.) occurring on non-calcareous soils with a mean pH of 5.1 and mean light conditions of 3.7 per cent full daylight. Represented in 20 per cent of the hornbeam/birch/sessile oak stands.
2. Dominated by dog's mercury (*Mercurialis perennis*), occurring on soils with a mean pH of 6.6, sometimes calcareous, and mean light conditions of 1.3 per cent full daylight. Represented in 30 per cent of the hornbeam/birch/sessile oak stands.
3. Sparse ground cover, occurring on soils with a mean pH of 5.5, and mean light conditions of 1.7 per cent full daylight. Represented in 26 per cent of the hornbeam/birch/sessile oak stands.

Consideration was now given to the question of whether all of these ground flora types needed to be represented in the new wood. There appeared to be a number of possibilities which included: (1) creating all three main ground flora communities; (2) creating one or two of the communities, chosen to be of greatest importance (e.g. conservation value); or (3) creating one main ground flora community, chosen to be that of poorest colonising ability, i.e. the least able to reconstitute itself.

Alternative (3) was attractive in that it was relatively simple, and might in time develop into a mosaic of ground flora types similar to those in Darenth Wood. Bramble would probably invade areas planted with dog's mercury if the conditions of soil pH and light described earlier in this section were provided.

Assessing the relative conservation value of the three ground flora types was difficult. On diversity grounds an analysis of the data indicated that the dog's mercury stands had an average of 4.8 species compared with 4.2 in the sparsely covered stands and 3.5 in the bramble stands. It was decided, therefore, that ground flora planting should consist largely of dog's mercury, planted into a hornbeam coppice providing a range of shade conditions from deep shade (about 1 per cent full daylight) to moderate shade (about 5 per cent full daylight). This should have created suitable conditions for the development of ground cover ranging from sparse to dense cover of dog's mercury and ultimately some bramble. Planting of other ground flora species (notably primary woodland species and local uncommon or rare species) could be carried out at the same time, transplanting from a nursery or another woodland. A range of soil conditions at the 'reception' site was required, the pH ranging from about 5 to 7. Calcareous soils should be represented but more acid conditions were also needed if the full range of ground flora types and primary woodland species found in the hornbeam coppice of Darenth Wood were to be established.

Experimental work

Darenth Wood was not available as a source of material, but an alternative woodland existed nearby in an area which was about to be quarried away. This woodland, although not identical to Darenth Wood, contained all of the main species of interest. A reception site for transplanted trees was allocated on an undisturbed quarry rim.

Twelve trees were transplanted, eight hornbeam, and an oak, ash, hazel and willow (Figure 19.1 and 19.2). Trees up to 8m high were moved whole; larger trees were cut back to coppice stools, partly to facilitate the physical problems of moving them and partly to reduce their water requirements on transplanting. The methods employed were deliberately crude, using only commonly available quarry machinery: a hydraulic back-hoe dug out the rootball and felled the tree; the tree was then coppiced; the stump loaded into a lorry and taken to the reception site where the procedure was reversed. Machine operators quickly became adept at these techniques and, with a 2km transfer, a rate of about 30 trees per day could have been sustained.

At the reception site, planting pits had already been prepared. A herringbone ditch pattern, connecting these pits, was also prepared, the idea being that a water bowser could discharge into the upper end of the system and thus water each tree very simply. In the event, the system was used once but was not required thereafter.

On the same site, one hundred commercial standard hornbeam were planted by conventional techniques, while 26 forest seedling hornbeam (kindly provided by the Nature Conservancy Council) were also established. Since 1978–79, when all these

Figure 19.1 Loading a cut stump using a sling.

Figure 19.2 Lifting out a semi-mature hornbeam specimen: the relatively small root ball obviated the need for slings

plantings were carried out, no maintenance of any description was undertaken.

An initial collection of ground flora was made in Spring 1979. The species chosen were *Anemone nemoroa, Mercurialis perennis, Euphorbia amygdaloides, Adoxa moschatellina, Hyacinthoides non-scripta, Luzula pilosa, Iris foetidissima, Arum maculatum, Carex sylvatica, Bromus ramosus* and *Brachypodium sylvaticum*. These were planted out in a nursery at Imperial College's field station at Silwood Park. Unlike the trees, the ground flora received close attention, and following good early establishment many established individuals were split. Such multiplication increased the numbers of individuals up to sixfold.

Results

Up to the termination of formal monitoring in 1980, all the mature tree transplants were surviving and had put on new growth. Leaf emergence tended to be later than for nearby undisturbed trees, leaf fall came earlier, and leaf sizes were reduced, indicating some transplanting stress. Since that time, casual inspection has shown that all trees have continued to thrive, although not until five years had elapsed was any useful degree of shade achieved. The forest seedlings likewise thrived,but the commercial hornbeam was not at all successful, with 20 per cent total failure and 50 per cent partial failure.

Ground flora propagation was extremely successful with at least75 per cent initial establishment rate. The subsequent growth rates of the woodland grasses (*Brachypodium sylvaticum* and *Bromus ramosus*) and the sedge (*Carex sylvatica*) were particularly good. Up to termination of the study, useful detailed information on plot design and planting density, to facilitate weed control, propagation and management, had been gained.

Discussion and conclusions

Although this chapter considers the specific work conducted for Darenth Wood, the lessons learned may have a more general applicability. The first and most important result is that it was technically and biologically practicable to transplant both canopy trees, and a variety of ground flora, by simple methods, and achieve establishment rates close to 100 per cent. No lengthy pre-preparation, nor unusual machinery, were required. A degree of specialist attention was, however, necessary for ground flora propagation.

It was not possible to answer the question 'would the ground flora have survived being transplanted from the nursery to the new "wood", and subsequently have thrived?' Since it took longer to achieve worthwhile shading than had been anticipated, a longer holding period in the nursery would have been essential.

The economic feasibility of the approach is also important. In 1988, assuming a transfer rate of 30 trees/day a cost of about £12 per tree moved is calculated. Ground flora propagation is not easy to cost, but pessimistically might be assumed to be equivalent. In other words, one tree plus ground flora within a 'radius of influence' might cost £20–25. At a tree planting density of say 500 per hectare, a capital expenditure of about £10,000 per hectare might be expected. This would be a high, but by no means unreasonable, expenditure.

Overall, it is considered that the work proved successful as far as it went, but much further study is essential before firm conclusions can be drawn.

References

Down, C.G., Morton, A.J., 1979, *Studies of the Re-creation of North Darenth Wood. First Report: Establishing Experimental Sites*, Imperial College Centre for Environmental Technology Library, London.

Down, C.G., Morton, A.J., 1979, *Studies of the Re-creation of Darenth Wood. Second Report*, Imperial College Centre for Environmental Technology Library, London.

Down, C.G., Morton, A.J., 1980, *Studies of the Re-creation of Darenth Wood. Third Report*, Imperial College Centre for Environmental Technology Library, London.

20

Soil transfer as a method of moving grassland and marshland vegetation

D.R. Helliwell

Introduction

There are several possible approaches to the transference of vegetation from one site to another. These include the use of seed, turves, or a thin layer of seed-rich soil (e.g. Miles, 1975; Wathern and Gilbert, 1978; Wells *et al.*, 1981). Movement of an intact soil profile, complete with surface vegetation, has also been used, and is described elsewhere in this volume by David Park. A similar approach has also been used by Penny Anderson (in this volume). This paper describes a different approach, which is likely to be less time-consuming and expensive than the movement of intact soil profiles.

The emphasis in this method is on retaining the characteristics of the soil. The vegetation is retained within the upper soil layer, but not as an intact layer of 'turf'. Theoretically, if the soil can be transferred in such a way that its physical, chemical and hydrological characteristics can be maintained, and an appropriate reservoir of plant propagules exists in the upper layer, then the vegetation which grows should be similar to that which existed previously. The way in which the new vegetation develops will depend primarily on the management to which it is subjected.

Methods

Under the direction of their land manager Barry Bransden, Greenham Construction Materials Ltd have for several years been restoring infilled gravel excavation areas to a high-grade agricultural land. Their method involves careful movement of soil in two or three layers, depending on the characteristics of the soil, to a depth of about 1 m. These layers are carefully removed by a hydraulic excavator, which stands on ground from which the soil has already been removed. The soil is removed and replaced in strips of about 6 m width, this being the width that can be worked by an excavator standing to one side of the strip. At no time is it necessary for excavators or dump trucks to travel on the soil.

At the start of the operation, a small quantity of the upper layer is removed and temporarily stored to expose the lower layer. The lower layer is then removed and transported by dump truck to a receptor site, where it is tipped out and levelled by the bucket of an excavator, over a similar area to that from which it was removed. The upper layer of soil from the next part of the donor site is then removed, transported and placed on top of the lower layer previously transported. This operation exposes the lower layer of soil on the second part of the donor area, which can then be removed. This stepped sequence of movements proceeds until all the soil has been removed from the donor area, when the stored topsoil is used to complete the final section. Three (or more) layers can be moved in a similar fashion, if appropriate amounts of each of the upper two (or more) layers are temporarily stored.

An example of the 'soil transfer' method used by Greenham Construction Materials Ltd is given below.

Donor site

The donor site in this case study was a 4.6ha area of hawthorn scrub, grassland, marshland and open water situated 2km west of Heathrow Airport and 26km west of Central London, which was due to be worked for gravel. Some 4,350m^2 of this included grassland or marshland vegetation of moderate to high floristic diversity, with a total of 85 species of vascular plant recorded on 1 and 2 October 1980. Over part of the area, 'dry' grassland occurred where the soil depth was about 400mm of well-structured loam/silty clay loam over gravel. 'Moist' grassland areas had a similar depth of soil of higher clay content (silty clay loam/silty clay) and were closer to the water-table. 'Marshy' vegetation was found on low-lying areas with up to 1.25m or more of silty loam/silty clay loam, with evidence of regular waterlogging.

Receptor site

The receptor site was a 40 × 100m rectangle within an area from which gravel had already been removed and then refilled with inert waste; mainly clay, with some bricks and concrete. The base of the receptor site was graded to provide for a higher finished level at the 'dry' end to that at the 'wet' end. Unwashed gravel was spread to a thickness of 400 mm over the base where the 'dry' and some of the 'moist' grassland was to be placed. The 'marshy' vegetation was transferred to the 'wet' end of the site.

As the level of the groundwater at the receptor site was likely to change in the future, as a result of gravel extraction activities in adjacent areas, the area was designed to be higher than the general groundwater level, and water was pumped into a 'French drain' of coarse aggregate around it in order to obtain a controllable level of groundwater.

Results

The results of this work have been summarised by Worthington and Helliwell (1987). At the receptor site there was a rapid re-growth of vegetation, with almost complete cover by the end of May 1981. There was an initial increase in species

Figure 20.1 The experiment at Hithermoor, 1980: front edge of reinstatement.

diversity in all three vegetation types (see Table 20.1). This increase has not been maintained in the 'dry' and 'wet' areas, but the 'moist' area is still significantly more diverse than it was in 1980.

Discussion

The decrease in species density in the 'dry' and 'wet' parts of the receptor area is almost certainly due to deficiencies in the water regime. The pump was, until the latter part of 1987, pumping water into the 'French drain' near to the 'dry' end of the site. Imperfect flow has resulted in the 'dry' end of the site being wetter than it should be, whereas the 'wet' end has frequently been too dry. This fault has, it is hoped, now been corrected, but it may be too late to recover the full diversity of species in the 'wet' area. It will be interesting to see what happens over the *next* seven years.

It had been predicted by some observers that such transference would result in a large increase in ruderal species and pernicious weeds such as creeping thistle (*Cirsium arvense*) and that many other species would be lost from the vegetation. This prediction has not turned out to be true. The vegetation is not identical in every aspect to the vegetation which existed previously, but it is very similar.

Table 20.1 Numbers of vascular plant species recorded, 1980–87

	'Wet area'		'Moist area'		'Dry area'	
	No. of species*	No. per m^2	No. of species	No. per m^2	No. of species	No. per m^2
1 Oct. 1980 (before transference)	54	10.7	47	10.4	43	15.9
2 Oct. 1981	61	16.0	48	13.9	41	17.0
1 Oct. 1985	44	10.7	46	14.1	33	10.3
1 Oct. 1986	49	11.8	56	14.6	45	12.6
1 Oct. 1987	39	9.1	59	13.7	44	8.6

*Total number of species recorded within the area, including those not recorded in any of the sample m^2 quadrats.

Figure 20.2 Hithermoor, 1981: looking west from the 'Dry' end of the receptor site.

Figure 20.3 Hithermoor, 1981: the 'Wet' end of the receptor site.

It is claimed by some observers that, even if the vegetation contains the same species in similar proportions, it can be no more than a facsimile of the previous vegetation, and that the link with the past has been broken. In certain cases this is clearly a valid point, particularly where studies of vegetation in relation to soils are currently being carried out. However, in other cases such a view might be regarded as being unduly purist and, as Jordan (in this volume) points out, there can be considerable heuristic value in habitat transference, reconstruction or repair. One major advantage of the method used in this instance is that it can be carried out reasonably quickly. It does not require large numbers of volunteers (or paid staff) to handle turves. Nor does it require any machinery which is not generally available throughout countries such as Britain and the rest of Europe.

References

Miles, J., 1975, 'Performance after six growing seasons of new species established from seed in Callunetum in north-east Scotland', *Journal of Ecology*, **63**, 891–901.

Wathern, P., Gilbert, O.L., 1978, 'Artificial diversification of grassland with native herbs', *Journal of Environmental Management*, 7, 29–42.

Wells, T.C.E., Bell, S., Frost, A. 1981, *Creating Attractive Grasslands Using Native Plant Species*, Nature Conservancy Council, Shrewsbury.

Worthington, T.R., Helliwell, D.R., 1987, 'Transference of semi-natural grassland and marshland onto newly created landfill', *Biological Conservation*, **41**, 301–11.

21

Relocating magnesian limestone grassland

D.G. Park

Introduction

The semi-natural grasslands characteristic of the magnesian limestone escarpment in Durham, England, are an important biogeographical bridge between the montane limestone grasslands to the north and the lowland calcareous grasslands to the south. Phytosociologically, the magnesian limestone grasslands form an association of the sub-alliance *Seslerio-Mesobromion* of the class *Festuca-Brometea*: this association, the *Seslerio-Helictotrichetum*, is a grassland type unique to the magnesian limestone of eastern Durham with no counterpart in Western Europe (Shimwell, 1968). Historically, it would appear that this type of grassland was once more widespread over eastern Durham, but agricultural losses have restricted it to small isolated areas, several of which are under threat from industrial development. One such area is Thrislington Plantation.

The planning application submitted in 1979 to extend the existing quarry at Thrislington met with considerable opposition on several accounts, a major one being that the proposed extension included part of the Thrislington Plantation Site of Special Scientific Interest. The Plantation is scheduled as a grade one site, and it is generally accepted that the area supports the most diverse, representative and extensive example of the magnesian limestone grassland remaining in northern England (Richardson, 1984). Following lengthy public inquiry, consent was given to the application, but with a condition requiring the operator to relocate the affected part of the Site of Special Scientific Interest to an adjacent site which would not be affected by future quarry development (Stanyon and Park, 1987).

In 1982, Steetley Quarry Products Limited embarked on a unique project in habitat management when the first area of semi-natural grassland was physically relocated. This chapter considers the work undertaken during the first two years of the project, the methods employed during relocation, and the results both in terms of the procedural difficulties encountered and the condition of the grassland in the years following relocation. Various aspects of grassland relocation are drawn together and discussed in the final part of the chapter.

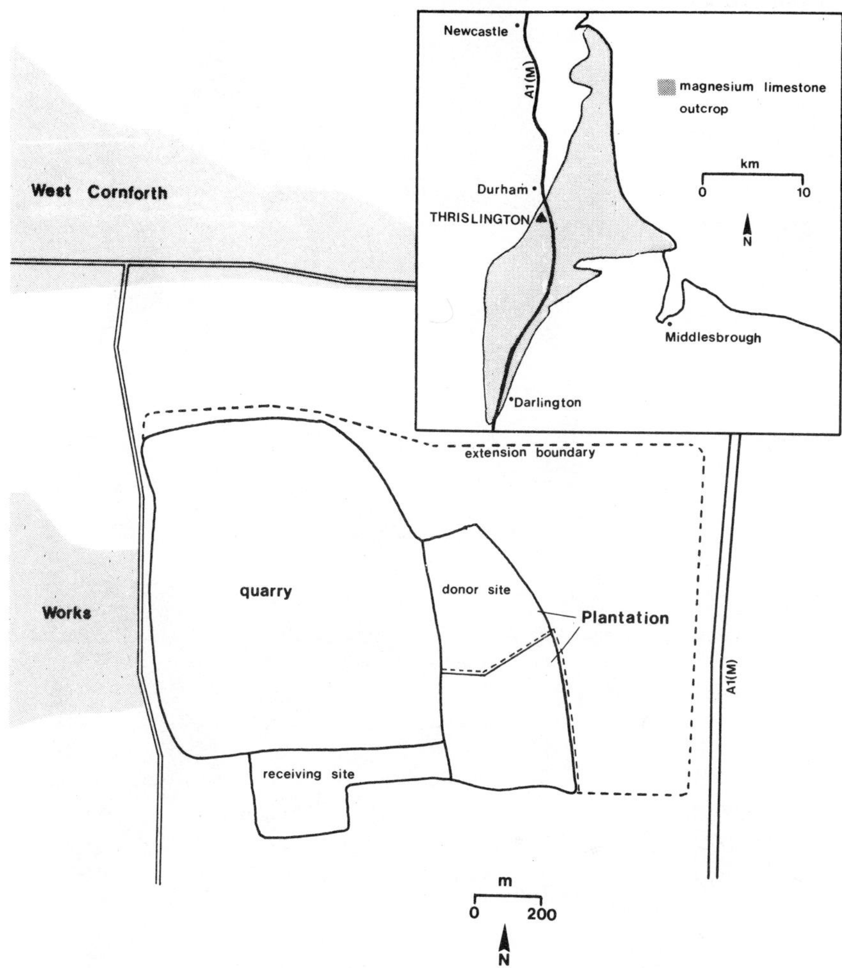

Figure 21.1 Location of Thrislington Plantation.

Site description

Thrislington Plantation (OS Ref. NZ 318 328) lies 12km to the south-east of the city of Durham, on the magnesian limestone escarpment which forms the western boundary of the East Durham Plateau (see Figure 21.1) The Plantation covers an area of 20.5 hectares and is broadly rectangular (700 x 350m) in outline, aligned north-south along its largest axis. The area is topographically heterogeneous, presenting slopes of different aspects and varying angles of repose; however there are tracts of relatively flat ground, particularly in the northern part.

The magnesian limestone bedrock (dolomite) is overlain by a layer of loose dolomitic-silcaceous material or overburden, and is of variable thickness (0.5 – 5.0m) and texture, ranging from a soft, powdery 'marl' through a very coarse 'gravel' to consolidated rock. The soil cover over much of the Plantation is typically a brown

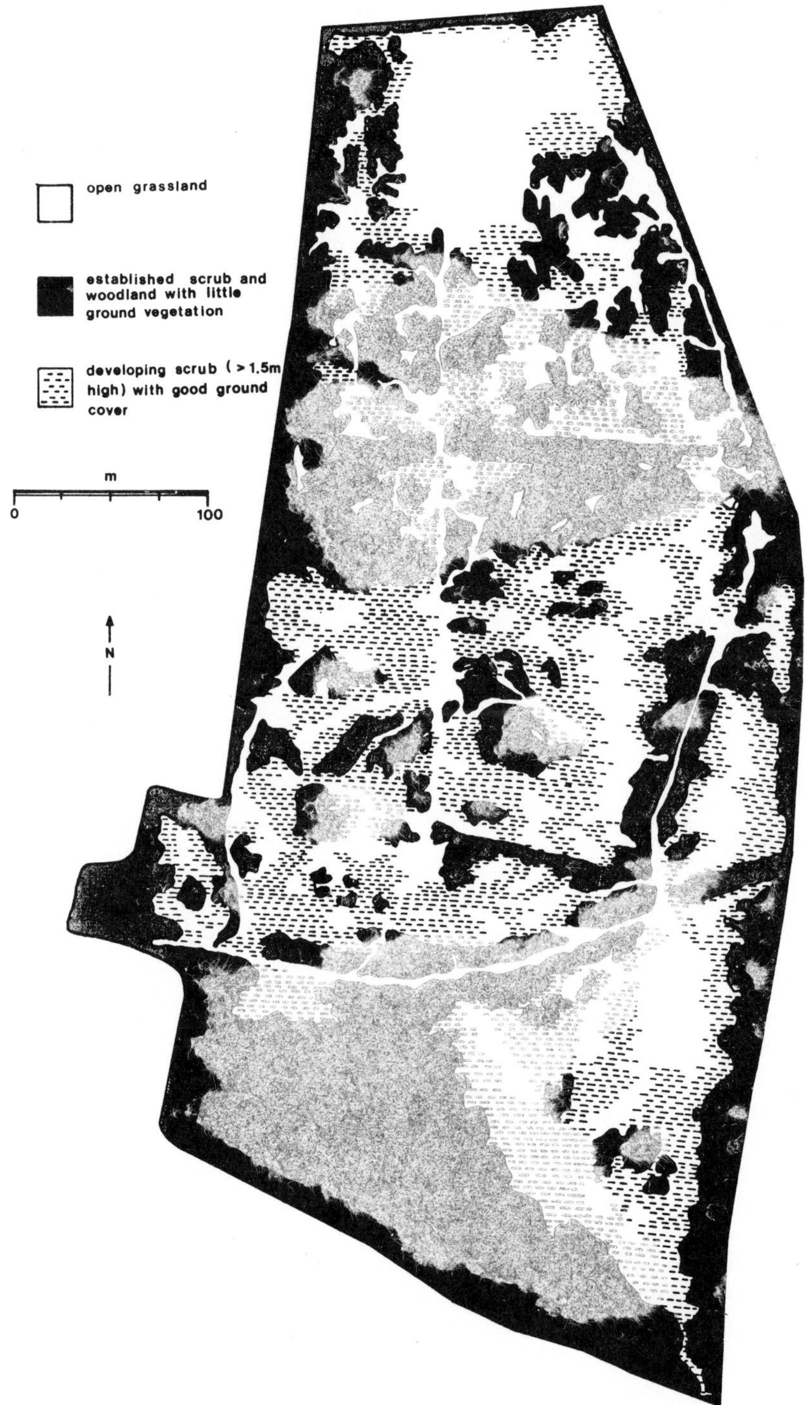

Figure 21.2 Distribution and development of scrub within Thrislington Plantation.

rendzina (the Elmton series of the Aberford Association) and is a shallow (100 – 300mm), slightly to moderately stony, clay-loam. However, there are isolated areas throughout the Plantation and more particularly in the extreme southern part which have deeper (550 – 850mm) soils similar to brown calcareous earths.

Although primarily important as a grassland site, the Plantation supports a widespread and diverse scrub vegetation (see Figure 21.2); the extreme southern part supports semi-mature secondary woodland and throughout the remainder of the Plantation there are areas of very dense shrub. These have particular impact on the relocation project.

A total of 8.5 hectares of the northern part of the plantation is affected by the turf transplantation operation (hereafter referred to as the donor site): the southern 12 hectares will remain *in situ* and are safeguarded from future mineral extraction. On the donor site, it will prove practical to transplant around 5.5 of the 8.5 hectares: the outstanding area is considered unsuitable for relocation as it has been altered by re-seeding (1.25 hectares) or is covered in dense scrub (1.75 hectares).

The grassland receiving site lies 800m to the south-west of the donor site and comprises two fields which were under arable cultivation until 1982 but have since lain fallow. The soil is generally shallow (300 ± 550mm) and past cultivation has brought fragments of dolomite to the surface. The underlying overburden varies in thickness and texture as previously described. The land slopes gently (c.3°) in a south-westerly direction and contrasts to the slope in the donor site (2° in a south-easterly direction).

Methods

Site preparation

The grassland receiving site is prepared before turf relocation by removing the soil to expose the overburden. As the overburden weathers easily, only an area sufficient to accommodate the expected amount of relocated grassland is exposed in any one year. The area stripped of soil is not levelled to any significant degree (i.e. using mechanised graders): the natural boundary between soil and overburden is followed as far as possible to produce a generally rolling appearance to the stripped area.

Preparation of the donor site requires the removal of trees and shrubs as near as possible to ground level. This is necessary to avoid the break-up of the turf, and to allow later steps in the relocation procedure to be carried out. Cut wood is manhandled from the grassland and stumps are treated with herbicide (trichopyr diluted in diesel oil).The ground immediately in front of the cleared grassland is prepared by removing *c.*500mm of soil, equivalent to the approximate thickness of the turves which will be taken. The overburden is normally exposed during this operation and is graded to produce a level surface.

Relocation procedure

Turf relocation is carried out by earth-moving machinery already in use within the quarry, but equipped with a specially designed transplant bucket. The actural procedure is straightforward, and has been developed through trial and error.

An articulated, wheeled, front-bucket off-loader (specifically a Caterpillar(R) 988) is used, weighing approximately 41,000kg and power-rated at 280kw (375 horsepower). The transplant bucket measures 4.75 × 1.75m and is equipped with a flat bottom, including reinforcing plates and wear-strips, and with a half-arrow cutting edge manufactured from toughened steel. The side-plates also have cutting edges.

Turf relocation incorporates three operations: first, during the pick-up operation, the machine is manoeuvred to present the cutting edge of the bucket parallel to the free edge of the grassland. The bucket is then raised, tilted until it hangs vertically and then lowered,slicing a turf approximately 1.9 – 2.0m, in from the free edge. The machine then reverses to allow the bucket to be lowered in front of the turf, and slowly advances into the turf with the side-plates cutting the remaining edge(s) of the turf. After pick-up, the machine carries the turf to the receiving area. Transfer is necessarily slow (approximately 16kmph) to avoid undue disturbance. During the set-down operation, the machine is manoeuvred to align the edge of the bucket parallel to the edge of the previously laid turf and to leave a gap of around 250mm. The bucket is then raised to a height of 1–1.5m and tilted until the edge is around 150mm above the overburden. The machine then advances to push the new turf up to the previously laid turf: the bucket is tilted and the machine slowly reverses, leaving the turf on the overburden. Finally, any gaps between the turves are infilled with soil. The relocation procedure, excluding infilling with soil, takes 12 – 20 minutes depending on site conditions.

Survey and recording

After the removal of scrub from the donor site, a vegetation survey is undertaken to identify the plant communities present within the cleared areas, using a classification of communities of the magnesian limestone grassland similar to that of Shimwell (1968). Within each delineated community stand, random quadrats are taken in which species presence and percentage cover are recorded. The survey is important to the programming of the sequential removal of grassland from the donor site in relation to the position of relocated areas in the receiving site.

Permanent quadrats are monitored to determine possible changes in the grassland following relocation. Quadrats were established in the donor site one or two years before relocation was to occur, but because of the difficulty of picking up a quadrat turf entire, additional quadrats were established in the receiving site immediately after relocation. Quadrats were established in 1982 and 1983, in a range of situations including areas of dense scrub clearance and soil infill areas. Thirty quadrats, each measuring 620 × 780mm, are sampled towards the end of June and July each year when species' presence and percentage cover are recorded. Cover is estimated using a non-repetitive point sampling method employing a frame of 20 points taken at three random positions within the quadrats.

Four turves removed from the Plantation in 1977, under the guidance of the Nature Conservancy Council, provide further information on possible changes in the grassland following relocation. The turves, each measuring 1200 × 600mm and 150mm deep, were relocated using an ordinary back-actor excavator and laid side by side in an exposed position on a waste mound. The turves were recorded for species presence in 1977 and 1978 but were not recorded again until 1982 and annually thereafter.

In 1982, an invertebrate survey of the donor site was undertaken by the Nature Conservancy Council to establish a baseline from which any changes in the fauna after relocation could be assessed. Samples were also taken from part of the Plantation which was not to be moved. The survey was repeated in 1985 when samples were also taken from the receptor site. Three sampling methods were employed. A grid of nine pitfall traps were set on each of the sampling sites: traps were set in May and emptied monthly until September. Sets of five samples obtained using a Dietrick pattern insect suction sampler were taken from each site on five occasions in each year. Finally, a lightweight sweep-net was used to obtain five sets of samples, each consisting of 20 sweeps, taken from each site on five occasions in each year. In this way it was hoped to obtain a sample of the representative species from the soil surface, the lower vegetation and the upper vegetation and air.

Results

Areas of grassland have been relocated over a period of seven years. During this time several constraints to the successful transfer of turf have been identified; some of those highlighted below may have wider applications for similar relocation schemes. This section is followed by an examination of the survey and monitoring work on the relocated grassland and finally two short sections examine the costs of the operation and the management of the relocated grassland.

Operational Constraints

TIMING OF OPERATIONS

Successful relocation of the grassland appears to depend to a large extent on the maintenance of the soil profile during pick-up and set-down operations. The set-down operation, if undertaken during summer months when the moisture content of the soil is low, was found to be unsuccessful. This happened because the upper 100–200mm of the soil profile, containing the highest density of grass and herbaceous plant roots, fractured over the lower profile material during set-down and formed voids, which later collapsed to give an irregular surface which was dangerous to walk over. These problems were avoided when the soil had a higher moisture content and as a consequence, relocation now occurs between October and March. While this period is acceptable ecologically, difficulties can occur in bad weather. Very wet conditions with standing surface water restricts the manoeuvrability of the machine, and relocation is not undertaken when there is snow cover. Severe ground frosts, however, help to maintain the integrity of the turf during relocation.

EQUIPMENT

Grassland relocation at Thrislington makes use of ordinary earth-moving equipment in a very precise operation. The main problems arise under wet conditions which cause wheel-slip and lateral movement of the machine, causing difficulty during pick-up and set-down of the turf.

The design of the transplant bucket was affected by several factors, including the need for the length of the bucket to be greater than the wheel-base of the machine, and to avoid undue stress on the arms during working.

GRASSLAND CONDITION

The presence of invasive scrub is a familiar problem in the management of calcicolous grasslands and poses additional constraints on relocation procedures. The need to remove trees and shrubs has already been referred to: however, due to the presence of tree and shrub roots within the turf area, a loss of turf amounting to 2–7 per cent of that picked up has been sustained as a result of the turf tearing and fragmenting. Where the scrub is dense, the ground vegetation is poorly developed and such areas do not relocate satisfactorily. Peripheral areas to densely scrubbed parts of the grassland which have a high presence of bryophyte species are prone to fragmentation on pick-up.

INTERSTITIAL SPACES BETWEEN TURVES

Although every effort is made during the set-down operation to abut turves together as closely as possible, interstitial spaces are always present. These arise along the short margins of the turf as a result of the design of the transplant bucket, which gives a minimum of 100mm: in practice, the space is generally of the order of 200–300mm. As the set-down procedure relies on gravity, there is a residual pull beween turf and bucket during reverting, leaving a gap of 200–300mm. In addition, small fragments from the turves may be lost in transit from donor to receiving sites.

Infilling the interstitial spaces with soils is a very necessary part of the relocation operation and gives a characteristic striped appearance (albeit temporary) to the transferred grassland. Initially, the infilled areas were seeded with a grass/herb seed mixture, but this is no longer undertaken as natural colonisation from the adjacent turves produces a complete vegetation cover within two to three years.

The establishement of 'weed' species either from the residual seed bank within the soil used for infilling, or from seed dispersal to the site, has caused some concern. Soil from the turf-receiving site was initially used but was found to have a high seed burden. A mixture of overburden and soil from the donor site is now used, which has a low seed burden and slows the rate of plant establishment from adjacent turves. However, 'weed' species do still establish: species such as *Veronica*, *Papaver*, *Sonchus* and *Brassica* appear in the summer following relocation, but become progressively more uncommon as the ground cover develops. However, two species, coltsfoot (*Tussilago farfara*) and creeping thistle (*Cirsium arvense*), are particularly troublesome as they are persistent and can spread to the turves from the infilled areas. Spot application of herbicide (glyphosate) with a weed-wiper is used to control these species.

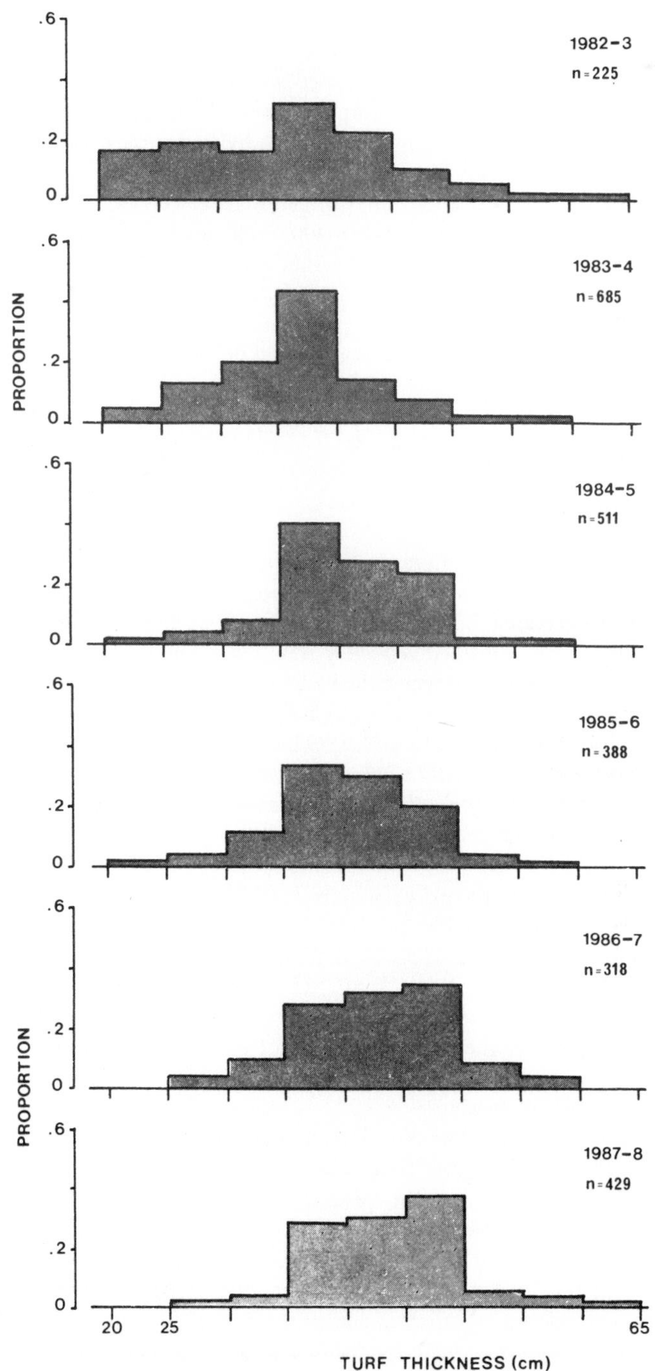

Figure 21.3 Variation in thickness of relocated turves.

IRREGULARITY IN GRASSLAND LEVEL

A noticeable feature of the relocated areas is the slightly uneven level of the grassland which at worst gives a somewhat 'stepped' surface making for difficult walking. Several factors combine to produce this uneveness. Variation in thickness occurs both between the leading and trailing edges of individual turves and between contiguous turves. As the turves are not double-handled, they are transposed during relocation such that the cut edges of contiguous turves do not correspond. The thickness of turf taken is usually between 400 – 500mm and this normally ensures that the entire soil profile is removed, together with a varying amount of overburden. Although the thickness range is from 250 – 750mm (see figure 21.3), the variation between adjacent turves is generally of the order of 50 – 150mm. This variation is due to uneven topography at the donor site and difficulties in machine operation (position of wheels in ruts, debris under bucket, etc.).

Monitoring of the relocated grassland

VEGETATION

The longest record of relocated grassland at Thrislington, and possibly elsewhere, is provided by the four turves taken in 1977. The number of species found in the individual turves has increased, but if the four turves are considered as a single unit the number of species has remained relatively stable (see Figure 21.4), This suggests that there has been internal movement of species between turves rather than new species colonising. The disturbed margins of adjacent turves may have provided the necessary opportunity for this to occur. Only two species (*Gentianella amarella*, *Knautia arvensis*) recorded in 1977 have not been found since. Some species (e.g. *Tussilogo farfara*, *Taraxacum officinale* agg., *Chamaenerion angustifolium*) have been able to establish along the exposed margins of the turves. The flux in species number is also affected by the appearance of four orchid species. The condition of the grassland remains compact and retains a strong resemblance to the grassland from which the turves were taken (see Table 21.1).

The findings from the permanent quadrat study can be summarised by reference to four representative quadrats(see Figure 21.5). Quadrat 15, established on infill material between turves, shows the relatively rapid colonisation of the bare soil within three years, due to the vegetative spread of species from adjacent turves. The increase in the number of species recorded during the second year results from the establishment of 'weed' species.

Two examples are shown from quadrats in grassland communities: quadrat 2 was established in the most common community type, the *Helictotrichon* (*Avenula*) sub-association (Shimwell, 1968), whilst quadrat 24 was established in an area of the typical *Seslerio-Helictotrichetum* grassland. In neither is there a marked change in the vegetation following relocation: the slight variation in cover from one year to the next may be typical of cyclical changes to be found in grassland systems. The change in the number of species recorded is due to the annual *Linum catharticum* and the orchid species *Gymnadenia conopsea*, *Dactylorhiza fuchsii*, *D. purpurella* and *Listera ovata*. Quadrat 7 is included by way of a contrast to the previous two quadrats: here, there is evidence of a directional change in the vegetation with the fairly rapid expansion

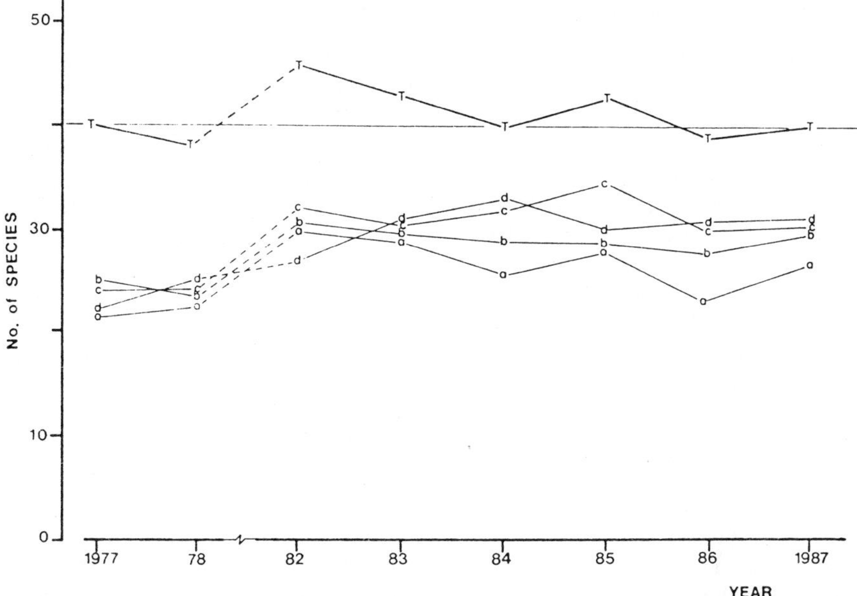

Figure 21.4 The number of species present in the four small turves (a–d inclusive) relocated in 1977, together with the total number of species (T) recorded from the turves.

of *Bromus erectus*. The spread of *B. erectus* into *Seslerio-Helictotrichon* grassland may result from the absence of grazing (Shimwell, 1968) rather than having been triggered by relocation.

INVERTEBRATES

Analysis of the samples collected during the surveys has not been completed and the results reported below are for the groups *Heteroptera*, *Homoptera*, *Coleoptera* (part), *Araneae*, *Opiliones* and *Pseudoscorpiones* only. Many species were present in very low numbers and it is difficult to draw conclusions for such species (Sheppard, 1988).

A total of 194 species were recorded from the relocated area: of these, 77 species were recorded both before and after relocation, 66 species were not found after relocation, and 51 species were new to the grassland after relocation (see Table 21.2). Of the 66 species apparently lost from the area, 34 were unique to the new site, not having been detected from the control area within the Plantation: recolonisation of the relocated area by these species may thus be difficult unless other reservoir populations occur nearby. Although no detailed assessment of the habitat requirements of those species lost has been made, many were associated with wetter grassland and sheltered scrubby situations. Of the 51 species recorded after relocation, 38 were unique to the site, not having been recorded from the control area. Again, detailed habitat assessments have not been made but many of the new arrivals are associated with dry, fairly tall grassland with a thick litter layer, or with dry gravelly soils with little vegetation.

Table 21.1 Estimates of species cover from the four turves relocated in 1977. Estimates are expressed as a percentage; the four small turves are T1–4 inclusive.

													Overall	
	S1	T1	S1	S2	T2	S2	S3	T3	S3	S4	T4	S4	T	Gr
Sesleria albicans	35.0	40.0	38.8	48.8	30.0	30.0	33.8	23.8	37.5	51.3	32.0	58.8	31.5	41.8
Sieglinglia decumbens	13.8	–	1.3	2.5	2.5	10.0	7.5	1.3	5.0	5.0	–	16.3	1.0	7.7
Brachypodium sylvaticum	–	–	–	–	–	–	–	–	–	–	6.3	–	1.5	–
Briza media	21.3	3.8	5.0	3.8	–	2.5	12.5	6.3	3.8	5.0	1.3	2.5	2.9	6.4
Festuca arundinacea	–	3.8	6.3	2.5	1.3	3.8	1.3	17.5	2.5	13.8	8.8	–	7.9	3.8
Festuca ovina	1.3	2.5	1.3	–	–	–	5.0	–	2.5	–	1.3	–	1.0	1.4
Carex flacca	13.8	8.8	25.0	21.3	5.0	11.3	20.0	2.5	8.8	12.5	20.0	15.0	9.0	16.0
Carex pulicaris	8.8	3.8	3.8	–	–	–	3.8	–	–	–	–	–	1.0	4.1
Avenula pratensis	–	–	–	–	3.8	1.3	–	2.5	2.5	–	–	–	2.2	0.6
Anthyllis vulneraria	–	–	–	–	–	–	2.5	–	–	–	2.5	1.3	–	0.4
Centaurea nigra	–	3.8	2.5	–	6.3	1.3	–	2.5	3.8	–	–	–	4.4	1.0
Galium verum	–	–	–	–	–	–	–	1.3	–	–	5.0	–	0.3	–
Helianthemum nummularium	–	–	–	12.5	28.8	15.3	–	2.5	8.8	–	–	–	7.8	4.6
Hypericum pulchrum	–	–	–	–	–	1.3	–	1.3	–	–	–	–	0.6	0.2
Leontodon hispidus	2.5	1.3	1.3	–	–	1.3	–	1.3	–	–	1.3	–	0.6	1.0
Lotus corniculatus	1.3	3.8	3.8	6.8	3.8	8.8	10.0	7.5	8.8	1.3	–	1.3	5.4	6.0
Pimpinella saxifraga	–	2.5	–	–	–	1.3	–	2.5	–	5.0	6.3	3.8	1.3	0.3
Plantago media	–	–	–	–	–	1.3	–	–	–	1.3	–	–	–	0.4
Sanguisorba minor	2.5	20.0	10.0	2.5	3.8	10.0	3.8	11.3	1.3	3.8	12.0	2.5	11.8	4.6
Ononis repens	–	–	–	–	12.5	–	–	16.3	15.0	–	–	–	4.1	1.9
Scabiosa columbaria	–	1.3	–	–	2.5	–	–	–	–	–	–	–	1.0	–
Trifolium repens	–	–	–	–	–	–	–	–	–	1.3	1.3	–	0.3	0.2
Vicia cracca	–	3.8	–	–	–	–	–	–	–	–	1.3	–	1.0	–

Note: Cover estimates from the grassland (S1–4 inclusively) on either side of the excavations left by the removal of the turves is also shown. The final two columns give mean cover values for the turves and grassland respectively. A non-repetitive point sampling method was used involving 4 frames each of 20 points placed at random positions within the turves of grassland. The cover values are for July 1987.

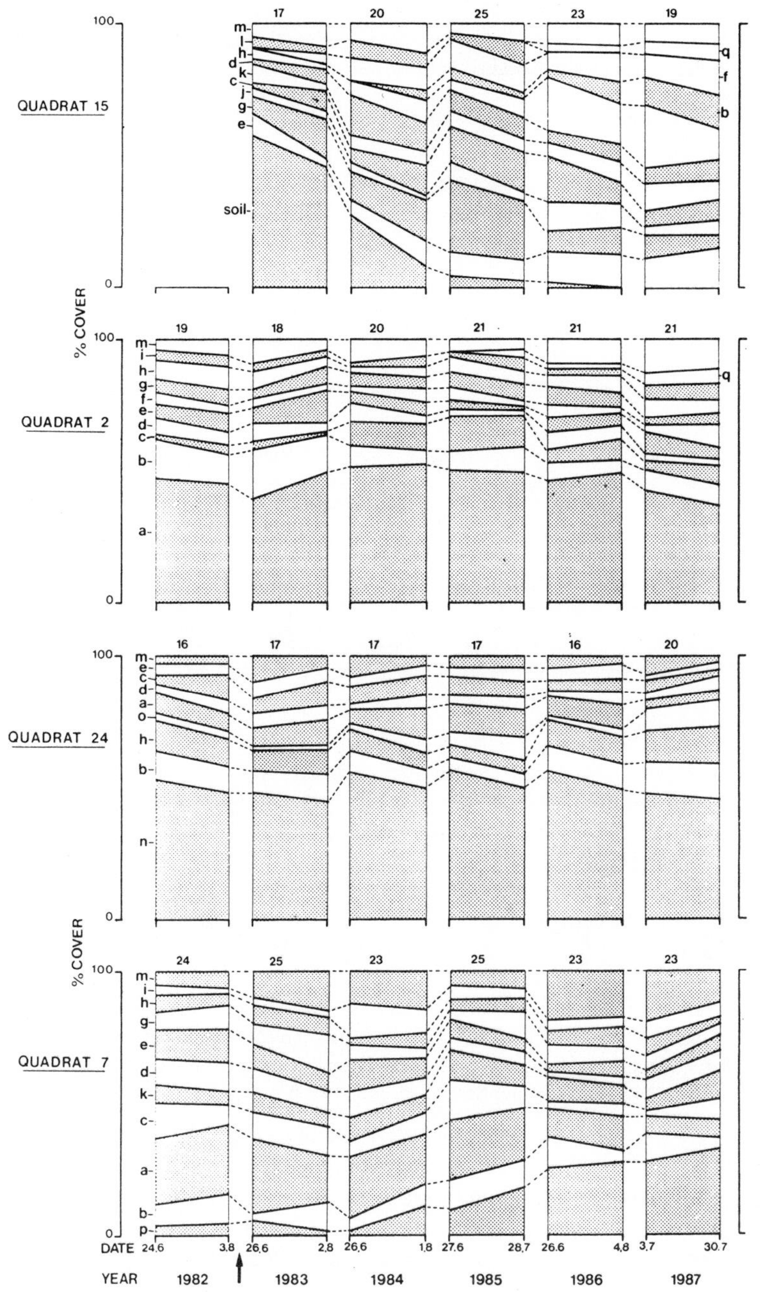

species recorded and the arrow indicates when relocation occurred.

Species Key *Helianthemum nummularium* (a), *Briza media* (b), *Sanguisorba minor* (c), *Lotus corniculatus* (d) *Festuca rubra* (e), *Brachypodium sylvaticum* (f), *Centaurea nigra* (g), *Carex flacca* (h), *Plantago lanceolata* (i), *Leontodon hispidus* (j), *Plantago media* (k), *Agrostis stolonifera* (l), *Sesleria albicans* (n), *Anthyllis vulneraria* (o), *Bromus erectus* (p), *Avenula pubescens* (q), miscellaneous (m).

Table 21.2 Changes in the invertebrate fauna of the relocated and control grassland areas detected during surveys undertaken in 1982 and 1985.

A. *Number of species recorded*	*1982 only*		*Both Surveys*		*1985 only*		*Difference*	*Total*
	No.	%	No.	%	No.	%		
Re-located grassland	66	34	77	40	51	26	−15	194
Control grassland	35	23	72	46	49	31	+14	156

B. *Species new to the relocated grassland*		*Predators*	*Herbivores*	*Detritivores*	*Total*
Unique		23	13	2	38
Control grassland	1982	2	-	-	2
	1984	2	1	1	4
	both years	7	-	-	7
Total		34	14	3	51

C. *Species lost from the relocated grassland*		*Predators*	*Herbivores*	*Detritivores*	*Total*
Unique		16	15	3	34
Control grassland	1982	11	5	-	16
	1984	7	1	-	8
	both years	4	3	1	8
Total		38	23	4	66

Note: Part A gives a summary of the number of species recorded and percentage of the total. Parts B and C highlight the species new to or lost from the relocated grassland respectively and indicate the numbers of species in each category recorded from the control grassland.

At the control site, 156 species were recorded, of which 72 species were found in the 1982 and 1985 surveys while 35 species were found only during the 1982 survey and 49 species were recorded only during the 1985 survey.

Costs

Costs quoted in Table 21.3 account only for those operations directly involved with grassland relocation: costs of site preparation (scrub-clearance, soil removal, etc.)

Table 21.3 Cost of the grassland relocation exercise undertaken at Thrislington Plantation.

	1982/3	1983/4	1984/5	1985/6	1986/7	1987/8
Rate of turf removal m^2 $hour^{-1}$	15.5	24.8	23.8	19.9	24.8	24.7
Average travel distance m	625	650	450	825	775	675
Cost (£) per $100m^2$	336	209	219	262	210	211

are excluded. The costings are based on current labour and machine hire charges, and the cost of the special bucket is not included.

The variation in costs (see Table 21.3) from year to year is attributable to weather conditions, the condition of the grassland, performance of the machine and the operators and the varying distance of travel between donor and receiving sites.

Management of relocated grassland

Management of the grassland during the first two years after relocation is confined to the levelling of the soil used to infill and the control of emergent weed species. Regrowth from the stumps of shrubs is cut and treated with herbicide. After two years, a flail-mower is used to cut the grassland annually: this is undertaken to check young saplings of *Crataegus mongyna* and the more vigorous grass species.

Discussion

Semi-natural grassland with high inherent biological interest is a fragile and diminishing resource: partly in response to this decline, some effort has been made to study the re-creation of grasslands through the use of appropriate seed mixtures (e.g. Wells *et al.*, 1981) or through the use of seed-rich topsoil and vegetation fragments (e.g. Helliwell, in this volume). However, such methods may not have been successful at Thrislington: the duplication of existing herb-rich swards with the attendant invertebrate fauna may be elusive. Moreover, seeds of many important species found in magnesian limestone grassland are difficult to collect and basic information relating to seed ripening, germination and establishment requirements is not available. The presence of tree and shrub roots in the grassland would make rotovation, soil-stripping and re-spreading very difficult.

The grassland relocation scheme adopted at Thrislington Plantation was conceptually simple, resembling the horticultural practice of turfing for ornamental or recreational purposes. However, there are important fundamental differences. Grassland relocation is broader in scope, dealing with a greater number of species (over 150 species of vascular plants at Thrislington), and encompassing an entire ecological system (flora and fauna): this is reflected in the scale of operations and the depth of soil taken. The Thrislington scheme is fortunate in three respects. Firstly, it is being undertaken by a company involved in the extractive industry which has all the equipment and expertise to carry through the scheme. Secondly, the nature of the grassland with shallow rendzina soils allows the removal of the entire soil profile. Thirdly, the area involved is relatively flat, without pronounced topographical features, such that matters of aspect and slope were not of overriding importance. However, the distance from donor to receiving site and the problems associated with scrub development are major considerations.

At the time of the public inquiry into the quarry extension, the only work which had been carried out on relocation in Britain was confined to the transplanting of small turves (less than $0.25m^2$) to act as nuclei for colonisation either of bare soil or existing, species-poor swards (e.g. Wathern and Gilbert, 1978; Gilbert and Wathern, 1980): also, some small experimental plots (15 and $60m^2$) had been moved mechanically (Down, 1982). The relocation of an entire sward consisting of several hectares was thought to be technically feasible, but the common belief held among conservationists was that relocation could not lead to the successful re-establishment of an old, species-rich grassland (Wells, 1981). How far, therefore, is the pessimistic view of the outcome of grassland relocation confirmed or refuted by the results at Thrislington?

Considering the grassland prior to and post-relocation, the most obvious change in the vegetation is the removal of scrub. However, in recognition of the importance of the grassland-scrub mosaic with respect to invertebrates and birds, provision will be made in the relocated area for the development of 'islands' of scrub which will be accompanied by peripheral tree and scrub planting. The relocation scheme has been successful in maintaining the botanical composition of the grassland, and none of the species recorded in preliminary vegetation surveys have been lost following turf removal. Moreover, there have been no instances of catastrophic changes in the vegetation, even from the few very shallow turves taken: species flower and set seed in the year immediately following relocation and continue to do so. The areas infilled with soil provide the opportunity for species new to the grassland to establish, but although several species have appeared, they generally do not persist.

The permanent quadrats did indicate that the number of species changed from one year to the next, though there was only one instance where a species was not subsequently recorded (*Tragapogon pratensis*). The intermittent appearance of annual species (Grubb 1976) and of flowering heads of orchids (Wells 1967 and 1981; Hutchings 1987) is a feature of calcerous grassland. Changes were also detected in the relative abundance of species as indicated by changes in plant cover, though these were slight and difficult to interpret. They may well have been caused by relocation but could equally be the result of intrinsic changes, reflecting the dynamics of the grassland system, or result from management practices (e.g. mowing), or simply reflect sampling error. Control quadrats were established and monitored within the Plantation but all have been affected by fire: additional quadrats have been established which will be monitored for at least four years before they are removed and should provide for a better understanding of the effects, if any, of relocation.

The results from the invertebrate surveys may substantiate the pessimistic views of relocation. However, the results need to be considered in the light of the changes in the control area, and low number of individuals sampled within species and the intensity of the survey. A large part of the change in the invertebrate fauna of the relocated area could have been caused by scrub removal alone, or through alterations in soil drainage, reducing moist-litter habitats in favour of dry grassland. While some species may simply not have survived during relocation, others may have been present in such low numbers that they were not found in 1985. The relocated grassland may have provided suitable habitats for species unable to colonise the undisturbed grassland, and may have affected other species through perhaps increased predator pressure of competition. Also, the 'new colonists' may well have been present on the site before relocation occurred, but were provided with a more favourable environment such that they increased in number at the expense of other species (Sheppard, 1988).

An important point is that the relocated and control areas experienced a fifty-four and sixty per cent change respectively in invertebrate fauna. While the explanations for the change may hold good for the receptor area, they are untenable for the donor site where habitat alteration was insignificant during the period 1982 – 1985. Clearly, other factors are at play in determining the insect fauna of the grassland. Additional surveys will be undertaken to determine whether the changes in the invertebrate fauna in both areas are transient features or symptomatic of a more permanent trend.

One of the main criticisms levelled at the Thrislington scheme concerns the disruption of the community matrix of the grassland. On the small scale, disruption arises because contiguous turves are transposed on set-down. On the large scale, disruption results from differences in shape and dimensions between the donor and receiving sites, such that the areas from which turves are taken are not always laid in correct juxtaposition. Disruption arising from the former cause is difficult to detect, but that from the latter cause is noticeable in one or two areas of relocated grassland. However, correct programming of turf removal in relation to the results of the preliminary vegetation surveys should ensure that such occurrences are minimised.

In conclusion, the Thrislington Scheme demonstrates that grassland relocation is both technically and biologically feasible. Although six years is a short period of time in the development of grasslands, the botanical results suggest a successful outcome, although more investigation of the invertebrate fauna is indicated. The grassland has proved to be more robust and resilient to the treatment it received during relocation than at first thought, and any misgivings about the operation are dispelled by the appearance of the grassland during the summer following relocation. Thrislington Plantation at present has the distinction of being the only mobile site of special scientific interest: the decision whether or not to confer this distinction on other sites is bound to generate much discussion.

Acknowledgements

I am grateful to Dr D. Sheppard of the Nature Conservancy Council for permission to use information from the invertebrate surveys. The success of the relocation operation owes much to the skill of the machine operators and I thank N. Liddle, R. Hull, G. Robson, J. Robinson, G. Robinson, N. McDonald, T. Timlin and T. Dolphin. I appreciate the assistance and advice given by members of staff at Steetley, Thrislington Works, in particular M. Turner, I.R. Congreve, G. Shepherd and W. Gillans. I thank A. Hamblin for the preparation of tables and J. Richards for typing the manuscript. This project is being undertaken by Steetley Quarry Products Ltd with advice from the Nature Conservancy Council.

References

Down, C.G., 1982, 'The re-creation of conservation value in mineral workings', in B.N.K. Davis (ed.), *Ecology of Quarries: The Importance of Natural Vegetation*, Institute of Terrestrial Ecology Symposium 11, Cambridge, pp. 62–4.

Gilbert, P.L., Wathern, P., 1980, 'The creation of flower rich swards on mineral workings', R.N. Humphries and T.T. Elkington (eds.), *Reclaiming Limestone and Fluorspar Workings for Wildlife*, Pergamon Press, London, pp. 197–207.

Grubb, P.J., 1976, 'A theoretical background to the conservation of ecologically distinct groups of annuals and biennials in the chalk grassland ecosystem, *Biological Conservation*, **10**, 53–76.

Hutchings, M.J., 1987, 'The population biology of the early spider orchid *Ophrys sphegoides*. I: A demographic study from 1975–1984', *Journal of Ecology*, **75**, 711–27.

Richardson, J.A., 1984, 'Some pioneer investigations with Professor Harrison on pit-heaps and magnesian limestone sites in County Durham', *Vasculum*, **69**, 25–35.

Sheppard, D.A., 1988, 'Changes in the fauna of magnesian limestone grassland after transplantation', in D.E. Wells (ed.), *Calcareous Grasslands – Ecology and Management*, British Ecological Society and Nature Conservancy Council joint symposium (in press).

Shimwell, D.W., 1968, 'The phytosociology of calcareous grasslands in the British Isles', unpublished Ph.D. thesis, University of Durham.

Stanyon, R.W., Park, D.G., 1987, 'Thrislington Quarry – case history of a compromise', *Transactions of the Institution of Mining and Metallurgy, Section A*, **96**, 88–90.

Wathern, P., Gilbert, O.L., 1978, 'Artificial diversification of grasslands with native herbs', *Journal of Environmental Management*, **7**, 29–42.

Wells, D.E., 1981, *Thrislington Plantation*, proof of evidence, (Document 27), public inquiry to the application to extend Dolomite Workings at Thrislington, West Cornforth, County Durham (1981), County Hall, Durham.

Wells, T.C.E., 1967, 'Changes in a population of *Spiranthes spiralis* at Knocking Hoe National Nature Reserve, Bedfordshire', *Journal of Ecology*, **55**, 83–99.

Wells, T.C.E., 1981, 'Population ecology of terrestrial orchids', in H. Synge (ed.), *The Biological Aspects of Rare Plant Conservation*, Wiley, Chichester, pp. 281–95.

Wells, T.C.E., Bell, S., Frost, A., 1981, *Creating Attractive Grasslands Using Native Plant Species*, Nature Conservancy Council, Shrewsbury.

SECTION 7

Diversifying and enhancing existing habitats

Many existing habitats contain features worth retaining, even though they may be considered somewhat degraded versions of their semi-natural counterparts. For such areas, which would be damaged by habitat 'creation' or transplantation operations to 'improve' them, some form of diversification or enhancement may be an acceptable compromise.

Habitats can be 'improved' in several different ways:

- by deliberately introducing new species into the existing community
- by artificially extending the habitat (e.g. by the flooding or inundation of an existing wetland area)
- by manipulating environmental conditions to favour some species (e.g. fertility, drainage)
- by appropriate management (e.g. by altering grazing regime intensity).

One of the cheapest of these alternatives is to introduce the 'missing' species by seeding and transplanting. Grasslands are popular and suitable subjects for this, particularly as mechanised methods are now being developed with the potential to diversify very large areas. One such method, slot-seeding, is described here by Wells *et al.*, who show that seedling establishment depends critically upon creating gaps of suitable size (in this case 10cm wide, herbicide treated strips) within the sward. Robust, pot-grown plants seem to require less ground preparation, but are more expensive to produce than seed. Competition-free areas can equally be used to introduce new, attractive and interesting species in other habitats such as woodlands or wetlands. This can be done directly or indirectly, using mowings, litter or thin layers of seed-rich soil in place of conventional seeds and transplants.

Another way to achieve diversification or enhancement is to extend the habitat so that it covers a wider range of conditions suitable for a greater number of species. Hill, for example, describes how wetland areas can be made more attractive to waders and wildlife by various management techniques, including extending lagoon areas and flooding low-lying pastures. The parallel case with grasslands might be to allow an improved pasture lying adjacent to a diverse, semi-natural sward to colonise naturally by relaxing the management regime. On arable farmland, considerable benefits to wildlife can also be achieved by reducing pesticide spraying at field margins, as demonstrated by Boatman *et al.*, allowing colonisation of these areas by species groups which would otherwise be narrowly confined to linear habitats between fields.

Other forms of management are more precisely designed to alter environmental conditions within the habitat: hence woodland canopies can be manipulated to adjust light levels; levels of fertility or drainage can be altered in grasslands, and water levels and salinity concentrations in wetlands can be controlled to suit particular groups of wildfowl and waders. In any of these cases management can be targeted at a particular wildlife group, or can aim simply to promote diversity. Gamebird management is a good example of the former objective, where the habitat requirements of the species are thoroughly established through longstanding research. However it is notable that in managing for gamebirds such as partridges and pheasants, habitat provision usually involves the enhancement of degraded farmland or woodland areas which also benefit other species of wildlife. In woodlands for example, Ludolf *et al.*, show that management for pheasants involving coppicing, skylighting and the creation and widening of rides, also increases the diversity of woodland plants and butterflies.

Enhancement and diversification procedures have ecological implications which may not always be fully understood. Waders and wildfowl do not necessarily increase their densities in response to extra habitat provision, and introduced herbs do not invariably spread and increase within the turf into which they are sown or planted. These and similar questions must be thoroughly examined if enhancement methods are to be considered practical alternatives to other forms of habitat reconstruction.

22

Diversifying grasslands by introducing seed and transplants into existing vegetation

T.C.E. Wells, Ruth Cox and *A. Frost*

Introduction

Since the publication of the booklet *Creating Attractive Grasslands Using Native Plant Species* (Wells, Bell and Frost, 1981) considerable interest has arisen in both the public and private sector in what has become known as habitat creation. This term encompasses a wide range of activities, from the digging of small ponds in field corners and in school gardens, to attempts at creating replicas of semi-natural grasslands and ancient woodlands. The most widespread method used in attempts to create grassland has been to sow mixtures of grasses and forbs selected for the particular soil/water conditions of the site (Wells, 1983) into prepared seed beds. Success with this approach has been variable (Wells, 1987; Gilbert, 1985); experience has shown that the key to success requires attention to four main points: (1) the careful selection of species suitable for the ecological conditions of the site (see Hodgson, in this volume), (2) a weed-free seed bed, (3) good seed bed preparation and (4) careful management of the site, especially in the first year after sowing. There is some evidence that soil nutrient levels may be an important factor in the successful establishment of diverse swards (Marrs and Gough, in this volume), but there are few published data to support the view that low soil nutrients are essential for success. Much more research is required to resolve this important point.

Over the past three to four years, interest has grown in enriching grass swards with wild flowers without destroying the existing sward. The reasons for not wanting to kill the sward and prepare a seed bed in the normal way are many: the cost of destroying the sward with a herbicide, cultivating the land and preparing a seed bed; the loss of use of the land while a newly sown grass/forb mixture is establishing – this is particularly inconvenient if the site is a public open space; the loss of grazing land for livestock for at least three months; and the adverse effect on the aesthetic appeal of the site of sprayed grassland – this is important if the grassland surrounds and sets off a building of historic importance, such as a stately home. There is also the objection that the few forbs present in the existing grassland will be killed unnecessarily. On slopes there is the added danger of soil erosion when the existing sward is destroyed.

Academic studies of species dynamics in permanent grasslands suggest that establishment of many species from seed is a rare event and considerable interest has been shown in the importance of 'gap size' since the publication of Dr P.J. Grubb's seminal paper 'The importance of the regeneration niche' (Grubb, 1977). The results from a number of studies (McConnaughay and Bazzaz (1987), Goldberg and Werner, (1983); Silvertown, (1980, 1981)) indicate that seed will not germinate and establish in an intact turf or in grassland with a litter-layer. Furthermore, germination is severely restricted or lowered beneath a leaf canopy (Silvertown, 1980). Thompson, Grime and Mason (1977) coined the term 'gap detection' to describe the mechanisms which enable seed to avoid germination until conditions are favourable for seedling survival.

The results obtained from these and other studies suggests that sowing seed of wild flowers into an established turf without any pre-treatment is unlikely to be successful. Dead plant material at the base of the sward may prevent the seed from reaching mineral soil and the seed will perish. Even if the seed germinates, competition from the established turf will be intense and may be fatal to the seedling. Attempts at providing gaps in the turf by harrowing and reducing competition by close mowing prior to sowing have been tried, but generally with little success. In an attempt to overcome some of these problems, experiments were made to introduce forbs into species-poor grassland using a slot-seeding machine. In another experiment, pot-grown transplants were introduced into a permanent grassland and their survival recorded. The results from these experiments are the subject of this chapter.

Slot-seeding

Slot-seeding was developed by agriculturalists and used principally to increase the productivity of pastures by introducing desirable species such as *Trifolium repens* into upland pastures (Boatman, Haggar and Squires, 1980). More recently it has been used to introduce seed of Italian ryegrass (*Lolium multiflorum*) and perennial ryegrass (*Lolium perenne*) into lowland pastures in an attempt to obtain production early in the year and to increase productivity in general (Haggar and Squires, 1982).

The principle of the method is simple and is described in detail by Squires, Haggar and Elliott (1979). Essentially it consists of spraying a band of herbicide, usually paraquat, to kill the existing sward and drilling the desired species or seeds mixture into a slit cut into the ground within the sprayed band, the whole process being done in a one-pass operation. The theory is that in the absence of competition, and with mineral soil exposed, the seed will germinate, establish and grow to a competitive size before the grass sward recovers and invades the sprayed area.

Description of machine

A one-pass seeder, based on the Stanhay precision drill[1] and a Massey Ferguson angled-steel tool bar, to which a polypropylene spray tank was attached, was used in all experiments (see Figure 22.1). Three slot-cutting units and three seeder units were spaced 50 cm apart on the tool bar. The slot-cutting unit consisted of a flat disc and a rigid coulter with a shaped blade.

Figure 22.1 Tractor with slot-seeder.

The coulter opens the slit cut by the disc, removing a narrow ribbon of turf and leaving a groove 5–15 mm deep and about 5 mm wide. The actual depth of the groove cut depends on soil type, moisture conditions and the type of sward, and particularly the amount of dead plant material (litter) present. Three individual seeder units were mounted on a sub-frame bolted behind the main frame. A common belt drive was used for all the seeder units. Seed, mixed with a filler of finely ground barley meal, was delivered from the hopper over a moving belt with holes punched in it at 3 cm intervals. Different seed rates and sizes of seed were accommodated by varying the ratios of the pulleys and by altering the proportions of seed and filler. A single spray nozzle was attached at the rear of each slot-cutting unit for applying the herbicide (paraquat) in a band about 10 cm wide over the slot area. The herbicide was fed, under pressure, from a 180 litre tank mounted on a frame above the seed hoppers. Paraquat was applied at the manufacturers' recommended rate of 5 litre/ha in 136 litres of water.

Experiments

Experiment 1

This experiment was designed to measure the establishment and subsequent development of 12 forbs (10 species and 2 cultivars of *Lotus corniculatus*) slot-seeded into

Table 22.1 Composition of nine amenity grass mixtures sown at Monks Wood on 5 September 1978

Species Number	Composition
1.	*Festuca rubra* ssp. *commutata* 'Cascade'
2.	*Festuca rubra* ssp. *commutata* 'Highlight'
3.	*Festuca rubra* ssp. *rubra* 'Rapid'
4.	*Festuca rubra* ssp. *rubra* 'Boreal'
5.	*Festuca rubra* ssp. *litoralis* 'Dawson'
6.	Johnson's J8 mixture
	Festuca longifolia 'Scaldis' 35%
	Festuca rubra ssp. *litoralis* 'Dawson' 15%
	Festuca rubra ssp. *rubra* 'Ruby' 25%
	Poa pratensis 'Prato' 10%
	Poa pratensis 'Parade' 5%
	Agrostis castellana 'Highland' 10%
7.	*Cynosurus cristatus*
8.	*Festuca rubra* ssp. *commutata* 'Waldorf'
9.	Mommersteeg's MM28 mixture
	Lolium perenne 'Stadion')
	Lolium perenne 'Grandstand') 20%
	Lolium perenne 'Majestic')
	Phleum pratense 'Eskimo' 40%
	Agrostis castellana 'Highland' 40%

Mixtures 5 and 7 sown at 14.0 gm m^{-2}, other plots at 23.0 gm m^{-2}.

nine different grassland types. The nine swards, consisting of popular amenity grass cultivars or mixtures of cultivars (see Table 22.1), were sown on a heavy clay soil at Monks Wood Experimental Station in September 1978. There were 4 replicates of each sward, arranged in a randomised block design. Plot size was 6 x 3 m and seed was sown at 23 g m^{-2}. Plots were cut regularly to establish a turf, and during 1984 moss was destroyed with 'lawn sand' and a selective herbicide (mecoprop) applied to control white clover and other dicotyledonous weeds.

Seed of 12 forbs (see Table 22.2), mixed with the appropriate amount of inert filler (barley meal), was slot-seeded singly on 17 October 1984. We aimed to drill the small-seeded species *Leucanthemum vulgare*, *Plantago media*, *Primula veris*, *Lotus corniculatus* (2 cultivars), *Ranunculus acris* and *Prunella vulgaris* at about 5 seed per cm and the large-seeded species, *Sanguisorba minor*, *Knautia arvensis*, *Centaurea nigra*, *Centaura scabiosa* and *Rhinanthus minor*, at about 1 seed per cm. The actual amounts drilled varied from 3.2 to 7.7 seeds per cm for the small-seeded species and from 0.2 to 0.9 seed per cm for the large-seeded species (see Table 22.2). Differences in the shape of the seed and texture of the seed coat, variations in the quality of the filler and jolting of the drill were probably responsible for the variations in the quantity of seed drilled.

Table 22.2 Sowing rates (seeds per cm length of row) of 12 forbs drilled at Monks Wood, 17 October 1984 using the slot-seeder and percentage establishment and survival, April 1985 and May 1986

Species	Seeds per cm	% establishment and and survival	
		1985	1986
Leucanthemum vulgare	6.87	10.6	9.0
Lotus corniculatus 'native'	7.74	3.7	3.0
Lotus corniculatus 'Leo'	6.82	7.1	6.4
Plantago media	4.64	28.4	19.1
Primula veris	5.10	44.7	40.2
Prunella vulgaris	5.41	27.3	24.4
Ranunculus acris	3.26	49.4	45.9
Centaurea nigra	0.97	20.9	22.3
Centaurea scabiosa	0.15	12.3	2.2
Knautia arvensis	0.25	16.8	20.2
Rhinanthus minor	0.38	28.1	24.4
Sanguisorba minor	0.53	19.5	16.4

The number of established plants in 1 metre lengths of row were counted in April 1985 and May 1986. Where plant density was too high to enable counting to be done with accuracy, a 20 cm strip was excavated whole and the individual plants washed out and counted. Results are expressed as percentage establishment. After arcsine transformation, data were subjected to ANOVA.

Experiment 2

The purpose of this experiment was to compare the establishment of mixtures of forbs slot-seeded in spring (April) with the same mixtures drilled in autumn (October). The site was an ancient meadow at Upwood, Cambridgeshire, situated on a heavy, water-logged clay soil, which had been sprayed and fertilised to improve it for agriculture in 1965. The vegetation consisted mostly of grasses, with occasional forbs such as *Ranunculus repens*, *R. ficaria* and *Cerastium fontanum*, which had survived the herbicide treatment. The site was grazed by cattle from mid-May to late October and is typical of the sort of site which is likely to be available for enrichment if the predicted changes in agriculture (ie 'set-aside') take place.

Four plots, each roughly rectangular in shape and 0.172 ha in extent, were marked out in October 1984. Plots 1 and 3 were slot-seeded on 11 October 1984, plots 2 and 4 on 4 April 1985. The composition of the seed mixture and the sowing rate is given in Table 22.3. The number of established plants in randomly selected 5 metre lengths of row were counted on 20–23 May 1985 (Plots 1 and 3), 3–4 July 1985 (Plots 2 and 4) and 9–13 June 1986 (all plots). Results are expressed as percentage establishment.

Table 22.3 Sowing rates (seeds per metre length of row) of 14 forbs drilled at Upwood, meadows on 12 October 1984 ('autumn sown') and 14 April 1985 ('spring sown) and percentage establishment and survival. Recorded May and July 1985 for the autumn and spring sown plots respectively and June 1986.

Species	Seeds sown per metre			% establishment and survival			
	Autumn	Spring		1985		1986	
	Plots 1 and 3	Plot 4	Plot 2	Autumn sown	Spring sown	Autumn sown	Spring sown
Achillea millefolium	20.8	20.8	20.8	0.5	0.3	0.5	0.4
*Centaurea nigra**	4.6	1.4	1.0	17.6	1.9	3.4	2.8
Galium verum	23.1	23.1	11.6	6.1	2.1	1.3	1.6
Leontodon hispidus	-	-	14.3	-	0	-	0
Leucanthemum vulgare	30.9	30.9	30.9	4.3	0.9	0.7	0.7
Lotus corniculatus	22.0	22.0	22.0	4.1	1.0	1.1	0.8
Lychnis flos-cuculi	19.4	19.4	19.4	4.4	1.1	2.6	2.1
Medicago lupulina	7.4	7.4	7.4	8.2	2.8	0	0
Plantago media	29.5	29.5	29.5	5.3	1.4	1.0	1.0
Primula veris	29.4	29.4	29.4	12.9	0.1	2.6	2.0
Prunella vulgaris	16.3	16.3	16.3	11.0	2.3	6.8	5.9
Ranunculus acris	9.2	9.2	9.2	54.9	3.3	28.3	22.9
Rhinanthus minor	4.9	4.9	4.9	0.6	0	0	0.2
Sanguisorba minor	6.5	6.5	4.3	1.1	2.1	0	0.3

* includes a proportion of *Centaurea scabiosa*

Experiment 3

The purpose of this experiment was to compare the establishment of mixtures of forbs slot-seeded in spring and autumn into a three-year-old ley, on a farm where sheep had been used to manage the sward. The site, at Cople, Bedfordshire, situated on a heavy clay soil, had been sown three years previously with a rye-grass/cocksfoot ley. Prior to slot-seeding, the grassland was grazed by 200 sheep for 5 days, which produced a short, but patchy sward, with tufts of grass and about 20 per cent bare ground. At the time of the autumn slot-seeding, the ground was hard and compacted, the coulter and discs having difficulty in penetrating the ground.

Four plots, each 188 m × 8 m were marked out in October 1984. Two plots, selected at random were slot-seeded and band-sprayed on 15 October 1984, the other two plots being similarly treated on 15 April 1985. The composition of the mixture sown is given in Table 22.4.

Results

Experiment 1

All 12 forbs established in all sward types but there were highly significant differences ($P<0.001$) between the establishment of the various species (see Table 22.5). Both cultivars of *Lotus corniculatus* germinated within a week of sowing, but nearly all of these seedlings were killed in the following three months by a combination of low temperatures and water-logged soil conditions. The plants of *Lotus* counted in 1985 grew from seed which had remained dormant during the winter of 1984 germinating in early spring 1985. The other 10 species germinated in March 1985. Germination of *Ranunculus acris* (49.4 per cent) and *Primula veris* (44.7 per cent) was excellent and all species except *Lotus corniculatus* had an establishment of greater than 10 per cent. By mid-June, 1985 rows of seedlings and young plants formed clear 'green lines' in the band-sprayed zone.

Comparison of establishment in 1985 and survival until May 1986 revealed that the only species to suffer any substantial mortalities were *Plantago media* and *Centaurea scabiosa*, the last named species disappearing completely by 1987, probably because it is unable to withstand prolonged winter flooding.

There was no significant effect of sward type or grass variety on initial establishment and survival up to two years after drilling of any of the 12 sown species. Cultivars differed, however, in their ability to recolonise the sprayed band. 'Dawson', 'Boreal', 'Waldorf', 'Cascade' and 'Highlight', all cultivars of *Festuca rubra*, were slow to fill the gap, whereas 'Rapid' and the mixtures MM28 and J8 had completely colonised the spray band after 2 years. The plots sown with *Cynosurus cristatus* had been invaded by the weed grass *Agrostis stolonifera*, which had quickly spread into the sprayed band.

Rhinanthus minor flowered in 1985 and seeded itself into the surrounding grassland. Seven species flowered in 1986 (*Leucanthemum vulgare*, *Lotus corniculatus* 'Leo' and 'Native', *Sanguisorba minor*, *Prunella vulgaris*, *Ranunculus acris*), the remainder in 1987. (*Centaurea scabiosa* died before it reached flowering size).

Table 22.4 Sowing rates (seeds per metre length of row) of 15 forbs drilled at Cople, on 15 October 1984 ('autumn sown') and 15 April 1985 ('spring sown') and percentage establishment and survival, 19 June 1986

Species	Seeds sown per metre		% establishment
	Autumn	Spring	
Achillea millefolium	104.5	98.1	0.2^{i}
Centaurea nigra	3.3	3.1	10.8^{b}
Centaurea scabiosa	1.0	0.9	0.3^{ij}
Galium verum	33.3	31.3	0.9^{g}
Knautia arvensis	4.0	3.7	7.7^{c}
Leontodon hispidus	14.3	13.4	4.0^{d}
Leucanthemum vulgare	31.1	29.2	1.7^{f}
Lotus corniculatus	22.2	20.8	0.8^{g}
Medicago lupulina	7.5	7.0	0.1^{j}
Plantago media	29.7	27.9	1.7^{e}
Primula veris	29.6	27.8	4.6^{d}
Prunella vulgaris	16.4	15.4	8.2^{c}
Ranunculus acris	9.3	8.7	16.9^{a}
Rhinanthus minor	5.0	4.7	0.3^{ij}
Sanguisorba minor	3.2	3.0	0.9^{h}
Total	314.4	295.0	

Means with the same superscript are not significantly different ($P<0.001$)

Experiment 2

All sown species except *Medicago lupulina* and *Leontodon hispidus* (only sown in Plot 2) established in both the autumn and spring drillings (see Table 22.3), but there was a trend for better establishment when species were sown in the autumn. (Mean establishment in autumn was 10.1 per cent, spring 3.7 per cent). In both autumn and spring sowings there was a great range in establishment, from 0.53 per cent (*Achillea millefolium*) to 54.9 per cent (*Ranunculus acris*) in autumn and from zero (*Medicago lupulina*) to 22.9 per cent *Ranunculus acris* in spring. However, the species rankings for both sowing dates remain almost the same, which suggests that for the range of species used, autumn sowing is preferable to spring. While it is only possible to speculate as to the cause for this seasonal difference, one important factor may be that the competition-free zone created by the spray band lasts for at least six months when applied in autumn, whereas when applied in spring, it persists for only about two months.

In 1986, six species flowered (*Centaurea nigra*, *Leucanthemum vulgare*, *Lotus corniculatus*, *Lychinis flos-cuculi*, *Ranunculus acris* and *Primula veris*); in 1987, in addition to the six species which flowered in 1986, the following four species also flowered: *Plantago media*, *Sanguisorba minor*, *Prunella vulgaris*, and *Rhinanthus minor*. Whether any of these plants actually produced seed for innoculating inter-row areas seems unlikely as the site was grazed heavily both during and after flowering. The benefit of grazing in restricting competition from the taller growing grasses is clear, but there is likely to be trade-off between the better vegetative growth of the forbs and the loss of seed for dispersal to other areas.

Experiment 3

All 15 sown species established but there was no significant difference in establishment between plots sown in autumn compared with those sown in spring (cf. Experiment 2). There were, however, highly significant differences ($P<0.001$) between the establishment of the sown species (see Table 22.4). As in previous experiments, *Ranunculus acris* had the best establishment, *Medicago lupulina* the worst, other species occupying roughly the same position in the rankings. Careful management of the site by the farmer with his sheep enabled more than half of the sown species to flower and set seed and it will be interesting to see if these species now spread in future years into the inter-row areas.

Pot-grown transplants

Native plants can be raised from seed and grown in pots quite easily using conventional horticultural methods. Details of how plants were raised at Monks Wood Experimental Station, where more than 20,000 have been grown since 1979, are given in Wells, Frost and Bell (1986).

Four to five month-old plants, with about four true leaves, are conveniently sized plants for inserting into established grasslands, using a bulb-planter or similar tool. In theory, the benefit of using young plants rather than sowing seed should lie in the ability of plantlets to compete effectively with established grasses, although clearly factors such as sward type, time of year of planting, weather conditions following planting and the management of the established sward, both before and after planting, will be important in affecting the degree of success achieved.

The use of pot-grown transplants for diversifying grasslands is of special interest where subtle differences of slope, aspect and drainage provide a range of habitat conditions which cannot be fully catered for by straightforward grass/forb mixtures. It is also more cost-effective to use pot-grown plants (a) where seed is scarce or expensive, (b) where species are known to be slow-growing and likely to be swamped at the seedling stage by surrounding vegetation, for example *Campanula rotundifolia*, *Thymus pulegioides* and (c) for species with seed dormancy mechanisms which prevent them from germinating immediately with other species in the sown mixture, thereby putting them at a competitive disadvantage when they do germinate, for example, *Geranium pratense*, *Silaum silaus* and *Filipendula vulgaris*.

As far as we are aware, there are no references in the literature to the survival through time of plantlets introduced into established grasslands. To obtain some preliminary information on this important aspect of habitat creation, we monitored the survival of 699 pot-grown plants inserted on 5–6 June 1984 into a short, previously mown permanent grassland at Monks Wood Experimental Station. The soil was a heavy calcareous clay, derived from the Boulder Clay, pH 5.8–6.6.

The 14 species used were distributed randomly in 70 rows, each row being 5 m long and 1 m wide. Planting density was two plants per m^2. The survival and state of the plants was recorded on 25 August 1984, 17 June 1985 and 7 July 1986 (see Table 22.5). Despite low rainfall in the two months following planting, the survival of all plants, except for *Ranunculus bulbosus*, exceeded 90 per cent. The grass was mown in September 1984 and again in late October 1984. By the following year, 426 (61 per cent) of the plants were still alive and a year later (1986), 320 (45.8 per cent) still survived.

Mortalities were not spread evenly among the 14 species planted. All plants of *Anthyllis vulneraria* were dead by June 1985; 5 per cent flowered in 1984, 5 per cent in 1985, the remaining plants dying before they could flower. Forty-seven per cent of the plants of *Ranunculus bulbosus* died in the first four months after planting, 17.8 per cent in 1985 and 22 per cent in 1986. Barling (1955) noted that *R. bulbosus* prefers free draining soils with a low water-table, quite the opposite of conditions

Table 22.5 Percentage survival and flowering behaviour of 14 species inserted into a permanent grassland at Monks Wood as 4½ month-old pot-grown plants. Planted 5–6 June 1984, recorded 25–26 August 1984, 17–20 June 1985 and 7–8 July 1986

	No. of pots inserted	% survival August 1984	% survival June 1985	% survival July 1986
Anthyllis vulneraria	42	95.2	4.6	0
Campanula glomerata	36	91.7	16.6	2.8
Centaurea nigra	51	100.0	82.7+	72.5+
Centaurea scabiosa	51	100.0	65.3	33.3
Leucanthemum vulgare	45	100.0	80.4+	66.7+
Knautia arvensis	43	97.7	84.1	84.1+
Leontodon hispidus	50	100.0	52.0	36.0
Lychnis flos-cuculi	52	96.2	84.6+	84.6+
Ononis spinosa	50	98.0+	42.0	16.0
Ranunculus acris	51	94.1	58.8+	17.6
Ranunculus bulbosus	45	53.3	35.6	13.3
Plantago media	55	98.2	74.1	52.7
Primula veris	52	100.0	96.1	86.5
Stachys officinalis	41	97.6	92.5	87.8+
Unidentified species*	35	0	0	0
Total	699	90.0	60.8	45.8

* 35 plants were recorded as missing, presumed dead, when recording took place in August 1984, but because of an error on the original planting plan could not be identified to the species level

\+ more than 50% of plants flowered

Table 22.6 The performance (mean height and rosette diameter) of 12 species inserted into a permanent grassland at Monks Wood as 4½ month-old pot-grown plants. Recorded 25 months after planting.

	Mean Height (cm) ± S.E.	Mean rosette diameter (cm) ± S.E.
Centaurea nigra	36.00 ± 2.92	18.66 ± 1.12
Centaurea scabiosa	25.80 ± 5.51	11.13 ± 1.58
Leucanthemum vulgare	54.96 ± 3.63	16.06 ± 1.81
Knautia arvensis	21.61 ± 2.64	13.64 ± 1.13
Leontodon hispidus	35.58 ± 3.32	15.94 ± 1.75
Lychnis flos-cuculi	46.91 ± 2.41	10.54 ± 0.84
Ononis spinosa	41.00 ± 4.80	18.66 ± 3.14
Plantago media	26.84 ± 2.31	14.0 ± 1.25
Primula veris	19.16 ± 1.11	13.08 ± 0.58
Ranunculus acris	59.44 ± 5.22	11.88 ± 2.11
Ranunculus bulbosus	36.83 ± 6.19	6.66 ± 1.99
Stachys officinalis	21.53 ± 2.42	9.47 ± 0.70

prevailing on the heavy clay soils at Monks Wood. Harper (1957) commented that *R. bulbosus* seldom persists in grassland which is cut for hay and silage and it seems probable that the combination of mowing twice a year and a water-logged soil did not provide suitable conditions for a species which renews itself each year with a corm, the size and performance of the corm being dependent on conditions the previous year.

Thirty-three per cent of *Campanula glomerata* plants flowered in the first year, but 75 per cent were dead by June 1985. A year later, a further 13.9 per cent had died and only a single plant survived until 1986. Another rosette hemicryptophyte, *Plantago media*, flowered in the first year (45.5 per cent) but by June 1985 23.6 per cent were dead. A further 21.8 per cent died the following year, although 45.5 per cent of surviving plants flowered in 1986. *Ranunculus acris* flowered in all three years (11.8, 54.9 and 17.6 per cent in 1984 to 1986) but mortalities of 35.3 per cent and 41.2 per cent in 1985 and 1986 reduced the population to 18 per cent of the original planting by 1987. However, new cohorts have established from seed shed in 1984 and 1985.

The performance of 12 of the 14 planted species, as measured by height of the inflorescence and diameter of the rosette (see Table 22.6) in July 1986 was good, indicating that the species were able to compete successfully with the established grasses.

Four species, *Stachys officinalis*, *Knautia arvensis*, *Lychnis flos-cuculi* and *Primula veris*, have survived exceptionally well, less than 20 per cent mortalities occurring since planting. Other species which have survived almost as well include *Centaurea nigra* (72.5 per cent), *Leucanthemum vulgare* (66.7 per cent), and *Plantago media* (52.7 per cent).

With the exception of *Ranunculus acris*, there is to date little sign of the established species spreading by seed, although considerable amounts of seed have been shed. These have either perished, or have joined the seed bank in the soil and may well germinate if the grassland is closely mown or scarified.

Discussion

This series of experiments has shown that it is possible to enrich established swards with forbs, introducing them either as seed, using the slot-seeder, or as plantlets, using a bulb-planter. Using the one-pass slot-seeder, the 10 cm wide strip of grassland killed by the herbicide paraquat provided a competition-free zone in which the sown forbs were able to germinate and grow to a competitive size before there was any regrowth of grasses. The length of time for which the sprayed band remained open, and free of new grass growth, varied according to when the herbicide was applied. With autumn (October) applications, the sprayed bands were visible for up to six months, whereas regrowth of grasses was much quicker following slot-seeding in spring (April) with the sprayed bands remaining open for only about two months. This may explain the slightly better establishment obtained with autumn slot-seeding at the Upwood site, compared with spring drilling, but the fact that no significant difference between season of drilling was obtained at the Cople site suggests that other site factors, such as sward composition and grazing regime, may be more important than time of year of slot-seeding. On the other hand, Haggar and Squires (1982) attributed most of the variation they encountered in the establishment of Italian ryegrass (*Lolium multiflorum*) in 1976 and 1977 to the weather conditions following slot-seeding (see Table 22.7). They concluded that mid-August to September was probably the safest time to slot-seed in the Midlands, the cooler temperature and heavy dews ensuring that the slots did not dry out too quickly. In our experiments, sowing in mid-October encouraged some species, notably *Lotus corniculatus*, to germinate and produce small seedlings which were unable to survive

Table 22.7 Number of plants of Italian Rye Grass (*Lolium multiflorum*) established after sowing at different months (plants m^{-1} length of row).

Month of sowing	Rainfall 1976 mm	6 weeks after sowing 1976	Rainfall 1977 mm	3 weeks after sowing 1977	6 weeks after sowing 1977
April	7.0	15	39.2	184	106
May	30.2	17	34.4	137	34
June	18.6	4	82.7	163	132
July	22.6	18	6.8	41	46
August	34.3	146	98.0	170	not recorded
Sept	84.0	106	10.1	73	not recorded
S.E.		6.9		13.2	15.0

Source: Data of Haggar and Squires (1982).

the rigours of winter, and it is clear that much more work is required to establish the optimum time for slot-seeding. In view of the range of weather conditions encountered in the British Isles it seems likely that the optimum time for slot-seeding will vary according to geographic location. A predictive model, based on evapotranspiration and day degrees above 5°C (see Figure 1 in Wells, Frost and Bell, 1986), may prove helpful.

Mortalities among seedlings were considerable and were caused by a number of factors, but we were not able to investigate these in any quantitative way. Damage from molluscs and other invertebrates occurred at all times of the year, but was particularly severe following prolonged rainfall. Squires (1976) applied slug pellets at 5 kg per ha in an attempt to reduce damage from slugs in a slot-seeding experiment, but did not report if this enhanced forb survival. In our trials, slug pellets were not used because of cost and the potential danger to other wildlife.

Pleurocarpous mosses, particularly *Calliergon cuspidatum*, *Brachythecium rutabulum* and *Campylium chryosophyllum*, rapidly colonised the sprayed bands, and birds searching for food among the mosses uprooted some seedlings.

The mean establishment of all pot-sown species 18 months after sowing was significantly higher at Monks Wood (19.5 per cent), than at Upwood (3.3 per cent) or Cople (3.9 per cent). Whether this was due to the higher seed rates used at Monks Wood (351 seed per metre) compared with 16 seeds per metre at Upwood and 20 at Cople cannot be ascertained as there were other differences between the sites which may have affected forb establishment. The nine swards at Monks Wood were managed by cutting twice a year, whereas the ley at Cople was grazed by sheep and the permanent grassland at Upwood by cattle. As a result, the 'texture' of the swards was quite different, the permanent pasture having a close-knit turf compared with the tufted and more open ley. At Monks Wood, the nine amenity mixtures differed considerably in texture, but there were no significant differences in forb establishment attributable to the various swards, which suggests that it is the creation of a competition-free gap in the turf that is important, not the species composition of the sward. However, sward composition may be important in the long term, when re-invasion of the sprayed band occurs and the different competitive abilities of the grasses becomes apparent.

All 16 species of forbs used in the three slot-seeding experiments established but there were highly significant differences (see Tables 22.2, 22.3 and 22.4) between species in percentage establishment and survival. If species are ranked according to their performance within individual experiments a clear pattern emerges of some species consistently doing well, e.g. *Ranunculus acris*, while others consistently perform badly, e.g. *Medicago lupulina*. Their behaviour overall is most easily demonstrated by calculating a 'mean rank' index from their performance in the various experiments. The following arranges them in descending order of success:

	Mean Rank Index
Ranunculus acris	1.0
Prunella vulgaris	2.5
Centaurea nigra	3.0
Primula veris	4.0

	Mean Rank Index
Lychnis flos-cuculi	4.0
Knautia arvensis	4.5
†*Galium verum*	6.3
Plantago media	6.7
Leucanthemum vulgare	8.0
Lotus corniculatus	8.2
Rhinanthus minor	8.2
Sanguisorba minor	8.5
†*Centaurea scabiosa*	10.5
†*Medicago lupulina*	11.0
**Leontodon hispidus*	11.0
**Achillea millefolium*	11.0

*Sown in only 1 experiment
†Sown in only 2 experiments

There is no relationship between rank order and seed size; small-seeded species such as *Lychnis* and *Primula* were more successful than large-seeded species such as *Knautia arvensis* and *Sanguisorba minor*, while some medium-sized species, such as *Ranunculus acris* and *Prunella vulgaris*, did best of all. We interpret the relative establishment and survival of species as expressing the effects of a combination of environmental factors, both biotic and abiotic, acting on the germinating seed and ensuing seedling, with those species most suited to the site conditions performing well. We suggest the term 'environmental sieving' to describe the phenomenon.

Those species which are not fitted to a particular environment will eventually die or perform badly, although this may take some time to become apparent. For example, both *Centaurea scabiosa* and *Medicago lupulina* established initially from seed but all young plants succumbed to prolonged winter wetness in 1987. In a previous experiment on the same clay soil, *Scabiosa columbaria* did well for three years, flowering in its second year, but the whole population died during the following winter, as a result of water-logging. These examples emphasise the need for carefully selecting species based on a knowledge of their ecological requirements – when these are not known, perhaps the best guide is to select species which grow and flourish in neighbouring grasslands.

The success of diversifying species-poor grasslands with seed or plantlets will eventually be measured by the ability of species to spread into the surrounding grassland or in the case of slot-seeding, into the inter-row areas. *Rhinanthus minor*, the only annual among the species tested so far, has already spread out of the rows into other areas, but it is too early to measure the success of other species in this respect as most have only flowered for the first time in 1987. Special management procedures, such as scarifying the turf with harrows in late autumn or early spring, or destroying patches of grass with herbicides, may be necessary to create suitable sites for regeneration. More work is required on defining the characteristics of the regeneration niche, *sensu* Grubb (1977), if we are to succeed in enriching existing grasslands, without resorting to sward destruction and total reseeding.

Acknowledgements

The authors wish to thank R. Plant for help with slot-seeding, Mr Charles Porter of Wood End Farm, Cople, and the Bedfordshire and Huntingdonshire Wildlife Trust for permission to work on their land at Cople and Upwood respectively.

Notes

1 Agriculturists use the Gibbs slot-seeder or the Hunter Rotary Strip Seeder, both of which are available in commerce, but because of financial constraints, we used the smaller, and cheaper Stanhay drill.

References

Barling, D.J., 1955, 'Some population studies in *Ranunculus bulbosus*', *Journal of Ecology*, **43**, 207–218.

Boatman, N.D., Haggar, R.J., Squires, N.R.W., 1980, 'Effects of band-spray-width and seed coating on the establishment of slot-seeded grass and clover', Proceedings of 1980 British Crop Protection Conference – Weeds, pp. 503–9.

Gilbert, O., 1985, 'A wild flower mix with a short life', *Landscape Design*, 47–49,**157**.

Goldberg, D.E., Werner, P.A., 1983, 'The effects of size of opening in vegetation and litter cover on seedling establishment of golden rods (*Solidago* spp.)', *Oecologia* (Berlin), **60**, 149–55.

Grubb, P.J., 1977, 'The maintenance of species-richness in plant communities: the importance of the regeneration niche', *Biological Reviews*, **52**, 107–45.

Haggar, R.J., Squires, N.R.W., 1982, 'Slot-seeding investigations. 2. Time of sowing, seed rate and row spacing of Italian ryegrass', *Grass and Forage Science*, **37**, 115–22.

Harper, J.L., 1957, 'Biological flora of the British Isles, *Ranunculus acris* L. *Ranunculus repens* L. *Ranunculus bulbosus* L., *Journal of Ecology*, **45**, 289–342.

McConnaughay, K.D.M., Bazzaz, F.A., 1987, 'The relationship between gap-size and performance of several colonising annuals', *Ecology*, **68**,(2), 411–16.

Squires, N.R.W., 1976, 'The use of band applications of three herbicides in the establishment of direct drilled grasses and legumes by the WRO one-pass sowing technique', Proceedings of 1976 British Crop Protection Conference – Weeds, pp. 591–96.

Squires, N.R.W., Haggar, R.J., Elliott, J.G., 1979, 'A one-pass seeder for introducing grasses, legumes and fodder crops into swards', *Journal of Agricultural Engineering Research*, **24**, 199–208.

Silvertown, J.W., 1980, 'Leaf-induced seed dormancy in the grassland flora', *New Phytologist*, **85**, 109–18.

Silvertown, J.W., 1981, 'Micro-spatial heterogeneity and seedling demography in species-rich grassland', *New Phytologist*, **88**, 117–28.

Thompson, K., Grime, J.P., Mason, G., 1977, 'Seed germination in response to diurnal fluctuations of temperature', *Nature*, **267**, 147–9.

Wells, T.C.E., 1983, 'The creation of species-rich grasslands', in A. Warren and F.B. Goldsmith (eds.), *Conservation in Perspective*, Wiley, Chichester, pp. 215–32.

Wells, T.C.E., 1987, 'The establishment of floral grasslands', *Acta Horticulturae*, **195**, 59–69.

Wells, T.C.E., Bell, S.A., Frost, A., 1981, *Creating Attractice Grasslands Using Native Plant Species*, Nature Conservancy Council, Shrewsbury.

Wells, T.C.E., Frost, A., Bell, S.A., 1986, *Wild Flower Grasslands from Crop-grown Seed and Hay-bales*, Nature Conservancy Council, Peterborough.

23

Modification of farming practice at field margins to encourage wildlife

N.D. Boatman, J.W. Dover, P.J. Wilson, M.B. Thomas, and *S.E. Cowgill*

Introduction

Until recently, arable farmland has been largely ignored as a wildlife habitat by ecologists in favour of more diverse, but less extensive, ecosystems. As a result, the effects of the dramatic changes in farming practice which have occurred over the last decades upon farmland ecology, and upon the individual species adapted to live in this environment, have gone largely unrecorded, except for those species considered to be serious weeds or pests. An exception to this generalisation is the Grey Partridge, *Perdix perdix*, which has been the subject of a long-term study in southern England, known as the Partridge Survival Project (Potts and Vickerman, 1974; Potts, 1986). Data from the Game Conservancy's National Game Census have shown that this species has suffered an 80 per cent decline since 1952. Detailed monitoring of a 62km^2 study area in West Sussex since 1968, backed up by experiments, has implicated increasing pesticide use as a major cause of this decline (Potts, 1986).

The pesticides were not directly toxic to partridges, but caused a reduction in numbers of the insects which form the major part of the diet of partridge chicks during their first few weeks of life. This has led to increased rates of chick mortality during this early phase when they are entirely dependent on insect food (Potts, 1986; Green *et al.*, 1987). The activity of pesticides on chick-food items may be directly insecticidal (Vickerman and Sunderland, 1977; Sotherton *et al.*, 1987; Sotherton and Moreby, 1988) or indirect, in the case of herbicides, by removal of the weed host plants of preferred chick-food insects (Southwood and Cross, 1969; Vickerman, 1974; Sotherton, 1982; Potts, 1986).

Broods of grey partridge chicks feed almost exclusively in cereals, and prefer the edges of fields (Green, 1984). From 1983 onwards, experiments have been carried out in which pesticide use has been modified over a 6m wide strip of crop at the edge of cereal fields. Greater numbers of preferred chick-food insects were found within these areas, with dramatic results in terms of improved grey partridge chick survival rates (Rands, 1985, 1986, in press; Sotherton *et al.*, 1985). Survival rates of pheasant chicks, which are similarly dependent on insect food in early life (Hill, 1985), were also increased (Rands, 1986). The implementation of this technique over approximately half the cereal acreage on the main study area (25 per cent of the total farm area) over four years (1983–6), has brought about an increase in spring pair density of grey partridges from four to twelve pairs per km^2 (Rands, in press).

Development of the 'conservation headland' concept

In early experiments, all pesticides (herbicides, fungicides and insecticides) were excluded from the 6 m wide headland after 1 January. Subsequent studies have led to a refining of the technique to produce a practical package of recommendations for use in commercial farm crops. The aim throughout has been to maximise the benefits while minimising the agricultural consequences. For example, screening of cereal fungicides in the laboratory (Sotherton *et al.*, 1987; Sotherton and Moreby, 1988), has shown only one chemical, pyrazophos, to have significant direct, toxicological, insecticidal activity. Consequently, recommendations now permit the use of most fungicides on what were formerly known as 'unsprayed headlands'. This misleading term has therefore now been changed to 'conservation headlands' (see Figure 23.1).

A survey in 1985 of the attitudes of farmers who subscribe to the Cereals and Gamebirds Research Project showed that, while most were reluctant to consider excluding herbicides completely from the headland area, only a few weed species were specifically identified as likely to make the idea unworkable in practice. The most frequently cited was cleavers (*Galium aparine*), others being mainly grass weeds including barren brome (*Bromus sterilis*), black-grass (*Alopecurus myosuroides*), wild-oats (*Avena* spp.), common couch (*Elymus repens*) and meadow-grasses (*Poa* spp.) (Boatman and Wilson, 1988). Bond (1987) also identified brome, cleavers, black-grass and wild-oats as being unacceptable in this context, and Roebuck (1987) considered grass weeds to be the dominant problem on arable headlands, with cleavers the only broadleaved species of importance. Recent research into the relative competitive abilities of different weed species supports this view (Wilson, 1986; Wilson and Wright, 1987).

Cleavers, barren brome, black-grass and rough meadow-grass are mainly problems of autumn-sown crops. Conversely, many of the species which have been identified as valuable host plants for chick-food insects, or whose seeds are eaten by gamebirds, are spring-germinating. Most are dicotyledons e.g. knotgrass (*Polygonum aviculare*), black-bindweed (*Bilderdykia convolvulus*), chickweed (*Stellaria media*), fat-hen (*Chenopodium album*), charlock (*Sinapis arvensis*) and hemp-nettles (*Galeopsis* spp.), though annual meadow-grass (*Poa annua*) is also considered useful. However, studies of the weed flora in headlands of autumn-sown cereal fields showed that, where herbicides were applied in autumn, weed populations were often sparse in the spring, and there was no significant difference in abundance of most 'desirable' species between headlands treated with herbicide in autumn only and those treated in autumn and spring (see Table 23.1). In some parts of the country, particularly where continuous growing of winter cereals is practised, the weed flora is predominantly composed of autumn-germinating species, and removal of these results in a virtually weed-free crop in the spring. Moreover, the increased availability and use of broad-spectrum herbicides with very long-lived residual activity in the soil often

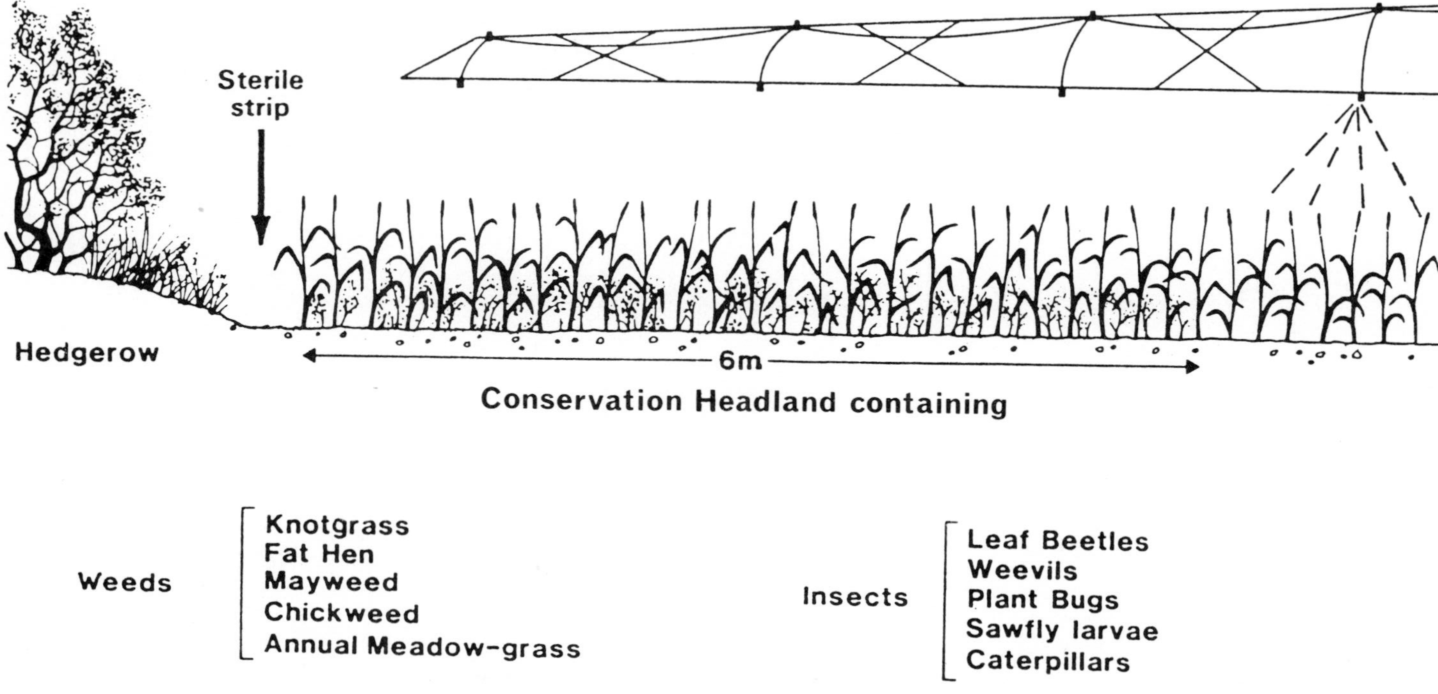

Figure 23.1 The relation of 'conservation headlands' to the field boundary and an arable crop.

Table 23.1 Mean numbers of weeds per m^2 in headlands of winter wheat fields receiving different herbicide treatments

Species	I No Herbicide	II Autumn[a] Herbicide	III Autumn[a] + Spring[b] Herbicide	P (Mann–Whitney U–Test) I vs II + III	II vs III
Mayweeds (*Matricaria* spp.)	19.6	0.0	0.1	<0.001	NS
Fluellens (*Kickxia* spp.)	15.0	0.0	0.0	<0.01	NS
Knotgrass (*Polygonum aviculare*)	14.5	11.5	1.2	<0.05	NS
Speedwells (*Veronica* spp.)	10.5	0.3	0.2	<0.01	NS
Field pansy (*Viola arvensis*)	9.8	0.4	0.4	<0.01	NS
Black bindweed (*Bilderdykia convolvulus*)	3.9	0.1	0.0	<0.01	NS
Charlock (*Sinapis arvensis*)	3.7	1.6	0.0	NS	<0.05
Shepherd's purse (*Capsella bursa-pastoris*)	3.0	0.1	0.0	<0.01	NS
Others	20.8	3.5	2.9		
Total Broadleaved	99.7	17.5	4.8	<0.001	NS
Meadow-grasses	8.3	3.7	2.8	NS	NS
Others	2.2	0.4	2.3		
Total grasses	10.5	4.1	5.1	NS	NS
Mean number of species	17.2	5.1	3.8	<0.001	NS

a pendimethalin or chlorsulfuron + metsulfuron-methyl
b fluroxypyr or ioxynil + bromoxynil

means that emergence of spring-germinating species is prevented even where they are represented in the seed bank.

In order to overcome these difficulties, selective control options are being developed which allow the less ecologically valuable and more agriculturally damaging weed species to be removed without affecting the other, more 'desirable' species (Boatman, 1987). This has allowed a modification of the earlier approach to herbicide use such that chemicals with broad-spectrum activity against dicotyledonous weeds can be excluded from conservation headlands. Guidelines are now available to subscribers, specifying which chemicals may be used at different times of year without affecting the benefits deriving from conservation headlands (see Table 23.2). Where a specific problem is not covered by these guidelines, individual advice is available.

Benefits to other forms of wildlife

It has been shown that, by implementation of the 'conservation headland' technique it has been possible to reverse the decline of native game species. This is achieved by protecting an entire food chain (see Figure 23.2). The pressure is removed from those non-target organisms whose populations are normally limited by agrochemicals. As a result additional resources are created which are then available for use by higher fauna. A general improvement in species diversity and abundance might, therefore, be expected to result. Recent research has shown that such trends do indeed occur in a number of taxa including butterflies (Rands and Sotherton, 1986; Dover, 1987), small mammals (Tew, 1988) and rarer components of the arable flora (Wilson, 1988). Increased abundance of common species may be a desirable end in itself if they are considered aesthetically pleasing (e.g. butterflies) or could themselves provide a resource utilised by rarer species (e.g. phytophagous insects as food for whitethroats). For other groups such as certain plants adapted to arable land, sympathetic headland management could make an important contribution to their conservation in their traditional habitat. Two examples will be considered in more detail.

Table 23.2 Outline recommendations for conservation headlands

	Autumn	*Spring*	
Insecticides	✓	×	
Fungicides	✓	✓	(most products)
Growth Regulators	✓	✓	
Herbicides (grass weeds)	✓	✓	
	(only selective graminicides)		
Herbicides (broadleaved weeds)	×	×	
	(except for selective control of certain species eg. cleavers)		

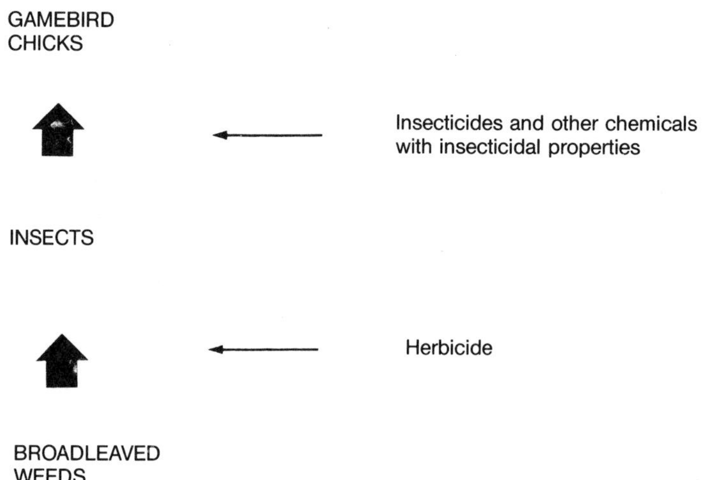

Figure 23.2 Diagrammatic food chain indicating sources of disruption.

Butterflies

Baseline monitoring of butterfly abundance over 'conservation' and fully sprayed headlands has been carried out on the main study farm in North Hampshire since 1984. The method employed a modified 'Pollard Walk' (Pollard *et al.*, 1975; Rands and Sotherton, 1986). Once a week during the summer, the recorder walked along field edges in the fully sprayed and conservation regimes that had been paired for habitat (e.g. grass bank, short hedge, etc.), aspect and wherever possible, crop. The data for the years 1984–7 are presented in Figure 23.3 and show quite conclusively that, when making within year comparisons, significantly more butterflies were seen over conservation headlands compared with field edges which received 'full' pesticide inputs.

Behavioural observations of adult butterflies were carried out in order to pinpoint the reason(s) for differentially higher butterfly numbers in the conservation regime. Butterflies were followed in field margins which had been paired as above. Activities such as flying, feeding, basking and interacting with other butterflies were recorded by rapid dictation into a tape recorder. The activity profiles of target species could then be compiled by going over the tapes with a stop-watch. The activity profile for females of the green-veined white butterfly (*Artogeia napi* L.) is presented in Figure 23.4.

It is evident that for many species of butterfly 'conservation headlands' represent a valuable resource, namely nectar, which may be in short supply in arable fields. An example is shown in Figure 23.5. Under the fully sprayed regime, the major activity of green-veined white females is flight; in the conservation regime the emphasis switches to feeding with almost all activity observed being carried out in the headland at the expense of the nectar-deficient hedgerow. One inference that can be

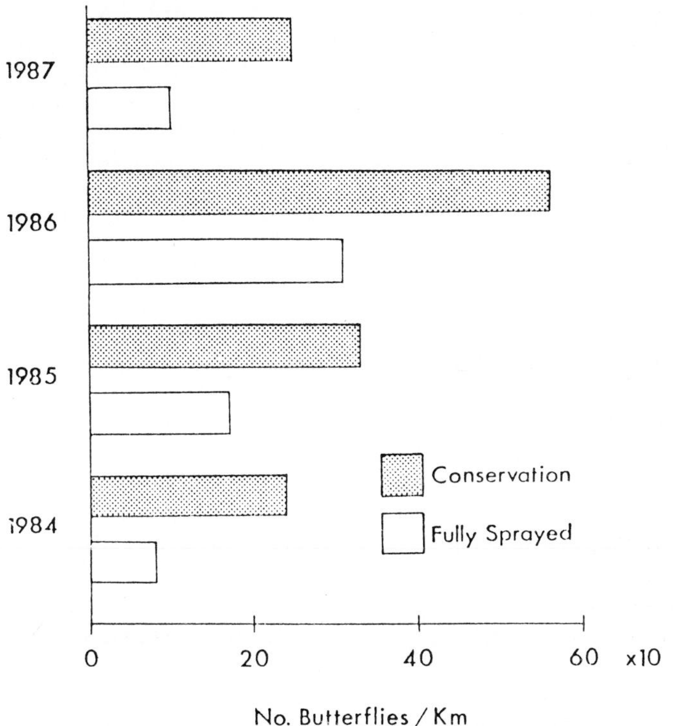

Figure 23.3 Butterflies seen over fully sprayed and conservation headlands 1984–7. Differences between the two regimes significant ($p < 0.001$) for each year. (1984 data from Rands and Sotherton, 1986).

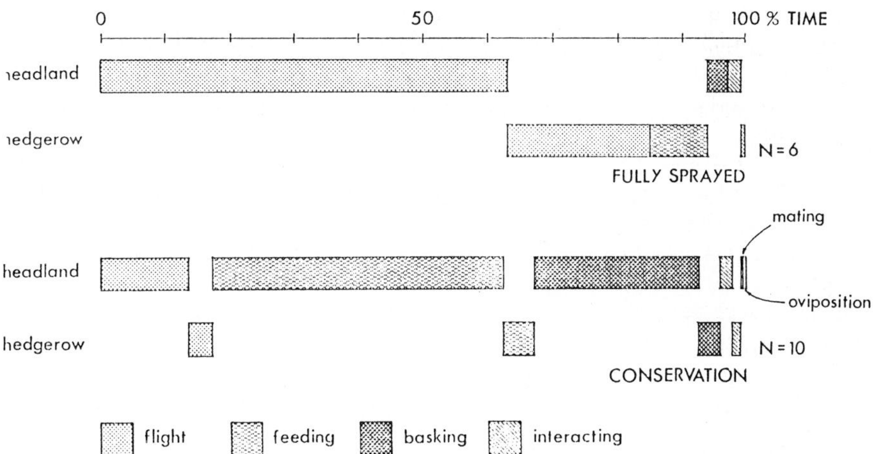

Figure 23.4 Activity profile for females of the green-veined white butterfly over fully sprayed and conservation headlands in 1986, showing the proportions of the various activities and distribution between hedgerow and headland. N = no. of butterflies observed.

drawn from the readiness of some butterfly species to desert the hedgerow in favour of a nectar-rich headland is that the farm hedgerow may not provide sufficient nectar for butterflies. Some research (Wiklund and Karlsson, 1984) suggests that nectar represents an important resource for ovipositing butterflies – an extension of the argument being that, in a habitat where nectar is in poor supply, females may not achieve the reproductive potential of those in nectar-rich habitats.

Flora

The accelerating rate of decline in recent decades of much of the flora traditionally associated with arable land has frequently been noted (Salisbury, 1961; Fryer and Chancellor, 1970; Chancellor, 1977). In some European countries, particularly in West Germany, this observed decline was acted upon by Governmental Agencies as long as ten years ago (Schumacher, 1987; Eggers, 1987). Serious attention was only focused on these species in Britian in 1986 when, in conjunction with the Botanical Survey of the British Isles (BSBI), the Nature Conservancy Council launched a country-wide survey of 25 arable weed species believed to be in decline. Preliminary results confirm that many of those plants once commonly associated with arable land, such as the corn buttercup (*Ranunculus arvensis*) and shepherd's needle (*Scandix pecten-veneris*), have experienced a catastrophic decline since the 1950s. Others, e.g. lamb's succory (*Arnoseris minima*) and thorow-wax (*Bupleurum rotundifolium*), are now extinct.

As part of the same project, the possibility of conserving some of the richest remaining sites by management agreement and legal measures was considered. This approach however, does not include any provision for the conservation of traditional weed communities over most of the countryside. Surveys conducted by the Cereals and Gamebirds Research Project in 1986 and 1987 have demonstrated the potential of conservation headlands as the foundation for a conservation strategy for rare weed communities. In the first year alone, 17 of the 25 species in the BSBI survey were discovered in conservation headlands (see Table 23.3; Wilson, 1988), and preliminary data from the second year indicate similar results. However, although herbicide use may be a major reason for changes in our weed communities, it is not the only one (Chancellor, 1980). Research is currently in progress to identify more precisely the reasons for the observed decline in many species, to provide a scientific basis for the management of rare weed communities. This research is at an early stage, but already it has been found that at many sites, even where herbicide use is restricted, rare species are confined to the edges and corners of fields where the effects of farming operations are often reduced in intensity. These results have important implications for current 'set-aside' farming schemes designed to reduce cereal production. Practices such as sowing grass strips or trees around the margins of fields could have disastrous consequences for the survival of remnant arable floras.

Field boundaries

So far the emphasis has been on manipulation of inputs into the cropped area of the field margin. However, as a habitat this region is strongly influenced by and insepa-

Table 23.3 Declining weed species recorded in a survey of conservation headlands on 17 farms, 1986

Species	Number of farms
Dense silky-bent (*Apera interrupta*)	1
Corn chamomile (*Anthemis arvensis*)	1
Dwarf spurge (*Euphorbia exigua*)	7
Broadleaved spurge (*Euphorbia platyphyllos*)	1
Dense-flowered fumitory (*Fumaria micrantha*)	3
Red hemp-nettle (*Galeopsis angustifolia*)	2
Sharp-leaved fluellen (*Kickxia elatine*)	7
Round-leaved fluellen (*Kickxia spuria*)	9
Venus's-looking-glass (*Legousia hybrida*)	7
Field gromwell (*Lithospermum arvense*)	2
Prickly poppy (*Papaver argemone*)	2
Rough poppy (*Papaver hybridum*)	4
Corn parsley (*Petroselinum segetum*)	1
Shepherd's-needle (*Scandix pecten-veneris*)	1
Night-flowering catchfly (*Silene noctiflora*)	3
Field woundwort (*Stachys arvensis*)	3
Narrow-fruited cornsalad (*Valerianella dentata*)	2

rable from the adjacent field boundary, be it hedgerow, grass bank, wood, wall, ditch or other structure. Moreover, modification of management practices on the cropped headland can also influence habitat quality within the field boundary itself.

Many existing farm hedgerows (and banks) are subject to potential damage from pesticide and fertiliser 'drift'. The concept of linking fragmented habitats such as woods by planting new hedgerows is currently popular. However these new hedges and their attendant ground flora are also subject to damage from 'bioactive' drift. Where new landscape structures such as hedges are grant-aided, and to a certain extent under the control of the grant-aiding body, damage by pesticide drift can be ameliorated by advising use of conservation headlands as buffer zones.

In the context of habitat linking in the arable environment, a newly planted or even well established hedge may represent only a narrow ribbon for animal dispersal. Such structures can be considerably widened during the active spring and summer season by the use of conservation headlands. For the nesting requirements of wild gamebirds, the Game Conservancy has recommended that hedgerows be some 3.3 m wide (Anon., 1986), an ideal which is unlikely to be realised on farms without a shooting interest. However, the wildlife corridor can be widened by 12 m if conservation headlands are employed on both sides of the boundary.

Hedgerows are an important nesting habitat for a number of bird species. Many studies have been carried out, and these have been recently reviewed by O'Connor (1987). The specific requirements of partridges have been discussed by Rands

(1987). Presently research is concentrating on the value and management of field boundaries for insects.

Butterflies

As part of the current programme of research into the ecology of butterflies on arable land, the impact of cutting regimes on field edge habitats (grass and hedge banks) is being examined from several standpoints. The questions being asked are: What is the influence of grass height on the oviposition preferences of butterflies? If the current pre and post-harvest tidying-up of banks and hedgerows is delayed, will this have an impact on the nectar resources available? Will a late spring/early summer cut extend the flowering period of some nectar species?

A further area of interest is the establishment of a good perennial flora in newly planted hedge banks. Currently, there is little advice on the management of the hedge-base flora, and that which is available restricts itself to weed control. Forestry companies frequently offer management services which include (say) a five-year management package of weed control to ensure good hedge establishment. In practice this often means strimming off well developed thistles and treating with a broad-spectrum herbicide.

The problems involved in new hedge planting, excluding establishment, are firstly, how to prevent the new hedge becoming a problem site for pernicious annual weeds such as barren brome, cleavers, etc., secondly, how to ensure that the hedge has the maximum wildlife benefit. An approach being examined currently is the seeding of a perennial grass and wild flower mixture into new hedge-bases and boundary banks (see Kaule and Krebs, in this volume). If a good perennial flora can be developed, the annual 'problem' species may not gain a foothold, and a good basal flora will be available for invertebrate colonisation/exploitation.

Polyphagous predators

Polyphagous predators may have an important role to play in maintaining cereal aphid populations below economic threshold levels (Carter and Sotherton, 1983; Edwards *et al.*, 1979). Sotherton (1984, 1985) has shown that field boundaries are of particular importance in providing overwintering refuges for many species in this group. Raised banks with rough grass cover at the base of hedges may support densities of Carabidae, Staphylinidae and spiders of up to 1,000 m^{-2} in the winter. The boundary/field area ratio however is usually small, and thus rapid colonisation of fields by the predators in the early spring, when their control potential is thought to be at its highest, is impaired (Wratten *et al.*, 1984; Coombes and Sotherton, 1986).

In order to increase predator numbers, we are experimenting with the creation of overwintering habitats on farmland which favour the development of high populations of predators. An aspect of this research is the reduction of field size experimentally by creating new within-field overwintering refuges. These take the form of raised banks (0.4 m $\times$ 1.5 m $\times$ 350 m) that penetrate the centres of two cereal

Table 23.4 Predator densities in autumn 1987 on raised banks positioned across cereal fields and sown with different grass seed mixtures. Treatments within a column with the same letter are not significantly different at the 5% level (Tukey's test)

Treatment	*Field 1* Predators m^{-1}	Difference	*Field 2* Predators m^{-1}	Difference
1. *Dactylis glomerata*	194.0	a	86.0	b
2. *Lolium perenne*	128.5	b	159.0	cd
3. *Holcus lanatus*	196.0	a	93.5	a
4. *Agrostis stolonifera*	134.5	b	51.5	c
5. 25% mix of 1–4	103.5	c	116.0	ef
6. 33% mix of 1–3	115.0	bc	74.0	de
7. Bare ground	31.5	d	46.0	g
8. Field	46.0	d	24.5	fg

fields. Sections of each bank have been sown (spring 1987) with various grass species in a randomised block design with six blocks per bank.

Although the grass establishment is at an early stage the results show that the raised banks support significantly higher densities of polyphagous predators than do the adjacent fields (see Table 23.4). With greater grass development differences between bank and field, and also between treatments within the bank, should be enhanced.

Future work will monitor the dispersal, distribution and predation rate of the predators in the crop in spring and summer. Also, it is intended to quantify the feeding rates and food items, reproductive condition, fat content and survival of the predators in order to gain an insight into the advantages, if any, of creating such overwintering sites for pest control.

Conclusion

In this paper we have summarised some of the recent and current research into field margin management for the benefit of wildlife. Although much remains to be achieved, a basis now exists for the development of a management strategy for field edges. In this context there is a need for the field margin to be considered as a single habitat and not in more discrete categories: i.e. the headland, the hedge, the hedgebank/grass bank, etc. This is particularly true when the question of grant-aid for a new hedge is considered. Currently, aid is available only for the planting of the hedge – not for the creation of the physical bank to put it on, nor for the seeding of a basal flora, nor for the protection of the new habitat from pesticide and fertiliser drift, though these features may be of the greatest importance to wildlife!

Wider opportunities are presented by the current debate on the reduction of the cereal surplus, with political initiatives on extensification and set-aside. It is vital that these initiatives take account of the requirements of farmland wildlife, otherwise such measures may be environmentally more harmful than beneficial.

References

Anonymous, 1986, *Advisory Booklet 4: The Grey Partridge*, The Game Conservancy, Fordingbridge, pp. 60.

Boatman, N.D., 1987, 'Selective grass weed control in cereal headlands to encourage game and wildlife', *1987 British Crop Protection Conference – Weeds*, **1**, 277–84.

Boatman, N.D., Wilson, P.J., 1988, 'Field edge management for game and wildlife conservation', *Aspects of Applied Biology*, **16**, 53–61.

Bond, S.D., 1987, 'Field margins: a farmer's view on management, in J.M. Way and P.W. Greig-Smith (eds.) *Field Margins*, BCPC Monograph No. 35, London, pp. 79–83.

Carter, N., Sotherton, N.W., 1983, 'The role of polyphagous predators in the control of cereal aphids', *10th International Congress of Plant Protection*, **2**, p.778.

Chancellor, R.J., 1977, 'A preliminary survey of arable weeds in Britain', *Weed Research*, **17**, 283–7.

Chancellor, R.J., 1980, 'New weeds for old in annual crops', in R.G. Hird, P.V. Biscoe and C. Dennis (eds.), *Opportunities for increasing crop yields*, Pitman, London, pp. 313–22.

Coombes, D.S., Sotherton N.W., 1986, 'The dispersion of polyphagous predators from their overwintering sites into cereal fields and factors affecting their distribution in the spring and summer', *Annals of Applied Biology*, **108**, 461–74.

Dover, J.W., 1987, 'The benefits of conservation headlands to butterflies on farmland', *Game Conservancy Annual Review*, **18**, 105–9.

Edwards, C.A., Sunderland, K.D., George, K.S., 1979, 'Studies on polyphagous predators of cereal aphids', *Journal of Applied Ecology*, **16**, 811–23.

Eggers, Th., 1987, 'Environmental impact of chemical weed control in arable fields in the Federal Repulic of Germany', *1987 British Crop Protection Conference – Weeds*, **1**, 267–75.

Fryer, J.D., Chancellor, R.J., 1970, 'Herbicides and our changing arable weeds', BSBI Conference Reports, 'The flora of a changing Britain', **11**, 105–18.

Green, R.E., 1984, 'The feeding ecology and survival of partridge chicks (*Alectoris rufa* and *Perdix perdix*) on arable farmland in East Anglia', *Journal of Applied Ecology*, **21**, 817–30.

Green, R.E., Rands, M.R.W., Moreby, S.J., 1987, 'Species differences in diet and the development of seed digestion in partridge chicks *Perdix perdix* and *Alectoris rufa*', *Ibis*, **129**, 511–14.

Hill, D.A., 1985, 'The feeding ecology and survival of pheasant chicks on arable farmland', *Journal of Applied Ecology*, **22**, 645–54.

O'Connor, B., 1987, 'Environmental interests of field margins for birds, in J.M. Way and P.W. Greig-Smith (eds.), BCPC Monograph No.35, London pp. 35–48.

Pollard, E., Elias, D.O., Skelton, M.J., Thomas, J.A., 1975, 'A method of assessing the abundance of butterflies in Monks Wood National Nature Reserve in 1973', *Entomologists's Gazette*, **26**, 79–88.

Potts, G.R., 1986, *The Partridge: Pesticides, Predation and Conservation*, Collins, London, pp. 274.

Potts, G.R. Vickerman, G.P., 1974, 'Studies on the cereal ecosystem', *Advances in Ecological Research*, **8**, 107–97.

Rands, M.R.W., 1985, 'Pesticide use on cereals and the survival of grey partridge chicks: a field experiment', *Journal of Applied Ecology*, **22**, 49–54.

Rands, M.R.W., 1986, 'The survival of gamebird chicks in relation to pesticide use on cereals', *Ibis*, **128**, 57–64.

Rands, M.R.W., 1987, 'Hedgerow management for the conservation of partridges *Perdix perdix* and *Alectoris rufa*', *Biological Conservation*, **40**, 127–39.

Rands, M.R.W., in press, 'The effects of pesticides on grey partridge chick survival and breeding density', *Journal of Applied Ecology*.

Rands, M.R.W., Sotherton, N.W., 1986, 'Pesticide use on cereal crops and changes in the abundance of butterflies on arable farmland', *Biological Conservation*, **36**, 71–82.

Roebuck, J.F., 1987, 'Agricultural problems of weeds on the crop headland', in J.M. Way and P.W. Greig-Smith, (eds.), *Field Margins*, BCPC Monograph No. 35, London, pp. 11–22.

Salisbury, E., 1961, *Weeds and Aliens*, Collins, London.

Schumacher, W., 1987, 'Measures taken to preserve arable weeds and their associated communities in central Europe', in J.M. Way and P.W. Greig-Smith (eds.), *Field Margins*, BCPC Monograph No. 35, London, pp. 109–12.

Sotherton, N.W., 1982, 'Effects of herbicides on the chrysomelid beetle *Gastrophysa polygoni* (L.) in laboratory and field', *Zeitschrift für angewandte Entomologie*, **94**, 446–51.

Sotherton, N.W., 1984, 'The distribution and abundance of predatory arthropods overwintering on farmland', *Annals of Applied Biology*, **105**, 423–9.

Sotherton, N.W., 1985, 'The distribution and abundance of predatory arthropods overwintering in field boundaries', *Annals of Applied Biology*, **106**, 17–21.

Sotherton, N.W., Moreby, S.J., 1988, 'The effects of foliar fungicides on beneficial arthropods in wheat fields', *Entomophaga*, **33**, 87–99.

Sotherton, N.W., Moreby, S.J., Langley, M.G., 1987, 'The effects of the foliar fungicide pyrazophos on beneficial arthropods in barley fields', *Annals of Applied Biology*, **111**, 75–87.

Sotherton, N.W., Rands, M.R.W., Moreby, S.J., 1985, 'Comparison of herbicide treated and untreated headlands on the survival of game and wildlife', *1985 British Crop Protection Conference – Weeds*, **3**, 991–8.

Southwood, T.R.E., Cross, D.J., 1969, 'The ecology of the partridge III, breeding success and abundance of insects in natural habitats', *Journal of Animal Ecology*, **38**, 497–509.

Tew, T., 1988, 'The effects of conservation headlands on small mammals', *Game Conservancy Annual Review*, **19**, 88–90.

Vickerman, G.P., 1974, 'Some effects of grass weed control on the arthropod fauna of cereals', *Proceedings of the 12th British Weed Control Conference*, **3**, 929–39.

Vickerman, G.P., Sunderland, K.D., 1977, 'Some effects of dimethoate on arthropods in winter wheat', *Journal of Applied Ecology*, **14**, 767–77.

Wiklund, C., Karlsson, B., 1984, 'Egg size variation in satyrid butterflies: adaptive vs historical, "Bauplan", and mechanistic explanations', *Oikos*, **43**, 391–400.

Wilson, B.J., 1986, 'Yield responses of winter cereals to the control of broad-leaved weeds', in *Proceedings of the European Weed Research Society Symposium on Economics Weed Control*, 1986, Stuttgart, pp.75–82.

Wilson, B.J., Wright, K.J., 1987, 'Variability in the growth of cleavers (*Galium aparine*) and their effect on wheat yield', in *Proceedings of the 1987 British Crop Protection Conference – Weeds*, **3**, 1051–7.

Wilson, P.J., 1988, 'The conservation of rare and vanishing arable weeds', *Game Conservancy Annual Review*, **19**, 80–3.

Wrattan, S.D., Bryan, K., Coombes, D., Sopp, P., 1984, 'Evaluation of polyphagous predators of aphids in arable crops', in *Proceedings of the 1984 British Crop Protection Conference – Pests and Diseases*, pp. 271–6.

24

Changes in the ground flora and butterfly populations of woodlands managed to encourage pheasants

I.C. Ludolf, P.A. Robertson and *M.I.A. Woodburn*

Introduction

The intimate patchwork of coppice and farmland that still characterises certain parts of lowland Britain is a reflection of the important historic role played by woodland management (Rackham, 1986).This landscape was widespread in the middle Ages and began to decline in the eighteenth century with the Enclosures Acts and a shift in the emphasis of woodland management from underwood to timber production (Warren, 1976). The loss slowed down in the latter half of the nineteenth century as farming became less profitable, so that almost all the woods present in lowland Britain in 1845 still existed in some form in 1945. Since the Second World War, however, the combined effects of replacing the existing trees with conifers and grubbing out for agriculture have led to the destruction of between a third and a half of the remaining ancient woodland (Rackham, 1986).

Many landowners consider management of their small farm woods uneconomic and have neglected them for many decades. As a result, lowland woods are currently shadier than at any time in the last millenium (Thomas and Webb, 1984). This continued neglect is likely to affect the woodland ground flora and may cause other changes to the historic character of the woodland and its associated wildlife (Peterken, 1981).

The value of small farm woodlands in terms of heritage, landscape and recreation as well as nature conservation is widely recognised (DART, 1983). However, apart from the few sites with statutory protection, their conservation is largely dependent on the efforts of the private landowner. An important incentive for such conservation is game production: 67 per cent and 56 per cent respectively of Timber Growers UK members responding to a survey stated that they retained and planted woodlands of less than 10 ha in size partly for their value to the pheasant shoot. The value of small woodlands for game was perceived by respondents to be second only to their value in the landscape and greater than their value in terms of timber, wildlife or shelter (Cobham Resource Consultants, 1983). The same survey revealed that many landowners undertook management of their small woods primarily to enhance the quality of the shoot.

This chapter examines the principle of woodland management for game and nature conservation and discusses the compatibility of the two interests. A case study into the effects of woodland management for game on the ground flora and butterflies within a shooting estate in Dorset is used as an illustration. Recommendations are then made for the compilation of long-term management plans for woodland which satisfy both game and nature conservation interests.

Principles of woodland management for game and nature conservation

Management to conserve wildlife

Woodlands provide important habitats for many forms of wildlife including epiphytes, invertebrates and mammals as well as higher plants. The conservation of ancient, semi-natural woodland is considered a top priority by Peterken (1981). These woods are known from map records to have been in continuous existence at least since 1650 and are likely to be much more than four centuries old. They appear little altered in structure or species composition by man and the soil structure, flora and fauna may bear some resemblance to that of the wildwood which existed on the site in prehistory. Nationally, about 17 per cent of Britain's existing woodland, or about 1.5 per cent of the land surface, is ancient, semi-natural.

Ancient woodland characteristically holds a ground flora of three types: shade tolerators able to grow beneath heavy tree and shrub canopies; marginals which persist in the seed bank of shaded woodland and germinate when the light intensity is increased; and ruderals which are short-lived species with strong powers of dispersal (Brown and Oosterhuis, 1981). The traditional management in the majority of ancient, semi-natural woods in lowland Britain was coppicing. This involves the cutting of blocks of woody growth on a rotation and results in a temporal and spatial patchwork of varying light and soil conditions (Rackham, 1986), allowing the perpetuation of ruderal, marginal and shade-tolerant woodland plants. In turn, the floral diversity created by coppicing benefits a range of animals, including small mammals and invertebrates (Warren, 1976).

Wide rides cut through woodland and maintained by regular management provide areas free from shrubs and trees which favour perennial forbs and grasses, including some marginal species common to the younger stages of the coppice cycle. In addition, species favouring more permanent open areas, such as pasture and meadowland are common on long-established rides and may encourage a fauna characteristic of unimproved grassland (Steel and Khan, 1987). Coppicing and ride maintenance are, therefore, considered complementary and optimal forms of conservation management for the majority of lowland woods.

Management increases the structural diversity of neglected woodland and consequently tends to increase the number of plants and animal species to be found within it. Species-richness of song birds and mammals have both been found to increase as the structural composition of woodland becomes more complex (Moss *et al.*, 1979; Staines, 1983). Ancient, semi-natural woodland which has remained unmanaged for decades or centuries may, however, hold species, such as some epiphytes and invertebrates, which are intolerant of disturbance (Hammond, 1974). Continued neglect must therefore be considered as an option in any long-term plan of woodland management for nature conservation.

Management for game

Pheasants tend to be birds of woodland or other areas of permanent cover in winter. They establish breeding territories along woodland edges in spring and typically move out to cereal fields in summer (Robertson, 1988). One of the aims of the Game Conservancy's 'Pheasants and Woodlands Project' has been to determine the habitat requirements of pheasants within woodland. Studies have shown that pheasant densities are highest in small woods with a high proportion of low, shrubby cover. Woodland edges which gradually reduce in height from the tree canopy through shrubs to open ground are particularly attractive. In other studies where birds were radio-tracked (see Figure 24.1), individuals were found to spend the majority of their time within 20 m of the woodland edge (Hill and Robertson, 1988; Robertson, 1988).

Management of existing woodlands to improve pheasant holding capacity should, therefore, aim to increase the length of edge per unit area by dividing up large blocks of trees with wide, sunny rides. In addition, dense, low cover should be promoted, particularly near the woodland edge, either by coppicing, small group fellings or the encouragement of shade-tolerant shrubs beneath the tree canopy.

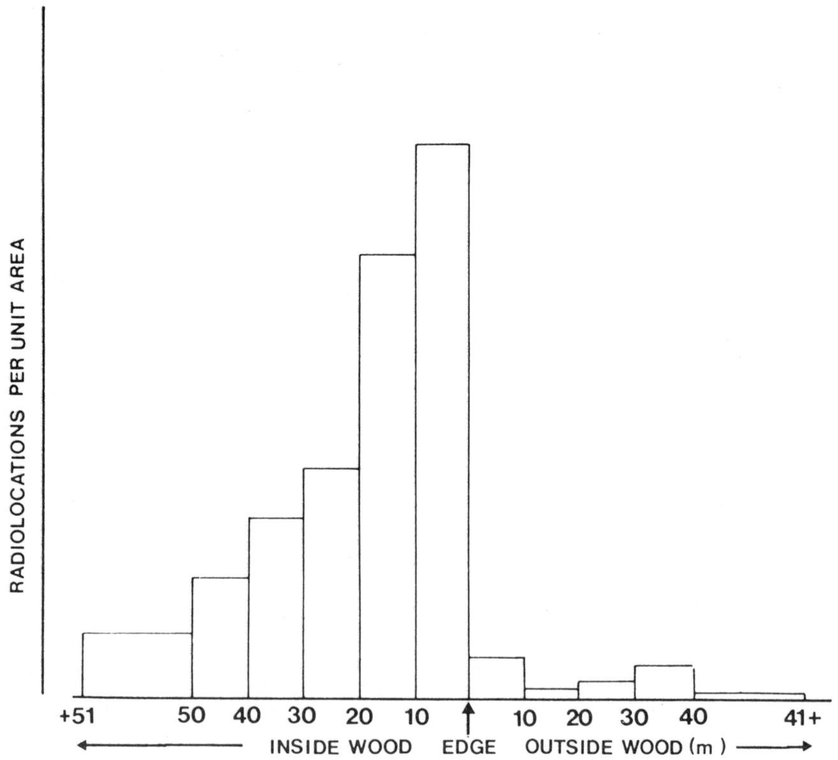

Figure 24.1 Distribution of pheasants in woodland. Position of radiolocations in relation to woodland edge from birds tracked in winter.

Such management is, on the whole, complementary with the aims of wildlife conservation.

Woodland management for game is also aimed at providing good sport by encouraging pheasants to fly high and straight. Pheasants have little stamina in flight and should take to the air about 40 m in front of the guns, rising steeply to reach maximum height while in shooting range. This is achieved by creating areas of impenetrable low cover or open ground termed 'flushing points'. The birds are driven by a line of beaters towards these points which are unattractive to running birds and encourage them to take to the air. They can be created by cutting a ride or small 'skylights' in the canopy of densely shaded woodland. The patches of undergrowth on an otherwise bare floor that are promoted by skylights encourage the birds to take wing in a staggered fashion, presenting themselves in a slow stream rather than a dense flock. The canopy at a 30° angle above the flushing points should be open to allow the birds to rise unimpeded and fly over the trees towards the guns (Gray, 1986). On a smaller scale, skylights create conditions similar to those resulting from the forestry practice of thinning a stand of trees grown as 'high forest'.

Methods

Study sites

The impact of various forms of woodland management for game on an ancient woodland ground flora and butterfly populations were examined in a small survey. The study was carried out in the summers of 1986 and 1987 in woodland on the Wimborne St.Giles estate in East Dorset, Southern England. The area is situated on acidic Bagshot sands with clay seams and was managed as hazel (*Corylus avellana*) coppice under pedunculate oak (*Quercus robur*) standards in the past (Thomas and Webb, 1984). This woodland is shown on maps dating from 1800 (Warren, 1976) and is thought to be ancient, semi-natural (Peterken, 1981).

The ground flora was examined by means of 30 sample sites chosen from woodland managed in 6 distinct ways (see Table 24.1; Figure 24.2).

UNMANAGED COPPICE

Five samples were allocated within areas predominantly of hazel with some ash (*Fraxinus excelsior*) and the occasional field maple (*Acer campestre*) stools under mature oak standards. These represented semi-natural woodland which had long been managed as coppice-with-standards. Having been coppiced at least 17 years previously, they were overgrown and shady, and provided little low cover for pheasants.

MANAGED COPPICE

The five sample areas selected were thought to be similar to the unmanaged coppice, except that the understorey had been cut within the last five years and the

Table 24.1 Sample sites for the plant surveys, all situated within ancient woodland on the Wimborne St. Giles Estate in East Dorset

Management category	Site No.	Description
Unmanaged coppice	1.	Hazel coppice with 30 years' regrowth under oak
	2.	Hazel coppice with 30 years' regrowth under oak
	3.	Hazel coppice with 17 years' regrowth under oak
	4.	Hazel coppice with 30+ years' regrowth under oak
	5.	Hazel coppice with 30 years' regrowth under oak
Managed coppice	6.	Hazel coppice with 1 years' regrowth under oak
	7.	Hazel coppice with 2 years' regrowth under oak
	8.	Hazel coppice with 3 years' regrowth under oak
	9.	Hazel coppice with 3 years' regrowth under oak
	10.	Hazel coppice with 3 years' regrowth under oak
High forest	11.	Mixed mature trees, heavy canopy, no understorey
	12.	Mixed mature trees, heavy canopy, no understorey
	13.	Mature oak, heavy canopy, no understorey
	14.	Mature oak and ash, heavy canopy, no understorey
	15.	Mature oak and ash, heavy canopy, no understorey
Ride	16.	15 year old, 50m wide ride, cut annually.
	17.	15 year old, 50m wide ride, cut annually.
	18.	15 year old, 50m wide ride, cut annually.
	19.	10 year old, 25m wide ride, cut annually.
	20.	10 year old, 25m wide ride, cut annually.
Young conifer	21.	4 year old Norway spruce, 1st rotation
	22.	7 year old Norway spruce/Douglas fir, 1st rotation
	23.	8 year old Douglas fir/Cypress, 2nd rotation
	24.	10 year old Douglas fir/Larch, 2nd rotation
	25.	10 year old European larch, 1st rotation
Mature conifer	26.	Douglas fir, canopy closed
	27.	Douglas fir/Lawson cypress, canopy closed
	28.	Scots pine, canopy closed
	29.	Scots pine/larch, canopy closed
	30.	Larch, canopy closed.

standards thinned where necessary to encourage healthy regrowth from the coppice stools. An important motive for coppicing was to create an attractive habitat for pheasants in winter.

HIGH FOREST

True skylights were uncommon on the Wimborne St. Giles estate and the effects of the practice on the ground flora had to be examined by surveying areas where man-

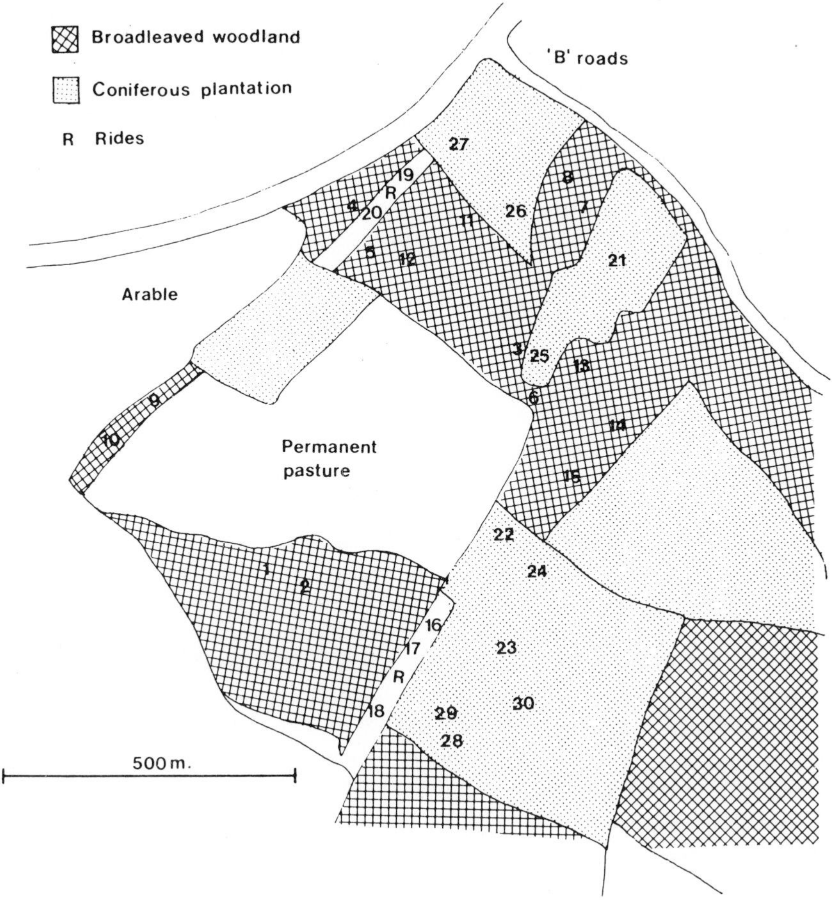

Figure 24.2 The study area on the Wimborne St. Giles Estate, Dorset, England showing the location of unmanaged coppice (numbers 1–5), managed coppice (6–10), high forest (11–15), ride (16–20), young conifer (21–25) and mature conifer (26–30) sample areas.

agement had created similar conditions. Of the five samples, sites 11 and 12 had been cleared of coppice and planted with a mixture of both native and exotic hard and soft woods about a century ago, and canopy closure was not complete. Site13 was located in a skylight created about twelve years ago by coppicing the hazel and leaving the mature oak standards. On sites 14 and 15 the tree canopy of stored ash coppice and oak standards was augmented by young trees which had grown up. The remaining hazel understorey had been coppiced two and three years ago respectively and showed few signs of regrowth.

RIDES

Five samples were located on rides cut 10 to 15 years previously to provide stands for the guns. Coppice regrowth, the establishment of woody species and the build-up of a tussocky sward on the rides were discouraged by annual mowing.

YOUNG AND MATURE CONIFERS

Traditional coppice-with-standards woodland had been cleared and stocked with conifers in the five young conifer and five mature conifer sample areas. Three areas of pre-thicket or thicket stage (young) conifers were first generation softwoods; the other two areas had been replanted after harvesting one conifer crop. No such distinction was made for the mature conifers as the recent history of the sites was not clear. The trees were planted as commercial crops.

Sampling methods

To sample the ground flora, twenty-five 0.25 m^2 quadrats were laid out on each sample area in May 1987, and again in August. Within each quadrat a record was made of each field and herb layer species present and a visual estimate of percentage cover made. The results were expressed as species density, total cover and diversity, the latter represented by the Shannon-Weiner Index (H) (Southwood, 1978). In addition, the contribution to the flora made by ancient woodland species was noted. Butterflies were studied by counting individuals seen along set transects within woodland managed in four different ways. Six of these transects were located in unmanaged coppice, eight in high forest areas, four on rides and four in mature conifer plantations. Transect counts were conducted in sunshine at approximately ten-day intervals during July and August, 1986 (Robertson, Woodburn and Hill, 1988). On each count the 22 transects, 200 m in length, were walked on the same day between 10.00 and 16.00 hours and the number of butterflies seen in an imaginary 5 × 5 m box immediately in front of the observer recorded (Pollard, 1977; Hall, 1981).

Results and discussion

Ground flora

The mean numbers of ground flora species found in the five sample plots of each management category in spring and summer are shown in Figure 24.3. Species number and diversity was relatively high in the managed coppice plots, where the amount of light reaching the woodland floor resulted in the germination and growth of marginal species (Brown and Oosterhuis, 1981). The number of ancient woodland indicator species (Peterken, 1981; Nature Conservancy Council, undated) found in unmanaged and recently managed coppice was, however, similar (see Figure 24.4). Many of these species are shade-tolerant and are able to thrive under a dense canopy, but tend to be unable to compete with vigorous, light-demanding species. They would therefore be expected to decline immediately after coppicing, increasing again as closed-canopy conditions begin to shade out many of the marginal species.

In most management categories a greater area of ground was covered by ground vegetation in summer, when a greater proportion of the incident light rays were

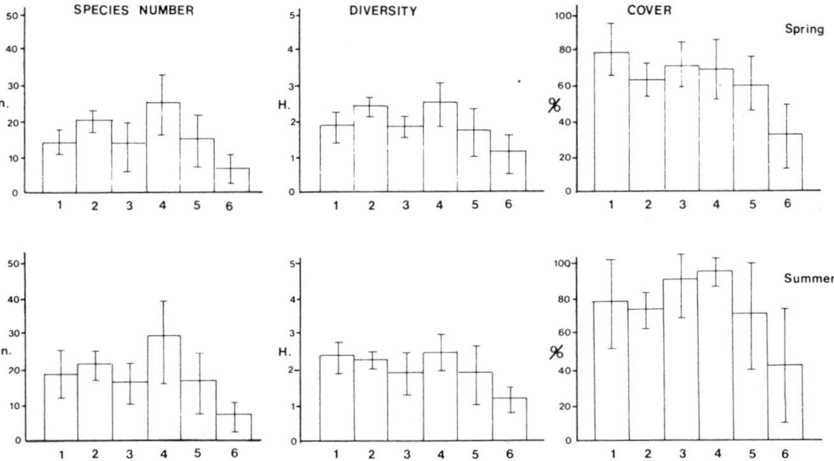

Figure 24.3 Characteristics of the ground flora during spring and summer in woodland managed for pheasants. The graphs show mean species number, mean diversity (H) and mean area of ground covered by vegetation (all with standard errors) for the 5 sample plots in unmanaged coppice (1), managed coppice (2), high forest (3), ride (4), young conifer (5) and mature conifer blocks (6).

vertical, than in spring (see Figure 24.3). The exception was the unmanaged coppice where it was likely that the amount of light reaching the woodland floor in spring was as great or greater than in summer, when both the trees and understorey were in full leaf. This suggests that plants undergoing their main growth period in summer are at a disadvantage under a dense canopy.

The management practices loosely termed 'high forest' let light through to the woodland floor, although this was insufficient to bring about vigorous growth of the coppice stools or woody seedlings. Herbivorous mammals, in particular roe deer (*Capreolus capreolus*), were attracted to the browse provided by growth in vegetation promoted by skylighting. The effect was particularly damaging in winter when green growth was absent and tree seedlings provided a staple food supply. Browsing pressure was concentrated by the small areas involved in 'high forest' management and, combined with the effects of low light levels in summer, killed much of the woody growth (Ratcliffe and Pepper, 1987). Hence, in the long term the woody understorey was lost from managed areas in favour of a dense ground flora (see Figure 24.3).

The high forest sample areas were characterised by an overwhelming dominance of a limited range of species. Bluebell (*Hyacinthoides non-scripta*) in spring was succeeded by bracken (*Pteridium aquilinum*) in summer in two of the sample sites while bramble (*Rubus fruticosus* agg.), soft rush (*Juncus effusus*), honeysuckle (*Lonicera periclymenum*), tufted hair grass (*Deschampsia cespitosa*) and ivy (*Hedera helix*) were extremely dominant in the other three areas. These highly competitive species appear to have thrived at the expense of ancient woodland plants which were less abundant in the high forest areas than in unmanaged coppice.

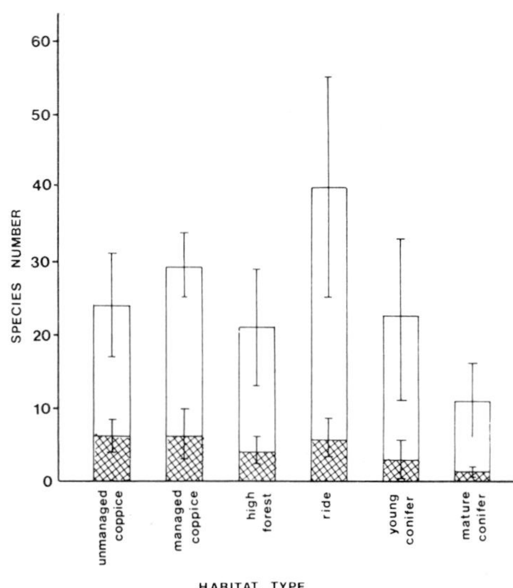

Figure 24.4 Mean number, with standard error, of ground flora species and ancient woodland indicator species (hatched) recorded in the five sample areas of each management category during spring and summer combined.

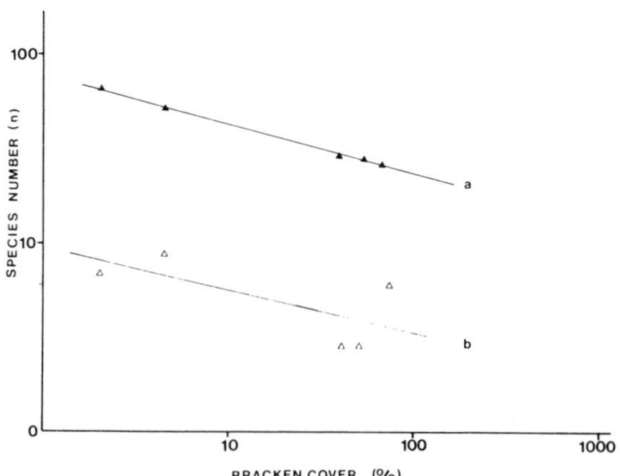

Figure 24.5 Relationship between bracken cover and species-richness on the rides, a: (solid triangles) correlation between bracken cover and total species number (r=0.99, t=38.7, p=⟨0.001), b: (open triangles) correlation of bracken cover and ancient woodland indicator species on each of the five sample areas (r=0.70, t=1.68, p=⟩0.10).

Table 24.2 Ancient woodland indicators found in each management category. The similarity between the ancient woodland flora in unmanaged coppice and other habitat categories is also shown

Species	Unmanaged coppice	Managed coppice	High forest	Ride	Young conifer	Mature conifer
Adoxa moschatellina	+	+	+	+	+	
Allium ursinum		+				
Anemone nemorosa	+	+	+	+	+	
Betonica officinalis	+			+		
Carex remota	+		+			
Carex sylvatica	+	+	+		+	
Conopodium majus		+				
Hyacinthoides non-scripta	+	+	+	+	+	+
Euphorbia amygdaloides	+			+	+	
Lamiastrum galeobdolon	+			+		
Galium odoratum	+			+		+
Holcus mollis						+
Iris foetidissima	+	+		+		
Lathyrus montanus				+		
Lysimachia nemorum	+	+	+	+	+	
Melica uniflora	+	+				
Milium effusum	+	+				
Oxalis acetosella				+	+	+
Potentilla sterilis	+	+	+	+		
Primula vulgaris		+				
Sanicula europaea				+	+	
Serratula tinctoria				+		
Solidago virgaurea				+		
Veronica montana	+	+	+			
Vicia sepium				+		
Total	13	14	9	17	10	3
Number of species common to unmanaged coppice (%)	100	79	100	65	90	33

More ground flora species were found on the rides than in other management categories. As in young coppice, the high light intensity encouraged the growth of marginal species while the practice of annual mowing probably increased species numbers by reducing the dominance of woody plants and tussock-forming grasses. Annual mowing, combined with the burrowing activities of animals, also helped to keep the sward open and the soil loose. This increased the germination rate of marginal species and allowed establishment of seed dispersed from other habitats (Grubb, 1977).

It can be seen that several ancient woodland indicator species were found on the rides but were not recorded in unmanaged or managed coppice samples (see Table 24.2). However, this difference is an artifact of the sampling technique, as often the species was present nearby but was not included in the sample. This phenomenon was less marked in the results of the high forest and conifer plots, as the species here were generally abundant throughout the estate and had a high chance of being recorded in other management categories.

On the rides, bracken tended to be highly dominant in summer: log-transformed values for the proportion of the sample area covered by bracken showed a strong negative correlation with species number ($r=-0.99$, $t=38.7$, $p=<0.001$; see Figure 24.5). The number of ancient woodland indicator species found on each study plot also declined but was not significantly correlated with bracken cover ($r=0.70$, $t=1.68$, $p=>0.1$). Autumn mowing of the rides removed much of the dead vegetation leaving a relatively open sward on bracken-dominated areas and providing space for vernal flowering plants to thrive. Many ancient woodland indicator species flower in early spring, a strategy which may have allowed co-existence with bracken.

Species-richness was high in first-rotation young conifers planted on ancient, semi-natural sites. This was due to the germination of marginal and ruderal species following clearance of the coppice-with-standards combined with the persistence of shade-tolerant species. Few species were found in the mature conifer stands, probably because of poor survival in the heavy shade characteristic of thicket-stage conifers. As the crop matures the canopy becomes more open and favourable for the growth of shade tolerant species. However, these species generally have short-lived seeds and are extremely slow to colonise from neighbouring areas (Brown and Oosterhuis, 1981). Of the three ancient woodland indicators found in mature conifer, *Hyacinthoides non-scripta* has a bulb and is able to tolerate unfavourable conditions for a short period of time; *Holcus mollis*, a marginal species, was found only in glades under Scots pine (*Pinus sylvestris*) while *Oxalis acetosella* is known for its shade-tolerance (Brown and Oosterhuis, 1981).

Conifers are thought to affect the ground flora in two ways: directly, by limiting the light available to lower strata of vegetation; and indirectly, by altering the leaf litter structure, soil microclimate and chemical characteristics of semi-natural woodland (Peterken, 1981; Anderson, 1987). The low species-richness recorded in second-rotation young conifer plots may well be due to these indirect effects which can influence the ground flora after the timber crop has been harvested. However, the study did not record species present as dormant seed, bulbs or other dormant stages, and these may have been present in the soil of conifer and high forest sample areas.

Butterflies

A total of 842 butterflies representing 21 different species were observed during the study. Numbers and species seen in each habitat category are shown in Table 24.3).

The results obtained from the unmanaged coppice areas can be considered controls against which the effects of management for game on butterfly numbers can be examined. Numbers here were low, at 2.4 individuals per kilometre, and the Speckled wood (*Pararge aegeria*), a species which is common in shaded woodland, predominated (Higgins and Hargreaves, 1983). Fewer species were seen in the

Table 24.3 The number of butterfly species and individuals seen in ancient woodland managed in four different ways. Results of transect counts conducted on the Wimborne St. Giles Estate in July and August, 1986

Species	Unmanaged coppice	High forest	Ride	Mature conifer
Aglais urticae (Small tortoiseshell)			×	
Aphantopus hyperantus (Ringlet)		×	×	
Argynnis paphia (Silver-washed fritillary)	×	×	×	
Artogeia napi (Green-veined white)		×		
Artogeia rapae (Small white)		×	×	
Clossiana euphrosyne (Pearl-bordered fritillary)		×	×	
Cynthia cardui (Painted lady)		×		
Gonepterix rhamni (Brimstone)		×	×	
Inachis io (Peacock)			×	
Lasiommata megera (Wall brown)		×	×	
Limenitis populi (White admiral)	×	×	×	×
Lycaena phlaeas (Small copper)			×	
Maniola jurtina (Meadow brown)	×	×	×	
Melanargia galathea (Marbled white)		×	×	
Ochlodes venatus faunus (Large skipper)		×	×	
Pararge aegeria (Speckled wood)	×	×	×	×
Pieris brassicae (Large white)		×	×	
Polygonia c–album (Comma)			×	
Polyommatus icarus (Common blue)			×	
Pyronia tithonus (Gatekeeper)		×	×	
Thymelicus flavus (Small skipper)		×	×	
No. species seen	4	16	19	2
No. km sampled	7.2	9.6	4.8	4.8
No. butterflies seen/km	2.4	25.4	89.4	5.2

mature conifer than in the unmanaged coppice and total butterfly numbers were also low, at 5.2 individuals per kilometre. These results may reflect the low light intensity here, and the absence of vegetative feeding or breeding sites beneath the canopy. Speckled woods were again the most common species in mature conifer.

Ten times the number of individuals representing 16 species were observed in high forest than in unmanaged coppice. However, the greatest diversity of butterflies was seen on the rides where an average of 89.4 individuals were counted along each kilometre of transect and a total of 19 species were recorded. Butterfly numbers and distribution therefore appear to respond proportionally to the increase in light intensity and diversity of the understorey vegetation brought about by woodland management for pheasants.

Compatibility of woodland management for pheasants and nature conservation

The conservation value of woodland benefits from pheasant management in two ways. Firstly, the retention of small farm woods, which are often valuable features in the landscape, is encouraged. These have a high edge-to-area ratio and pheasants, which are found near woodland edges, reach higher densities than in larger woods. Secondly, management of neglected woods is recommended. The value of such management for both the shoot and nature conservation interests are outlined below.

Neglected woodland can be of great value to a limited range of wildlife species such as epiphytes and species which live on dead wood. However, the canopy in these woods tends to be dense, shading out the lower strata of vegetation. In consequence the ground flora is represented largely by shade-tolerant species; few song bird species breed and butterflies are uncommon. Unmanaged coppice-with-standards tends to be open and draughty near the ground and is unattractive to pheasants. Such woodland can be made more attractive to pheasants by planting conifers or shade-tolerant shrubs, by creating skylights and rides or by resuming the coppice cycle.

A common method of creating low cover for pheasants is by planting shade-tolerant shrubs beneath the lightly thinned canopy of otherwise unmanaged woodland. The effects of this practice on wildlife were not examined in the case study, but it seems likely that such shrubs will cast heavy shade and reduce the abundance and diversity of both plants and butterflies beneath. The shrubs used are usually non-native evergreens, and are thought to be of limited value as food plants for our indigenous fauna (Southwood, 1961). They may, however, improve the value of the habitat for song birds which require structural diversity in woodland.

The establishment of conifers on ancient woodland sites may be a financially attractive method of improving the pheasant shoot. Young conifers provide low, dense cover giving shelter and protection from predators to ground-feeding birds while thicket stage stands, particularly of larch, form attractive roosting sites. Strategically placed blocks of conifers can, therefore, be of short-term benefit to the shoot. Such benefits are absent from the post-thicket stages, which provide little low cover, few or no sheltered roosting sites and create a dense tree canopy for over half the life-span of the crop.

In the above study both the species-richness and number of ancient woodland indicator plants found in semi-natural woodland were markedly reduced by growing a single crop of conifers. The effects on the ground flora are evident long after felling of the trees and loss of shade-tolerant species may be permanent. Butterfly numbers and diversity in mature conifer plantations are also low, although the creation of rides and maintenance of an open canopy throughout the rotation will ameliorate some of these harmful effects.

Skylights open the tree and shrub canopy, provide flushing points for the shoot, allow light to penetrate and promote growth of ground layer vegetation. Pheasants may obtain short-term benefit from the extra cover resulting from the practice although the aerial parts of much of the natural field and herb layer of woodland dies in autumn and any remains are flattened by winter rains and snow. Bramble, persisting later than most species, provides good shelter for pheasants but is detrimental to the shoot where it occurs in large, dense thickets from which birds cannot be driven. The abundant ground flora within skylights bears storage organs, seeds and fruit, providing food for pheasants and wildlife. Further, the combination of increased light intensity and abundance of the field layer leads to far higher numbers of butterflies here than in unmanaged woodland. However, the areas involved are generally small, providing insufficient light for vigorous regeneration of coppice or tree seedlings, encouraging intense browsing damage by deer and favouring the development of a dense and species-poor field layer. Hence, the long-term effect of this management technique is to reduce the structural and plant species diversity of the woodland.

In the first two years after cutting, a block of coppice-with-standards provides a sheltered, open area which may be valuable to the shoot as a flushing point or stand for guns. The boundaries of the cut area may also act as woodland edges, providing favoured areas for pheasants in winter. Marginal and ruderal species of the ground flora are stimulated to grow and reproduce in the early years of the coppice cycle. The combination of sunny conditions and a diverse ground flora creates conditions attractive to butterflies (Robertson, Woodburn & Hill, 1988).

Coppice with three to six-year-old regrowth forms dense, low cover which is highly attractive to pheasants in winter. It also provides a breeding habitat for a number of song bird species and cover for small mammals. By about six years after cutting the regrowth from coppice tends to cast dense shade, reducing the floral diversity and cover provided by the field layer. The woodland provides less shelter for pheasants than younger stages of coppice, but it may provide a valuable roosting area particularly if it is close to younger coppice.

Wide, sunny rides provide flushing points, sites for gun stands and increase the length of the woodland's edge where the highest densities of pheasants are found. The ground flora on well maintained rides within ancient, semi-natural woodland can be extremely species-rich while a wide range of butterfly species can be attracted. Other groups of invertebrates, small mammals including bats, amphibians and reptiles, also benefit from wide, sunny rides.

The features required by a shoot can be created and maintained in existing woodland by a carefully planned coppice cycle operated around a network of permanent rides. Thus, low cover and feeding areas can be provided by the young stages of the coppice cycle; roosting sites by older coppice; shrubby woodland edges by boundaries between coppice blocks and ride sides; and gun stands by wide, carefully sited rides. This management will provide woodland with a heterogeneous structure which will benefit the ground flora and encourage a wide range of woodland fauna. Species requiring undisturbed and overmature woodland can also benefit where the centre of large woodland blocks, unattractive to pheasants, are left unmanaged.

Commercially viable timber production is difficult to achieve in small farm woodlands as felling and extraction operations are rarely cost-effective on a small scale. Problems are frequently exacerbated by the limited access available to these woodlands. Standard forestry techniques such as clear-felling and replanting with a conifer nurse crop which must be removed before it impedes the growth of the hardwood main crop are not, therefore, ideal. In contrast, coppicing can frequently be carried out by farm staff in winter when agricultural work is slack and the cordwood can be extracted at the landowner's convenience: this is frequently at harvest time when stubbles allow easy access by farm vehicles to the woodland. Further, good quality timber trees can be grown at wide spacing in tree shelters among the coppice regrowth which acts as a nurse and encourages the development of straight, clean trunks. These trees will increase the capital and landscape value of the woodland in the long term.

The benefits to the pheasant shoot may provide the incentive to landowners for working small farm woodlands which have not been considered worthy of management for decades. As well as benefiting a wide range of wildlife groups, this management will ensure the conservation of the woodland and its intrinsic amenity and landscape value. In addition, timber trees may be introduced which increase the capital value of the woodland and help to ensure its long-term survival.

Acknowledgements

We would like to thank Lord Shaftsbury for permission to work on his estate and Don Ford and Harry Teesdale for their co-operation with the field work. Dr G.R. Potts, Mark Anderson and Clive Bealey all gave valued advice during preparation of the paper. The work was supported by a grant from the Forestry Commission.

References

Anderson, M., 1986, 'Conserving soil fertility: application of some recent findings', in R. Davies (ed.), *Proc. ICF Conference: Forestry's Social and Environmental Benefits*, Institute of Chartered Foresters, Edinburgh.

Brown, A.F.H., Oosterhuis, L., 1981, 'The role of buried seed in coppice woods', *Biological Conservation*, **21**, 19–38.

Cobham Resource Consultants, 1983, 'Countryside sports, their economic significance', *The Standing Conference on Countryside Sports*, Reading University.

Dartington Amenity Research Trust (DART), 1983, *Small Woods on Farms*. Countryside Commission Report, CCP 143.

Gray, N., 1986, *Woodland Management for Pheasants and Wildlife*, David and Charles, Vermont, USA.

Grubb, P.J., 1977, 'The maintenance of species richness in plant communities: the importance of the regeneration niche', *Biological Reviews*, **52**, 107–45.

Hall, M., 1981, *The Butterfly Monitoring Scheme*, Institute of Terrestrial Ecology, Monks Wood.

Hammond, P.M., 1974, 'Changes in the coleoptera fauna', in D. Hawksworth (ed.), *The Changing Flora and Fauna of Britain*, Academic Press, London, pp.333–67.

Higgins, L., Hargreaves, B., 1983, *The Butterflies of Britain and Western Europe*, Collins, England.

Hill, D.A., Robertson, P.A., 1988, *The Pheasant: Ecology, Management and Conservation*, Blackwell Scientific Publications, Oxford.

Moss, D., Taylor, P.M., Easterbee, N., 1979, 'The effects on song bird populations of upland afforestation with spruce', *Forestry*, **52**, 129–50.

Nature Conservancy Council, undated, *Ancient Woodland Survey -- South Region: Ancient Woodland Vascular Plants*, NCC, South Region.

Peterken, G.F., 1981, *Woodland Conservation and Management*, Chapman & Hall, London.

Pollard, E., 1977, 'A method of assessing changes in the abundance of butterflies', *Biological Conservation*, **24**, 317–28.

Rackham, O., 1986, *History of the Countryside*, Dent, London.

Ratcliffe, P.R., Pepper, H.W., 1987, 'The impact of roe deer, rabbits and grey squirrels on the management of broadleaved woodland', in *Oxford Press Institute Occasional Papers* No. **34**, National Hardwoods programme: report of the seventh meeting, pp. 39–50.

Robertson, P.A., 1988, 'Pheasant management in small broadleaved woodlands', in D.C. Jardine (ed.), *Wildlife management in forests*, Proc. ICF discussion meeting, April 1987, Institute of Chartered Foresters, Edinburgh, pp.25–33.

Robertson, P.A., Woodburn, M.I.A., Hill, D.A., 1988, 'The effects of woodland management for pheasants on the abundance of butterflies', *Biological Conservation*, **45**, 1–9.

Southwood, T.R.E., 1961, 'The number of insect species associated with trees', *Journal of Animal* Ecology, **30**, 1–8.

Southwood, T.R.E., 1978, *Ecological methods: with Particular Reference to Insect Populations*, Chapman & Hall, London.

Staines, B.W., 1983, 'Mammals and forestry' in E.Harris (ed.), *Forestry and Conservation*, Royal Forestry Society of England, Wales and Northern Ireland, Tring.

Steel, C., Khan, 1987, *The Management of Rides and Open Spaces*, Forestry Commission West England, Bristol.

Thomas, J., Webb, N., 1984, *Butterflies of Dorset*, Dorset Natural History Society.

Warren, M.S., 1976, 'The Dorset woodlands – their history and conservation', unpublished M.Sc. thesis, University College, London.

25

Manipulating water habitats to optimise wader and wildfowl populations

D.A. Hill

Introduction

The artificial creation of wetlands for waders and wildfowl has been a key feature of RSPB management over the past few decades. Initial successes, measured in terms of increases in species diversity and abundance, particularly in the case of rare species, has led to the increased use of this management technique. However, little is known of the sustainability of such systems, or of the underlying principles which drive them. In particular it is not known whether such systems are optimal nor whether they can sustain high levels of invertebrates indefinitely.

The aim of this type of conservation management is to provide a resource which is limiting at a particular time of year, such as food supplies or nest sites in the spring. The response of individuals to this resource will determine the success of the management technique. Density dependent regulating factors can allow a species to respond favourably to an increase in abundance of such a resource, whereas density independent factors are unlikely to be significantly influenced by management. Consequently an understanding of the dynamics of populations is important since a management technique aimed at increasing numbers at one stage in the life cycle of the species concerned may, in the long term, be detrimental if this stage is followed by one which acts in a strongly density dependent way. For example, we might increase wader breeding populations by creating favourable nesting habitat, but higher predation on clutches at high density would counteract some of the gains obtained by the management technique.

Conservation management usually attempts to push a species to a new, higher stable equilibrium density, although little has been documented on stability properties of simple wetlands as examined in this chapter. These systems often support large populations of opportunistic invertebrates which are eaten by wildfowl and waders. This chapter therefore aims to document changes in species abundance and diversity following hydrological management on five RSPB reserves: Havergate Island and Minsmere in Suffolk, Titchwell in Norfolk, Blacktoft Sands in Humberside and Elmley in Kent, as a first attempt towards understanding how such systems operate. The management techniques most commonly used to create feeding and nesting habitats for waders and wildfowl can be conveniently categorised as (1) lagoon creation by topsoil removal, (2) flooding of low-lying pasture and (3) reduction of salinity. In order to identify differences due to the three management categories the analyses presented are divided accordingly. General principles regarding responses of birds to these management categories are discussed.

Study areas

Havergate Island, grid ref: TM 425496, lies 3 km from the mouth of the river Ore estuary in Suffolk and is a low island of 108 ha protected from the North Sea by a shingle bank.

Titchwell Marsh is situated on the Norfolk coast, grid ref: TF 749436, covering 170 ha, comprising both tidal and freshwater reedbeds, a sea aster (*Aster tripolium*) saltmarsh, brackish and freshwater pools, sand dunes and a shingle beach.

Minsmere lies 21 km north of Havergate, on the low Suffolk coast, grid ref: TM 452680. The 592 ha site is an area of shallow brackish water, mud and islands inside a shingle beach with extensive reedbeds and meres, heathland and deciduous woodland.

Blacktoft Sands in Humberside, grid ref: SE 843232, is situated at the confluence of the rivers Ouse and Trent on the inner Humber estuary and is a 186 ha site consisting of tidal reedbed fringed by saltmarsh with an area of shallow brackish water lagoons.

Elmley Marshes, grid ref: TQ 926705, forms a 282 ha site on the extensive North Kent Marshes. Elmley is a coastal grazing marsh with fresh water fleets and shallow floods, and is bordered by saltmarsh on the north side of the Swale estuary.

Methods of management

Lagoon creation by excavation

At Minsmere, a series of large coastal lagoons were created artificially by the excavation of the surface of dry, poor grassland using bulldozers. The whole area (20.5 ha) is now known as 'The Scrape'. The Scrape is separated from the sea by a double sea-wall, water levels and salinity being manipulated by a sluice system. The lagoon area is dotted with islands suitable for nesting birds and these were formed by heaping material and covering with a shingle top. The shallow flooding of low areas of mud with sea water in the spring and summer helps to curtail succession.

At Blacktoft Sands areas of rough grass were converted to lagoons with islands suitable for breeding birds by raising and lowering water levels and by excavation, totalling an area of 24 ha. A bund embankment was bulldozed and scraped out, the scrape then being flooded by pumping water into the lagoon. Preferred levels within the separate lagoons were then maintained for different wader and wildfowl species.

Flooding of low lying pasture

At Elmley Marshes, existing low lying pastures were flooded to give areas of shallow water amounting to 33 ha. These were created by constructing dams and pumping water directly onto the land. Reservoirs were built and the water used to maintain water levels in the flood zones throughout the year. Vegetation died off and the sediment of fine silt within the lagoons was inhabited by large numbers of few invertebrates such as Chironomidae and Tubificidae (Reuth, 1987).

Reduction of salinity levels

Havergate Island is surrounded by an artificial embankment originally constructed to protect summer grazing. Presently 47 ha of the total 108 ha is managed as island and shallow water lagoons which exhibit wide seasonal variability in salinity. A sluice system is used to flush hypersaline water out of the lagoons and reduce salinity, increasing invertebrate numbers as a food supply for waders, principally avocets. Ditch clearance is carried out to maintain an adequate unhindered water flow. Hypersaline conditions above $65gl^{-1}$ are hostile environments for invertebrates eaten by avocet chicks, principally *Nereis diversicolor*, *Corophium volutator* and *Chironomus salinarius* (Mason, 1976). Lagoons were created in the late 1940s and early 1950s by scraping, with islands formed by heaping material scraped from surrounding areas, by deposition of soil excavated by machine or by flooding over low-lying areas with pockets of raised ground.

In a total of 36 ha of saltmarsh and tidal reedbed at Titchwell 15ha were transformed into freshwater, freshwater marsh and brackish marsh habitats. During a seven-year period sea-walls were constructed and established, and within the three new habitats salinity and water levels were regulated using a series of specially constructed dams, pipes, ditches and sluices.

Methods of analysis

Monthly counts of wildfowl and waders were collated and analysed as maxima per year in the case of wintering or passage abundances, together with breeding population size. Counts for individual species and communities were converted to densities using the amount of extant habitat, which changed in relation to the progress of management.

More intensively collected long-term data on avocet populations at Havergate and Minsmere are presented as a key factor analysis (Podoler and Rogers, 1975; Hill, 1988) in order to determine the individual mortality which explains most variation in total mortality and to identify the mortality most responsible for population regulation. Individual mortalities identified were (a) inability to realise maximum egg production, (b) egg loss, (c) chick loss and (d) overwinter loss (Hill, in press).

Results

Lagoon creation by topsoil removal

MINSMERE

The 20.5 ha scrape at Minsmere was created specifically for breeding avocets and terns *Sterna sandvicensis*, *S. hirundo* and *S. albifrons*. The progression in size of habitat following successive stages of management is shown in Figure 25.1.

The number of breeding avocets increased from 2 in 1963 to a peak of 138 in 1983 with a subsequent decline to 80 adults in 1986 (see Figure 25.2). The number of young produced significantly declined during the period 1963–86 ($t_{slope} = -4.31$, $r=-0.68$, $P<0.001$). Key factor analysis showed total losses to be highest in 1977 and 1984 due mainly to high egg and chick loss. Chick loss explained most variance in total loss and was hence the key factor ($t_{slope}=9.5$, $n=0.91$, $P<0.001$). The strongest density dependent regulating factor in the Minsmere avocet population was the loss occurring from the autumn population in year t and the breeding population in year $t+2$ (termed overwinter loss for convenience) since most avocets return to breed at their natal site in their second year of age. During the period 1963–76 the population was in an 'increase phase' and the dynamics were different from the period 1977–86 in which the population was in 'plateau phase'. During the increase phase birds responded to the increased availability of breeding space because of undercompensatory density dependent overwinter loss, (slope of log mortality on log autumn population density = 0.28 ± 0.06). During the plateau phase overwinter density dependence was almost perfectly compensatory (slope = 0.95 ± 0.25). This corresponded to the period when no new habitat was created.

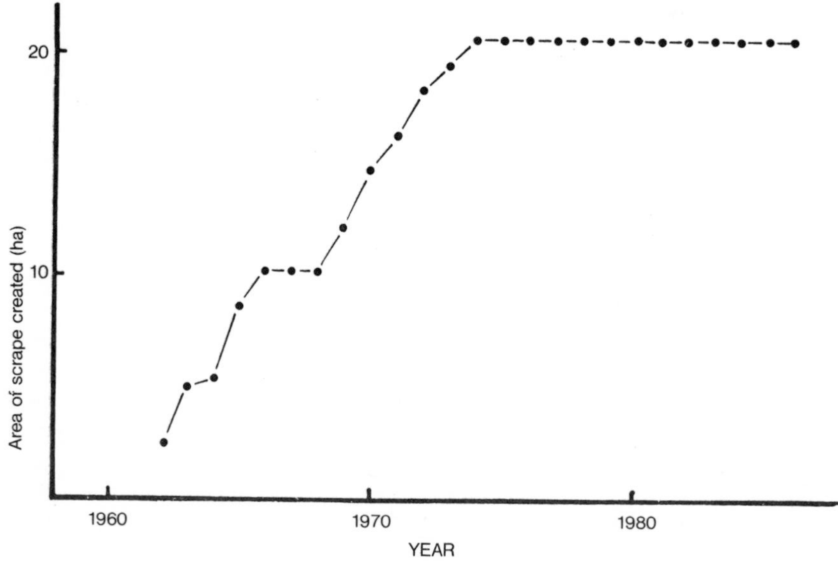

Figure 25.1 Increase in size of the scrape at Minsmere following management.

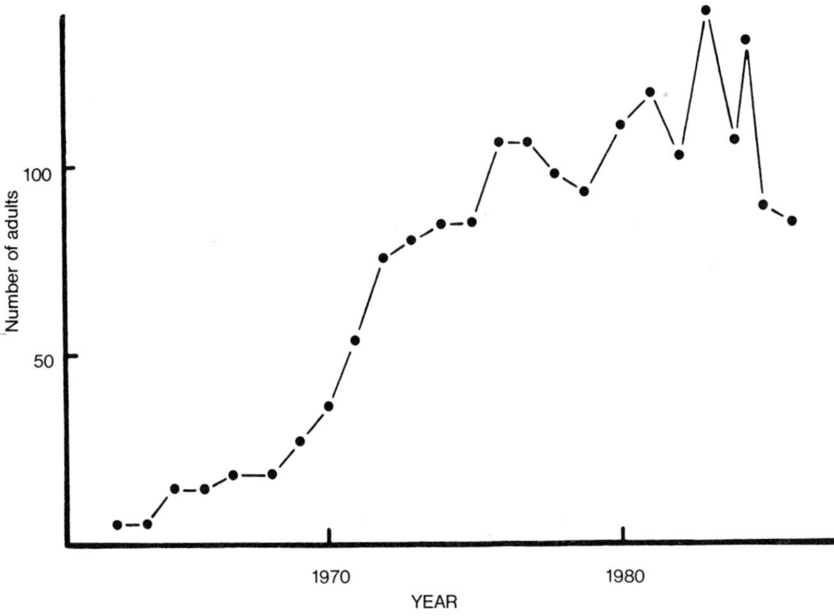

Figure 25.2 The number of breeding avocets at Minsmere, 1963–86.

Hence analysis of the population dynamics indicated regulation by lack of breeding space.

Following the creation of the scrape the maximum numbers of other waders using it (i.e. generally on passage) increased, but after 1969, when much of the creation was complete, numbers fluctuated about a mean (± s.e) of 3,536 ± 650 with no apparent trend. Total wader density on passage declined during 1963–86 (slope = -23.9 ± 9.3, $r=-0.50$, $n=22$, $P<0.01$), although wintering wildfowl density showed no trend. Densities of total breeding waders and wildfowl also declined however (slope waders = -0.17 ± 0.03, $r=-0.71$, $n=22$, $P<0.001$; slope wildfowl = -0.46 ± 0.09, $r=-0.76$, $n=22$, $P<0.001$).

BLACKTOFT SANDS

The increase in the area of new habitat created is shown in Figure 25.3 over the period 1977–87.

The number of breeding pairs of waders (all species) slightly increased during the period 1974–87 whereas the number of breeding pairs of wildfowl (all species) increased significantly (slope=1.8 ± 0.3, $r=0.88$, $n=14$, $P<0.001$) (see Figure 25.4). Maximum numbers of waders and wildfowl increased significantly during the study period (slope waders = 396 ± 67, $r=0.86$, $n=14$, $P<0.001$; slope wildfowl = 250 ± 68, $r=0.73$, $n=14$, $P<0.003$) (see Figure 25.5a) but declined when expressed as density (slope waders = -75 ± 26, $r=0.71$, $n=10$, $P<0.02$; slope wildfowl = -92 ± 28, $r=-0.76$, $n=10$, $P<0.01$) (see Figure 25.5b). Of particular importance was the

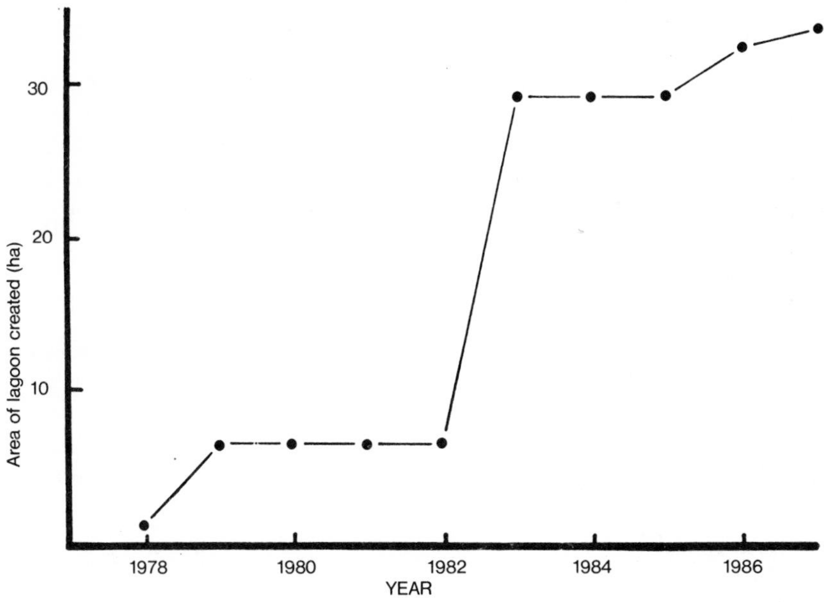

Figure 25.3 Increase in the area of lagoons created at Blacktoft Sands.

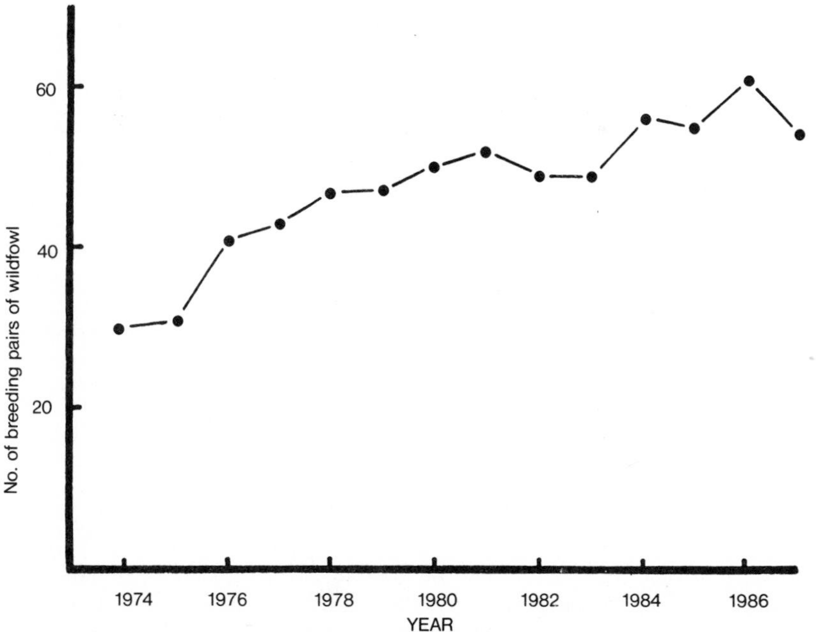

Figure 25.4 The number of breeding pairs of wildfowl (all species) at Blacktoft Sands, 1974–87.

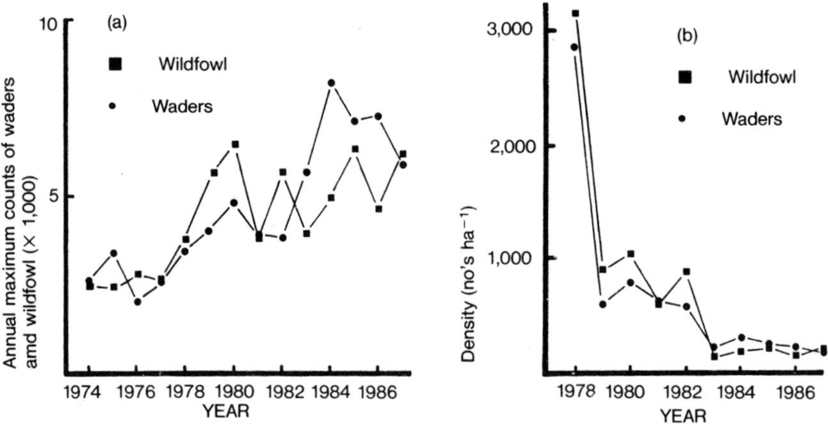

Figure 25.5 Annual maximum counts (a) and density (b) of waders and wildfowl at Blacktoft Sands, 1974–87 and 1978–87.

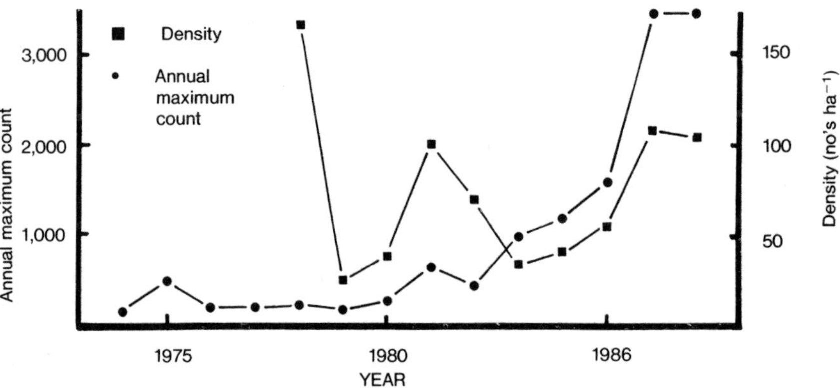

Figure 25.6 The number and density of golden plover using the Blacktoft Sands lagoons on passage migration.

exponential increase in golden plover using the lagoons on passage although density showed no trend (see Figure 25.6). The mean number of species of waders counted on passage increased significantly from 26.4 ± 0.3 before management to 33.0 ± 0.7 after the lagoon habitat was created ($P<0.001$).

Flooding of low-lying pasture

The increase in area of habitat created by flooding low-lying pasture at Elmley is shown in Figure 25.7 for the period 1975–86. Maximum numbers of both waders and wildfowl increased up to 1979 and 1980, but wader numbers declined during the period 1980–6 (see Figure 25.8). In particular the density of gadwall increased at Elmley following the creation of lagoons by controlled flooding (slope = 0.07 ±

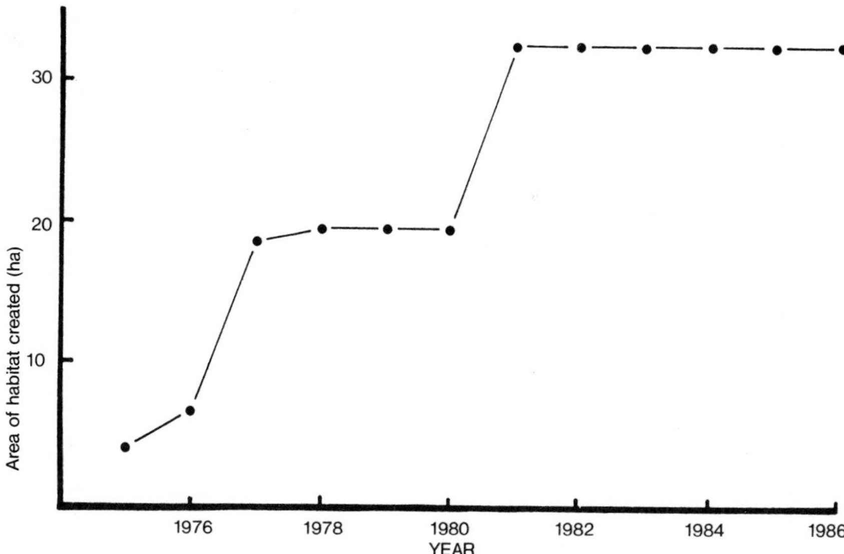

Figure 25.7 Increase in the area of habitat created by flooding low lying pasture at Elmley, 1975–86.

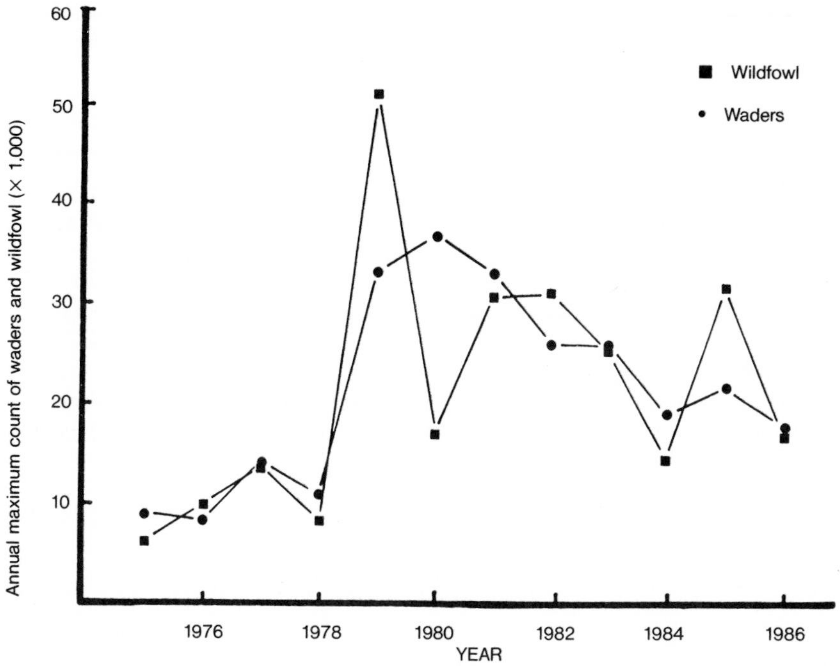

Figure 25.8 Annual maximum counts of waders and wildfowl at Elmley, 1975–86.

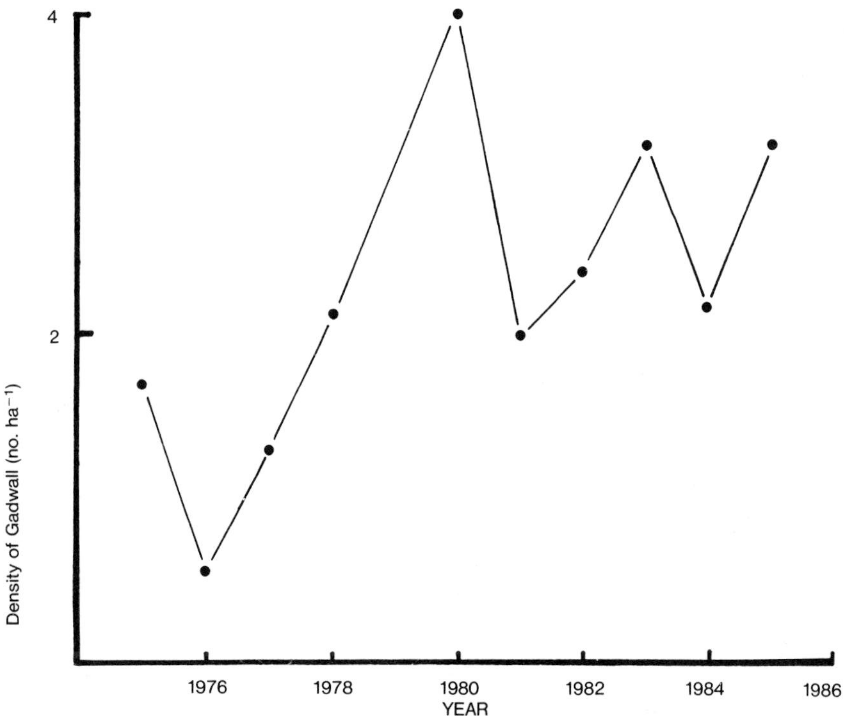

Figure 25.9 The density of gadwall using flooded lagoons at Elmley, 1975–86.

0.02, r=0.62, n=12, P<0.05) (see Figure 25.9). Passage and winter numbers of redshank and spotted redshank also increased during the period 1975–86 (slope redshank = 122 ± 40, r=0.69, n=12, P<0.01; slope spotted redshank = 10.2 ± 1.9, r=0.86, n=12, P<0.001) (see Figure 25.10), although when expressed as density redshank, but not spotted redshank, showed a decline (slope redshank = −4.1 ± 1.8, r=−0.59, n=12, P<0.05). Densities of all waders but not wildfowl declined during the period 1975–86 (slope waders = −38 ± 16, r=−0.60, n=12, P<0.05). The number of species of both waders and wildfowl increased significantly during 1975–86 (slope waders = 0.74 ± 0.18, n=12, r=0.79, P<0.001; slope wildfowl = 1.05 ± 0.33, r=0.71, n=12, P<0.001).

Reduction of salinity

HAVERGATE

Following the creation of the first lagoon numbers of breeding avocets increased from 8 in 1947 to a peak of 264 in 1985 (see Figure 25.11). During the period 1968–86 (most creation had ceased by 1968) the rate of growth was less than in preceeding years, and numbers fluctuated around 220 birds. The number of young

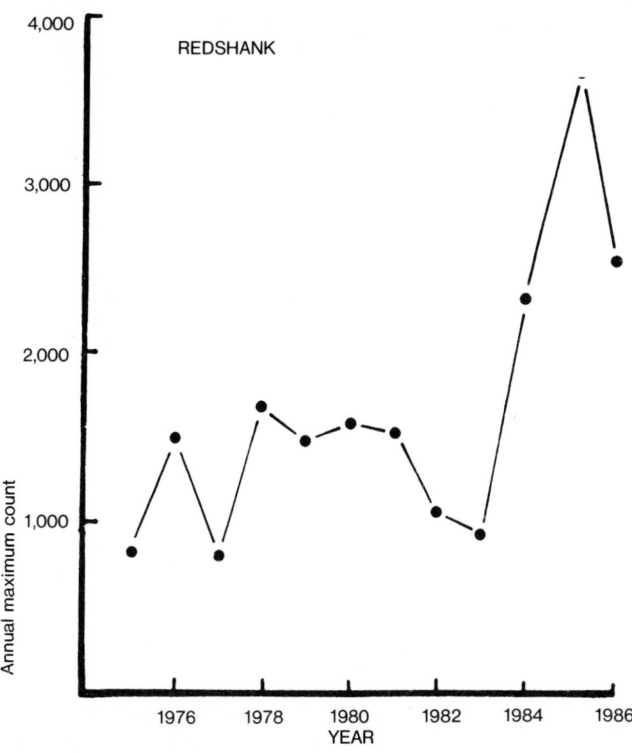

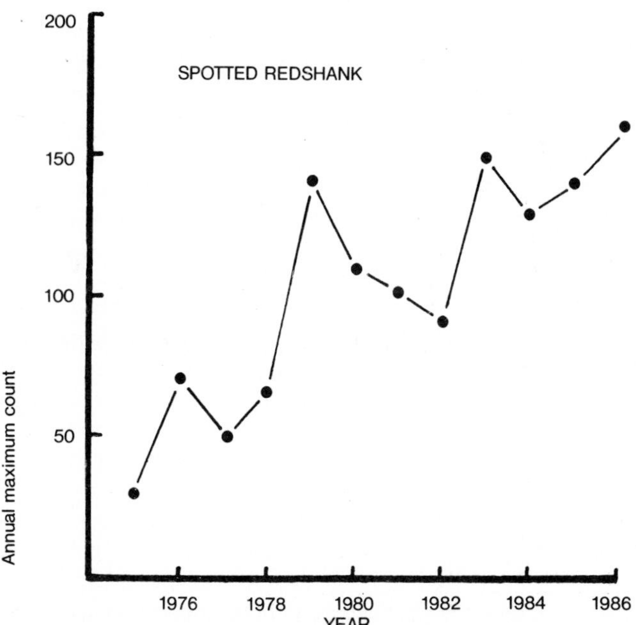

Figure 25.10 The number of redshank and spotted redshank using flooded lagoons at Elmley, 1975–86.

Figure 25.11 The number of breeding avocets at Havergate, 1947–86.

fledged per pair declined during the period 1947–86 (t_{slope}=−5.16, r=−0.64, P<0.001). Chick loss was identified as the key factor (t_{slope}=14.9, n=40, r=0.93, P<0.001). Overwinter loss, i.e. the loss occurring from the autumn population in year t and the breeding population in year t+2 was significantly correlated with autumn population size in year t (t_{slope}=6.60, r=0.74, n=38, P<0.001). During the 'increase phase' (1947–69) the slope of density dependent overwinter loss was undercompensatory (slope = 0.39), but almost perfectly density dependent (slope = 0.93) during the 'plateau phase' (1969–86). Consequently resources at the breeding site, principally breeding space, are most likely limiting.

Siltation of drains and ditches has resulted in a lower frequency of dilution of water within the Havergate lagoon complex. This has led to the progressive increase in salinity which is known to have a deleterious effect on invertebrates *Nereis diversicolor*, *Corophium volutator* and *Chironomus salinarius*. Research conducted by the University of Essex during 1978–81 showed that salinity peaks in relation to water temperature. Salinity reached 72.4 gl^{-1} in one year. Following the clearance of ditches between 1972–7 and then the building and operation of the more extensive sluice system, mean salinity in all lagoons declined in June (t_{slope}=−4.48, r=0.83, n=10, P<0.001), July (t_{slope}=−3.52, r=−0.78, n=10, P<0.01) and August (t_{slope}=−3.72, r=0.86, n=7, P<0.01), and has since been maintained at a lower level due to sluice control. During 1977–86 inclusive avocet chick mortality also declined significantly ($t_{slope}=-2.28$, $n=10$, $P<0.01$).

Generally no major changes in winter and passage abundances of wildfowl or waders have taken place following the hydrological control of lagoons and ditch clearance in the 1970s, although the number of avocets staying to winter has significantly increased.

TITCHWELL

The historical changes at Titchwell and the effects on vegetation are documented in Table 25.1. Following the conversion of saltmarsh and tidal reedbed to non-tidal marsh and freshwater reedbed habitats the annual maximum number of wildfowl increased at the site (see Table 25.2) particularly gadwall and tufted duck. Numbers of waders on passage, particularly four species (black-tailed godwit, curlew sandpiper, ruff and little stint) which use the non-tidal marsh, also increased significantly (see Table 25.3).

Total numbers of species of passage waders and wintering wildfowl, and the number of species of breeding wildfowl (but not waders) breeding at Titchwell also increased significantly following management in 1979 (see Table 25.2). The increase in number of wildfowl species was due largely to breeding by gadwall, shoveler and tufted duck following management.

One of the main objectives of habitat conversion was to provide breeding and rearing habitat for avocets. Indeed, after 1979 the number of avocets during breeding and autumn increased exponentially (see Figure 25.12).

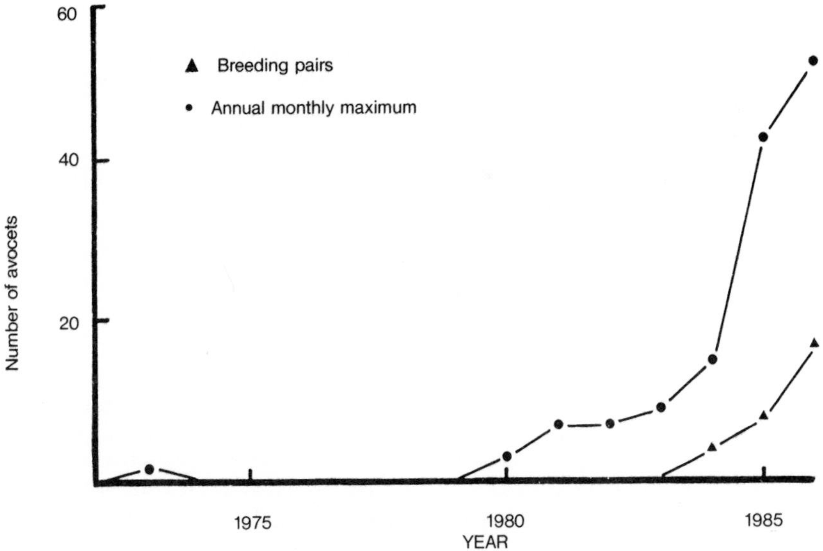

Figure 25.12 The number of breeding avocets at Titchwell, 1979–86.

Table 25.1 Stages in the conversion of saltmarsh to freshwater and brackish marshes at Titchwell together with ecological consequences

Date	Marsh history	Ecological characteristics
Freshwater marsh		
1974–8	Intermittent tidal flooding	Saltmarsh vegetation developed from *Asteretum* community to a *Puccinellia* dominated community
1979–80	Flooded with freshwater	Die-back of saltmarsh vegetation. Creation of thick organic layer
1981–6	Freshwater levels controlled	Freshwater mudflat fauna develops. Vegetation associated with freshwater marshes colonises islands
Brackish marsh		
1974–6	Natural tidal flooding	Saltmarsh vegetation maintained
1976–9	Intermittent controlled flooding with saltwater	Saltmarsh vegetation begins to die
1979–80	Flooded by seawater except Sept. and Oct.	Die-back of saltmarsh vegetation complete. Exposed mud flats and shallow water and island formation. Increase in mud-dwelling invertebrates. Increased feeding by waders
1981–3	Freshwater added to create brackish water (8–15 gl^{-1})	Increased diversity of mud-dwelling invertebrates. Arrival of *Corophium* sp. Increased use by spoonbills and avocets
1984–6	Salinity increased to 15–30 gl^{-1}	Increase in *Corophium* sp. Avocets begin to breed

Discussion

The three management techniques presented in this paper, i.e. creation of artificial brackish-freshwater lagoons by (1) excavation, (2) flooding of low lying pasture and (3) reduction of salinity in established sites, had markedly similar effects on wader and wildfowl use of the sites so created. In general, across all sites except Havergate, management was followed by (1) increases in peak numbers of waders and wildfowl in winter or on spring and autumn passage migration, (2) increases in number of species of waders and wildfowl visiting the site, principally on passage, (3) increases in the numbers of waders and wildfowl breeding on the site and (4) decreases in the

Table 25.2 Numbers (mean ± s.e) of waders and wildfowl before and after saltmarsh conversion to freshwater reedbed, freshwater and brackish marsh at Titchwell

Numbers of waders and wildfowl	Before (1973–9)	After (1980–6)
Maximum numbers all waders	48.7 ± 2.6	33.7 ± 1.3
Maximum numbers all wildfowl	6.9 ± 1.1	44.4 ± 6.7
Maximum species number–waders	25.0 ± 0.6	31.3 ± 0.6
Maximum species number–wildfowl	13.6 ± 1.5	20.4 ± 0.9
Number species breeding waders	3.3 ± 0.2	3.9 ± 0.3
Number species breeding wildfowl	2.0 ± 0.3	7.7 ± 0.4

Table 25.3 Increase in mean numbers per year of four species of waders counted feeding on passage migration at Titchwell following management

Wader species	Before 1973–9	After 1980–4
Black-tailed godwit	0.4	30.6
Curlew Sandpiper	13.0	28.2
Ruff	2.3	45.0
Little Stint	1.1	28.4

densities of passage and breeding waders and wildfowl. At Havergate the reductions in salinity following sluice operation and ditch clearance only benefited avocet chick production and had no subsequent effect on numbers of other waders and wildfowl at any time of the year.

The finding that densities declined with the progression of hydrological management and lagoon creation has important implications for future conservation management. The aim of such management from the point of view of creating wader habitat has been to increase numbers and species diversity of birds through habitat manipulation. Little regard has been paid to management in order to increase density. Such criteria for assessment of conservation 'quality' would be mediated if the site for manipulation contained other important considerations such as fragility, history, natural features etc., as used by the Nature Conservancy Council in their evaluation procedure. However, brackish-freshwater lagoons do attract large numbers of birds even though density generally declines over the time period that management is in operation.

What explanations can we pose for the declines in density expressed by these analyses, and can we draw conclusions that might help us in planning future management on coastal sites? The slope of the relationship of density regressed on available habitat would be expected to be 0 if birds responded to an increase in such habitat by exploiting it evenly. However, the decline in density is most likely attributed to an edge effect such that only the shoreline is used to any extent, and the centre of the lagoon is used significantly less. This would make sense for waders which can only exploit water to a certain depth, but most of the wildfowl counted comprise dabbling ducks which do not have such constraints. This explanation would also account for the decline in breeding density if expressed as birds breeding on land per unit area of water. This would suggest that shallow irregular lagoons are better than deeper circular ones.

The increase in species-richness is probably a consequence of the provision of a greater number of niches with the progression of development and macrophytic/invertebrate colonisation of the lagoons.

There is a body of information relating the number of individual waders and wildfowl to the size of a refuge or estuary, largely because of the recent threats posed to estuaries by land-use alternatives and energy production. Most attention has focused on the winter period since this is the time when British estuaries are vitally important for continental breeding waders and wildfowl during passage migration. Food seems to be the major determinant of shorebird distribution, and prey density and wader density are usually highly positively correlated (Prater, 1981). Spacing behaviour, often irrespective of aggressive interactions, causes individuals of many species of wader to feed in areas of lower prey density where high prey density areas are being exploited. For example, while redshank prefer to settle in places with the highest density of *Corophium* sp., they will avoid places of high bird density because of the additional loss of feeding time incurred by mutual interference (Goss-Custard, 1970, 1976, 1977). To some extent the behaviour of the invertebrates on which they feed determines this spacing behaviour in that at high predator density the prey withdraw into burrows and hence become less available to the birds. Feeding rates of oystercatchers and curlews (Zwarts, 1978) have also been shown to decline as the local density of birds increased.

Prater (1981) showed that the density of shore waders and all waders combined was significantly negatively correlated with estuary size, although when simply expressed as numbers of individual species against estuary size the relationship was positive. However, these results again suggest that the increase in numbers of individuals of certain species does not have as steep a slope as the increase in size of estuary would indicate if exploitation was even per unit area.

A closer look at the dynamics of one selected species, the avocet, breeding at Havergate and Minsmere, showed how overwinter density dependence regulates the size of breeding populations, most probably in relation to the amount of breeding space. On more recently created sites such as Titchwell, recolonisation is followed by an increase phase until the habitat becomes 'saturated'. The management techniques described in this chapter all aim to increase breeding and wintering numbers. However, management techniques which aim to increase productivity of a population which has a strong density dependence operating on numbers in spring, will not be successful in increasing breeding populations.

The principles of creating and managing lagoons on RSPB reserves has often been approached in an *ad hoc* way. The way in which these systems operate, bearing in mind the opportunistic invertebrate species which colonise them, is very little understood. An experimental approach, using (1) flooding regime, (2) vegetation management, (3) physical dimensions and (4) salinity control, as the manipulated variables, would help to isolate the most important factors to bird density. These experiments are presently being planned.

References

Goss-Custard, J, 1970, 'The responses of redshank to spatial variations in the density of their prey', *Journal of Animal Ecology*, **39**, 91–113.

Goss-Custard, J, 1976, 'Variation in the dispersion of redshank *Tringa totanus* on their winter feeding grounds', *Ibis*, **119**, 257–63.

Goss-Custard, J, 1977, 'The ecology of the Wash 3. Density related behaviour and the possible effects of the loss of feeding grounds on wading birds (*Charadrii*)', *Journal of Applied Ecology*, **14**, 721–39.

Hill, D.A., 1988, 'Population dynamics of avocets (*Recurvirostra avocetta*) breeding in Britain', *Journal of Animal Ecology*, **57**, 669–83.

Mason, C., 1976, 'Invertebrate populations and biomass over four years in a coastal, saline lagoon', *Hydrobiologia*, **133**, 21–9.

Podoler, H., Rogers, D., 1975, 'A new method for the identification of key factors from life table data', *Journal of Animal Ecology*, **44**, 85–114.

Prater, A., 1981, *Estuary Birds of Britain and Ireland*, T & A.D. Poyser, Calton.

Reuth, C., 1987, 'Invertebrate survey of Elmley Marshes, Kent', report to the RSPB, Industrial Applied Biology Group, University of Essex.

Zwarts, L., 1978, 'Intra- and inter-specific competition for space in estuarine bird species in a one-prey situation', *Proceedings of International Ornithological Congress*, Berlin.

Appendix 25.1: Wildfowl Species*

Common Name	*Scientific Name*
Mallard	*Anas platyrhynchos*
Teal	*Anas crecca*
Garganey	*Anas querquedula*
Gadwall	*Anas strepera*
Wigeon	*Anas penelope*
Pintail	*Anas acuta*
Shoveler	*Anas clypeata*
Scaup	*Aythya marila*
Tufted Duck	*Aythya fuligula*
Pochard	*Aythya ferina*
Goldeneye	*Bucephala clangula*
Long Tailed Duck	*Clangula hyemalis*
Velvet Scoter	*Melanitta fusca*
Surf Scoter	*Melanitta perspicillata*
Common Scoter	*Melanitta nigra*
Eider	*Somateria mollissima*
Red Breasted Merganser	*Mergus serrator*
Goosander	*Mergus merganser*
Smew	*Mergus albellus*
Common Shelduck	*Tadorna tadorna*
Mandarin Duck	*Aix galericulata*
Red-crested Pochard	*Netta rufina*
Ferruginous Duck	*Aythya nyroca*
Blue-winged Teal	*Anas discors*
Green-winged Teal	*Anas c. carolinensis*
Ruddy Shelduck	*Tadorna ferruginea*
Chiloe Wigeon	*Anas sibilatrix*
American Wigeon	*Anas americana*
Bahama Pintail	*Anas bahamensis*
Chilean Pintail	*Anas georgica spinicauda*
Cape Shelduck	*Tadorna cana*
Greylag Goose	*Anser anser*
White-fronted Goose	*Anser albifrons*
Lesser White Fronted Goose	*Anser erythropus*
Barnacle Goose	*Branta leucopsis*
Pink-footed Goose	*Anser brachyrhynchus*
Brent Goose	*Branta bernicla*
Bean Goose	*Anser fabalis*
Canada Goose	*Branta canadensis*
Snow Goose	*Anser caerulescens*
Red Breasted Goose	*Branta ruficollis*
Egyptian Goose	*Alopochen aegyptiacus*
Bar Headed Goose	*Anser indicus*
Emperor Goose	*Anser canagicus*
Greylag × Canada Hybrid	
Barnacle × Canada Hybrid	

* Possible escapes included

Appendix 25.1 Cont: Wildfowl Species

Common Name	*Scientific Name*
Snow × Canada Hybrid	
Mute Swan	*Cygnus olor*
Whooper Swan	*Olor cygnus*
Bewick Swan	*Olor columbianus bewickii*
Black Swan	*Cygnus atratus*

Wader Species

Oystercatcher	*Haematopus ostralegus*
Lapwing	*Vanellus vanellus*
Ringed Plover	*Charadrius hiaticula*
Lesser Ringed Plover	*Charadrius dubius*
Grey Plover	*Pluvialis squatarola*
Golden Plover	*Pluvialis apricaria*
Lesser Golden Plover	*Pluvialis dominica*
Dotterel	*Eudromias morinellus*
Turnstone	*Arenaria interpres*
Snipe	*Gallinago gallinago*
Jack Snipe	*Lymnocryptes minimus*
Woodcock	*Scolopax rusticola*
Curlew	*Numenius arquata*
Whimbrel	*Numenius phaeopus*
Black-tailed Godwit	*Limosa limosa*
Bar-tailed Godwit	*Limosa lapponica*
Green Sandpiper	*Tringa ochropus*
Wood Sandpiper	*Tringa glareola*
Common Sandpiper	*Actitis hypoleucos*
Redshank	*Tringa totanus*
Spotted Redshank	*Tringa erythropus*
Greenshank	*Tringa nebularia*
Knot	*Calidris canutus*
Purple Sandpiper	*Calidris maritima*
Little Stint	*Calidris minuta*
Dunlin	*Calidris alpina*
Curlew Sandpiper	*Calidris ferruginea*
Sanderling	*Calidris alba*
Ruff	*Philomachus pugnax*
Avocet	*Recurvirostra avosetta*
Temmincks Stint	*Calidris temminckii*
Red-necked Stint	*Calidris ruficollis*
Black-winged Stilt	*Himantopus himantopus*
Kentish Plover	*Charadrius alexandrinus*
Sociable Plover	*Vanellus gregarius*
Stone Curlew	*Burhinus oedicnemus*
Pectoral Sandpiper	*Calidris melanotos*
Broad-billed Sandpiper	*Limicola falcinellus*
Marsh Sandpiper	*Tringa stagnatilis*
Stilt Sandpiper	*Micropalama himantopus*

Appendix 25.1 Cont: Wader Species

Common Name	*Scientific Name*
Semi-palmated Sandpiper	*Calidris pusilla*
Sharp-tailed Sandpiper	*Calidris acuminata*
Buff-breasted Sandpiper	*Tryngites subruficollis*
White-rumped Sandpiper	*Calidris fuscicollis*
Terek Sandpiper	*Xenus cinereus*
Bairds Sandpiper	*Calidris bairdii*
Spotted Sandpiper	*Actitis macularia*
Greater Yellowlegs	*Tringa melanoleuca*
Lesser Yellowlegs	*Tringa flavipes*
Grey Phalarope	*Phalaropus fulicarius*
Red-necked Phalarope	*Phalaropus lobatus*
Wilsons Phalarope	*Phalaropus tricolor*
Hudsonian Godwit	*Limosa haemastica*
Long-billed Dowitcher	*Limnodromuscolopaceus*
Black-winged Pratincole	*Glareola nordmanni*
Collared Praticole	*Glareola pratincola*
Spoonbill	*Platalea leucorodia*
Crane	*Grus grus*
Crowned Crane	*Balearica pavonina*
Flamingo	*Pheonicopterus* spp.
Sacred Ibis	*Threskiornis aethiopicus*
Glossy Ibis	*Plegadis falcinellus*

SECTION 8

Conclusions

Losses of semi-natural habitat sustained in developed countries in the past forty years are a stark reminder of our commitment to creating living space, producing raw and manufactured materials, and growing food and timber. It might also be observed that the benefits of this technological achievement, and the confidence engendered by it, have had psychological effects extending well beyond the built and cultivated environment. One is the popular belief that semi-natural communities, too, can be assembled or reassembled by some benign, quasi-industrial process or husbandry almost within our grasp.

Development, however, has other more positive consequences. Improved standards of living and better educational opportunities increase our appreciation of the wider environment and focus attention on the damage being done to it. But while aesthetic, educational and scientific interest in semi-natural areas flourishes as never before, there is an ever-present optimism, rooted in technology, that such communities can easily be manipulated, rescued from destruction, or 'engineered' to order. Lying between these two positions, habitat reconstruction is a test of the paradox.

Attitudes towards habitat reconstruction rapidly polarise into a position taken by the restoration pragmatist or 'engineer' on the one hand, and the nature conservationist on the other. While horticulturalists and landscape designers are more concerned with the visual impact of habitats than their scientific authenticity, conservationists seek to avoid damage or alteration to existing semi-natural communities, and are committed to largely passive methods of restoration, such as natural colonisation or low-intensity management practices. The contrast of these ethical stances is most acute when a relocation scheme is suggested as an alternative to some high-quality habitat being damaged by development. To conservationists, habitat transfer is an unnatural event which diminishes or degrades the original habitat. Developers, in turn, will point to their technical skills and quote examples of successful transfer operations as an argument to justify future removals.

Ground rules can only be established for habitat reconstruction once the objectives for each project are clearly stated and an appropriate situation can be found. There is general acceptance that 'creating' new habitats within the urban environment poses no major ethical problems, and that a close copy of the original semi-natural community is rarely essential for objectives which include education, recreation and visual enjoyment.

Towns are traditional places for habitat reconstruction. A newer arena is the countryside, explored here by several different contributors. Their accounts show that small reductions in management intensity can quickly increase the wildlife diversity of areas of commercially motivated farming and forestry. Arable areas can be 'improved' by extending margins (if necessary by replacing them where they have been lost), and grasslands can be actively diversified by developments such as slot-seeding. Modified forestry practices such as retaining 'permanent' areas of forest, staggering felling coupes, heavy thinning and ride-widening have similarly tangible wildlife benefits.

One test of the effectiveness of different methods of habitat reconstruction is to compare their relative implementation costs and maintenance implications. Hence a cheap solution, popular with conservationists, is to rely on natural colonisation and minimal management to slowly convert a new site into a plausible habitat stereotype. However, as a practical method this has a number of attendant ecological snags, even if the site in question already adjoins good habitat. Soil conditions may be too fertile or otherwise unsuitable; seedbanks of the preferred plant species may not be present; and species with poor dispersal mechanisms may be unable to colonise effectively or at all. The problems of laissez-faire restoration must be faced and solved.

Another inexpensive technique is to sow a commercial wild flower seed mixture costing perhaps £1,000 ha^{-1}, although maintenance costs will be somewhat higher than for agricultural leys or grass in city parks, particularly when traditional methods of hay cutting and close grazing are implemented. In the case of meadows re-created from seed, it is not only maintenance but careful species choice which is critical if disappointment is to be avoided.

Other habitats are much more expensive to re-create, particularly ponds and lagoons which require heavy earth-moving machinery, special materials and often careful hydrological management, but they tend to give excellent results relative to their area. Aquatic systems also colonise quickly with plants, invertebrates and birds, and soon become a valuable educational resource which is easy to justify. In contrast, scrub and woodland habitats are very long-term, needing a canopy to be established before other ecosystem components can be introduced. The additional planting costs make them more expensive than meadows, but management in the long-term will be cheaper.

Compared with habitat 'creation', habitat transfers produce results faithful to the original community and are generally more successful. They give longer species lists, the soil layers used are the correct medium for the plants and other organisms they contain, and recovery of the transferred vegetation is often remarkably rapid. However the cost will be at least an order of magnitude greater than for conventional sowing, even when transport distances between the 'donor' and 'receptor' sites are minimal. Subtle changes will also occur at the new site owing to altered environmental conditions, while management must be continued as on the 'donor' area. 'Donor' sites also imply sacrifice, raise ethical questions and can involve lengthy impact assessment procedures.

Diversification and enhancement methods offer a compromise between those of habitat 'creation' and transfer, as well as some compromise in cost. The technique of introducing seed and transplants into gaps in existing vegetation shows promise and in some situations lends itself to mechanisation. Similarly, methods which rely on positive management to create the right conditions for greater community diversity are inherently attractive if the opportunity cost is not appreciably higher than conventional practice. However it is likely that techniques which reduce the use of pesticides and fertilisers on agricultural land, or open up forest canopies, will need further underpinning by compensation agreements to make them acceptable outside the public sector.

Finally, perhaps the most fascinating aspect of habitat reconstruction is the challenge that it presents to our understanding of ecology. Restorationists are continually thwarted by imprecise information, from the autecological requirements of individual species to the functioning of whole communities. For good habitat design we need to know much more about critical population sizes of the species concerned; the dispersal ability of organisms; the role of isolation and connectivity in landscapes; the importance of fertility in determining community diversity; and the relevance of gap size and disturbance in maintaining species-rich communities, to name only a few critical areas. Habitat reconstruction provides us with both the tools and the laboratory for the analysis.

Species Index

NOTE: All species have been referred to by their Latin name only.

Subject Index